ADVENTURES IN THE ANTHROPOCENE

Gaia Vince is a journalist and broadcaster specialising in science and the environment. She has been the front editor of the journal *Nature Climate Change*, the news editor of *Nature* and online editor of *New Scientist*. Her work has appeared in the *Guardian, The Times, Science, Scientific American, Australian Geographic* and the *Australian*. She has a regular column, Smart Planet, on BBC Online, and devises and presents programmes about the Anthropocene for BBC radio. She blogs at WanderingGaia.com and tweets at @WanderingGaia.

GAIA VINCE

ADVENTURES IN THE ANTHROPOCENE

A JOURNEY TO THE HEART OF THE PLANET WE MADE

Chatto & Windus
LONDON

Published by Chatto & Windus

2 4 6 8 10 9 7 5 3

Copyright © Gaia Vince 2014

Gaia Vince has asserted her right under the Copyright, Designs
and Patents Act 1988 to be identified as the author of this work

First published in Great Britain by
Chatto & Windus
Random House, 20 Vauxhall Bridge Road,
London SW1V 2SA

www.randomhouse.co.uk

Addresses for companies within The Random House Group Limited can be found at:
www.randomhouse.co.uk/offices.htm

The Random House Group Limited Reg. No. 954009

A CIP catalogue record for this book
is available from the British Library

ISBN 9780701187347 (hardback)
ISBN 9780701187354 (Trade Paperback)

Map © Jane Randfield
Geological Timescale © Francisco Izzo / Nautilus
All photographs © Nick Pattinson

Penguin Random House is committed to a sustainable future for
our business, our readers and our planet. This book is made from
Forest Stewardship Council® certified paper.

Typeset in Minion Pro by Palimpsest Book Production Limited,
Falkirk, Stirlingshire

Printed and bound in Great Britain by
Clays Ltd, St Ives plc

For Nick

CONTENTS

GEOLOGICAL TIMESCALE

CENOZOIC
65.5 mya to Present

- ANTHROPOCENE
- HOLOCENE
- PLEISTOCENE
- PLIOCENE
- MIOCENE
- OLIQOCENE
- EOCENE
- PALEOCENE

MESOZOIC
251.0 to 65.5 mya

- CREATACEOUS
- JURASSIC
- TRIASSIC

PALEOZOIC
542.0 to 251.0 mya

- PERMIAN
- CARBONIFEROUS
- DEVONIAN
- SILURIAN
- ORDOVICIAN
- CAMBRIAN

PROTEROZOIC 2500 to 542.0 mya

ARCHEON 4000 to 2500 mya

HADEAN 4600 to 4000 mya

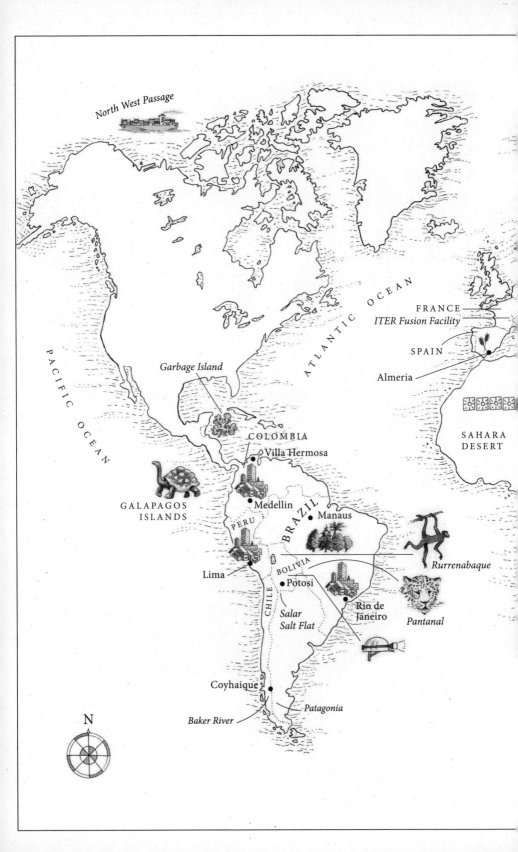

North West Passage

ATLANTIC OCEAN

FRANCE
ITER Fusion Facility

SPAIN

Almería

SAHARA
DESERT

PACIFIC OCEAN

Garbage Island

COLOMBIA
Villa Hermosa

GALAPAGOS
ISLANDS

Medellín

PERU

BRAZIL

Manaus

Rurrenabaque

Lima

BOLIVIA

Potosí

Pantanal

Salar
Salt Flat

Río de
Janeiro

Coyhaique

Patagonia

Baker River

N

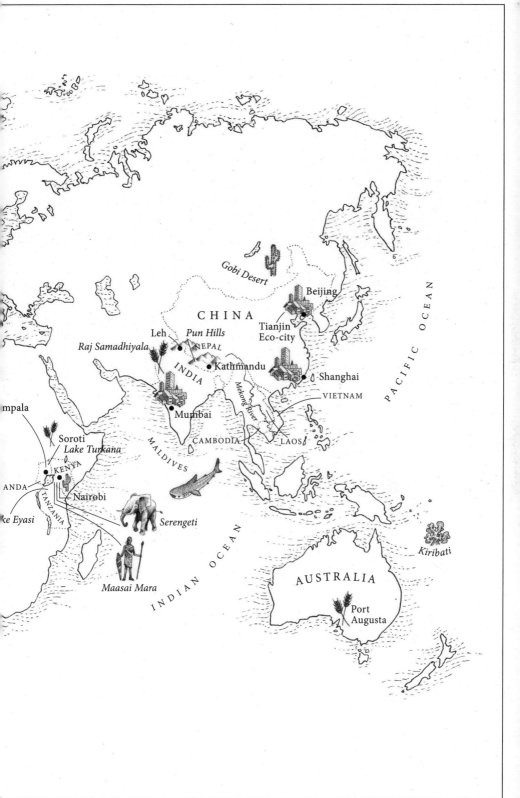

INTRODUCTION:
THE HUMAN PLANET

F our and a half billion years ago, out of the dirty halo of cosmic dust
left over from the creation of our sun, a spinning clump of minerals
coalesced. Earth was born, the third rock from the sun. Soon after, a
big rock crashed into our planet, shaving a huge chunk off, forming the moon
and knocking our world on to a tilted axis. The tilt gave us seasons and
currents and the moon brought ocean tides. These helped provide
the conditions for life, which first emerged some 4 billion years ago. Over the
next 3.5 billion years, the planet swung in and out of extreme glaciations.
When the last of these ended, there was an explosion of complex multicellular
life forms.

The rest is history, tattooed into the planet's skin in three-dimensional
fossil portraits of fantastical creatures, such as long-necked dinosaurs and
lizard birds, huge insects and alien fish. The emergence of life on Earth
fundamentally changed the physics of the planet.[1] Plants sped up the slow

breakdown of rocks with their roots, helping erode channels down which rainfall coursed, creating rivers. Photosynthesis transformed the chemistry of the atmosphere and oceans, imbued the Earth system with chemical energy, and altered the global climate. Animals ate the plants, modifying again the Earth's chemistry.

In return, the physical planet dictated the biology of Earth. Life evolves in response to geological, physical and chemical conditions. In the past 500 million years, there have been five mass extinctions triggered by supervolcanic eruptions, asteroid impacts and other enormous planetary events that dramatically altered the climate.[2] After each of these, the survivors regrouped, proliferated and evolved. The diversity of plants, animals, fungi, bacteria and other life on Earth is richer now than at any point in time.[3]

And us? Anatomically modern humans didn't arrive until nearly 200,000 years ago and it was touch and go whether we would survive. But something pulled us through, the something that differentiated us from the other species in this shared biosphere and made us so successful that we now rule our world: the human brain. We're more intelligent and use tools better than the other animals. And humans can make and control fire. Ever since the first human lit the first spark, our destiny as the most powerful species was assured. Having this external source of energy, which we could move wherever we chose, gave us power over the landscape, protection from other animals, allowed us to cook our food, keep warm and, ultimately, take over the world.

For thousands of years, humans shared the planet with Neanderthals and our other cousin species. A supervolcanic eruption at Toba in Indonesia 74,000 years ago nearly wiped us all out – the human population shrank to a few thousand. But, by 35,000 years ago, truly modern humans, indistinguishable from people alive today and littering caves and rocks with signs of their culture, had emerged and migrated out of Africa. Thus began the heroic ascent of man.

In the Stone Age, our impact as a species on the planet was limited to some extinctions – particularly of large mammals – and some local landscape *changes, such as the burning of forests. Technologies were primitive and minimal, and were fashioned entirely from renewable materials. Over the following centuries, our impact grew. Farming was invented around 10,000 years ago (about 300 generations ago; world population: 1 million), transforming some regional landscapes as human-bred plant varieties replaced wild flora. Around 5,500 years ago (world population: 5 million), cities were built and the first great civilisations emerged. The Industrial Revolution in Europe and North America, which replaced the labour of humans and beasts with machines, started having a measurably global impact about 150 years ago (world population: 1 billion), as large volumes of carbon dioxide from fossil fuels were released into the atmosphere.*

Nothing, however, compares to the scale and speed of our planetary impact since World War Two, driven by population expansion, globalisation, mass production, technological and communications revolutions, improved farming methods and medical advances. Known as the Great Acceleration, this rapid increase in human activity can be seen across a vast range of things, from the number of cars to water use.[4] It took 50,000 years for humans to reach a population of 1 billion, but just the last ten years to add the latest billion.

This rapid transformation spurred social and economic development – a century ago, life expectancy in Europe was less than fifty years, now it's around eighty years. But the Great Acceleration has been a filthy undertaking. Pea-souper smogs shrouded cities like London killing thousands, acid rain poisoned rivers, lakes and soils, eroding buildings and monuments, refrigerant chemicals ate away at the protective ozone layer, and carbon dioxide emissions changed in the global climate and acidified the oceans. Our voracious plundering of the natural world has led to massive deforestation, a surge in extinctions and destroyed ecosystems. It has produced a deluge of waste that will take centuries to degrade. In a single lifetime we've become a phenomenal global force and there is no sign of a slowdown – in fact, our extraordinary impact on the planet is only increasing.

Meanwhile, our closest relative, the chimpanzee, is living much as he did

50,000 years ago. Humans are the only creatures to have cumulative culture, allowing us to build on the past rather than continually reinvent the wheel. But, as we fumble about on Earth's surface, hostage to the whims of our phenomenally powerful brains, humanity is undertaking a brave experiment in remodelling the physical and biological world. We have the power to dramatically shift the fortunes of every species, including our own. Great changes are already being wrought. The same ingenuity that allows us to live longer and more comfortably than ever before is transforming Earth beyond anything our species has experienced before. It's a thrilling but uncertain time to be alive. Welcome to the Anthropocene: the Age of Man.

We live in epoch-making times. Literally. The changes humans have made in recent decades have been on such a scale that they have altered our world beyond anything it has experienced in its 4.5 billion-year history. Our planet is crossing a geological boundary and we humans are the change-makers.

Millions of years from now, a stripe in the accumulated layers of rock on Earth's surface will reveal our human fingerprint just as we can see evidence of dinosaurs in rocks of the Jurassic, or the explosion of life that marks the Cambrian or the glacial retreat scars of the Holocene. Our influence will show up as a mass of species going extinct, changes in the chemistry of the oceans, the loss of forests and the growth of deserts, the damming of rivers, the retreat of glaciers and the sinking of islands. Geologists of the far future will note in the fossil records the extinctions of various animals and the abundance of domesticates, the chemical fingerprint of artificial materials, such as aluminium drinks cans and plastic carrier bags, and the footprint of projects like the Syncrude mine in the Athabasca oil sands of north-eastern Canada, which moves 30 billion tonnes of earth each year, twice the amount of sediment that flows down all the rivers in the world in that time.

Geologists are calling this new epoch the Anthropocene, recognising that humanity has become a geophysical force on a par with the earth-shattering asteroids and planet-cloaking volcanoes that defined past eras.[5]

Earth is now a human planet. We decide whether a forest stands or is razed, whether pandas survive or go extinct, how and where a river flows, even the temperature of the atmosphere. We are now the most numerous big animal on Earth, and the next in line are the animals we have created through breeding to feed and serve us. Four-tenths of the planet's land surface is used to grow our food. Three-quarters of the world's fresh water is controlled by us. It is an extraordinary time. In the tropics, coral reefs are disappearing, ice is melting at the poles, and the oceans are emptying of fish because of us. Entire islands are vanishing under rising seas, just as naked new land appears in the Arctic.

During my career as a science journalist, it became my business to take special interest in reports on how the biosphere was changing. There was no shortage of research. Study after study came my way, describing changes in butterfly migrations, glacier melt rate, ocean nitrogen levels, wildfire frequency . . . all united by a common theme: the impact of humans. Scientists I spoke to described the many and varied ways humans were affecting the natural world, even when it came to seemingly impervious physical phenomena like weather and earthquakes and ocean currents. And their predictions were of bigger changes to come. Climate scientists tracking global warming told of deadly droughts, heatwaves and metres of sea-level rise. Conservation biologists were describing biodiversity collapse to the extent of a mass extinction, marine biologists were talking of 'islands of plastic garbage' in the oceans, space scientists were holding conferences on what to do about all the junk up there threatening our satellites, ecologists were describing deforestation of the last intact rainforests, agro-economists were warning about deserts spreading across the last fertile soils. Every new study seemed to hammer home how much our world was changing – it was becoming a different planet. Humanity was shaking up our world, and as I and others reported these stories, people around the world were left in no doubt about the environmental crises we were responsible for.[6] It was profoundly worrying and often overwhelming.

As I followed the latest research, I heard plenty of dire predictions about our future on Earth. But at the same time I was also writing about our triumphs, the genius of humans, our inventions and discoveries, about how scientists were finding new ways to improve plants, stave off disease, transport electricity and make entirely new materials. We are an incredible force of nature. Humans have the power to heat the planet further or to cool it right down, to eliminate species and to engineer entirely new ones, to resculpt the terrestrial surface and to determine its biology. No part of this planet is untouched by human influence – we have transcended natural cycles, altered the physical, chemical and biological processes of the planet. We can create new life in a test tube, bring extinct species back from the dead, grow new body parts from cells or build mechanical replacements. We have invented robots to be our slaves, computers to extend our brains, and a new ecosystem of networks with which to communicate. We have shifted our own evolutionary pathway with medical advances that save those who would naturally die in infancy. We have surmounted the limitations that restrict other species by creating artificial environments and external sources of energy. A 72-year-old man now has the same chance of dying as a 30-year-old caveman. We are supernatural: we can fly without wings and dive without gills, we can survive killer diseases and be resuscitated after death. We are the only species to leave the planet and visit our moon.

The realisation that we wield such planetary power requires a quite extraordinary shift in perception, fundamentally toppling the scientific, cultural and religious philosophies that define our place in the world, in time and in relation to all other known life. Up until the Middle Ages, man was believed to be at the centre of the universe. Then came Nicolaus Copernicus in the sixteenth century, who put Earth in its place as just another planet revolving around the sun. By the nineteenth century, Charles Darwin had reduced man to just another species – a twig on the grand tree of life. But now, the paradigm has shifted again: man is no longer just another species. We are the first to knowingly reshape the living earth's biology and chemistry. We have become the masters of our planet and integral to the destiny of life on Earth.

The last time our planet entered a new geological age was around 10,000 years ago and it had a profound effect on the survival and success of our species. As the last ice age ended, a new epoch of global warming called the Holocene began. Ice sheets retreated to the poles and the tropics became wetter. People came out of their caves and began taking advantage of the new conditions: grasses proliferated and those with nutritious seeds, like wheat and barley, could be farmed. Around the world, people began settling in larger communities and processing food rather than simply hunting and gathering. This stability led to the development of culture and civilisations – our species became more populous and so successful we spread across six continents. The impacts of the Anthropocene will be just as profound.

It was Nobel laureate Paul Crutzen who came up with the term Anthropocene. The Dutch chemist was sitting in a scientific conference, he told me, when it occurred to him that all the biophysical changes that researchers were discussing 'meant we weren't in the Holocene any more. The planet had changed too much from what would be considered normal for the Holocene.' Crutzen made a case for the Anthropocene in an article in *Nature* in 2002, and over the past decade the term has gained use in the scientific community.[7] Now, the British Geological Society is beginning the slow process of formally deciding and listing this new epoch, based on the changes humans are making to the biosphere that will be preserved in the geology, chemistry and biology of our planet for thousands or millions of years.[8] These include land-use changes, such as the conversion of forest to farmland, and radioactive fallout particles. Boundaries between geological times are fuzzy and often span thousands of years, as scientists try to calculate them from stripes in rocks around the world. The geologists will have to decide when the epoch started – was it thousands of years ago with the advent of farming, a few generations ago with the industrial revolution, or the 1950s with the Great Acceleration? That decision will depend on which marker the geologists use to define the Anthropocene: the atomic tests of 1949, say, or the rise in atmospheric carbon dioxide concentration around 150 years ago.

But while the geologists battle the conceptual difficulty of palaeontological dating for an era whose palaeontology and geology are still being created, the Anthropocene has escaped the confines of academia and been embraced by a far broader section of society. The idea that humanity is having a truly planetary effect has aroused the interest of artists and poets, sociologists and conservationists, politicians and lawyers. Scientists are using the term to describe multifaceted changes to our planet and its life. And it is in the spirit of this broader definition – and the growing consensus that we are now crossing the boundary into the Anthropocene epoch – that I write this book.

So how can we recognise the Anthropocene – what are the signs that we're entering a new geological age? In the atmosphere, carbon dioxide levels are almost 50% higher than the Holocene mean – our industrial and domestic emissions of greenhouse gases are warming the atmosphere, changing the climate and disrupting weather patterns across the globe.[9] The impacts of climate change are planetary and affect all life on Earth to some degree. The atmosphere has also newly become a repository for a range of other chemicals. Mountains are losing glaciers that have covered them for many thousands of years, causing them to crumble faster – and they are also being hacked at by miners. Rivers are rerouted, dammed, drained and exhibiting a dramatic reduction in sediment flow. Farmlands have appeared out of the natural landscape and there has been an explosion in the amount of available nitrogen on the planet because of the fertilisers we're adding. This nitrogen has increased crop yields, which has allowed human populations to soar, doubling in the last fifty years, with far-reaching consequences for the entire planet. The oceans are becoming more acidic as they dissolve our carbon dioxide emissions from the atmosphere, and they are becoming less biodiverse as coral dies and we lose fish through overfishing, pollution and warming waters. The Arctic is melting and coast-lines are eroding as storms increase in frequency and violence, the sea level rises and protective sediments, mangroves and wetlands vanish.

Deserts are spreading across savannahs, forests are drying and being logged. Wildlife is being hunted and dying because of habitat loss, climate change and species invasions, pushing the planet towards the sixth mass extinction in its history. Meanwhile, we are causing the proliferation of our domesticated species and indiscriminately scattering others around the globe. We are disembowelling the Earth through mining, drilling and other extractions, littering the planet with novel compounds and materials, devices and objects, that could never have occurred naturally. And we are building enormous steel, concrete and glass cities that light up the night sky and are visible from space.

And what about the impacts of our changed planet on us? After all, we've evolved and adapted to a life in the Holocene, and the new changes have occured very rapidly. The transformations we have made to our planet have been key in enabling us to become this superspecies – and they have also been a consequence of our extraordinary ascent. In changing the Earth we have been able to thrive, to live longer and healthier, in better comfort even in greater numbers than ever before. However, for now, at least, humans are still of nature – we evolved on this living planet, we are made of cells, we breathe air, drink water and eat protein. We rely on the biological, chemical and physical parts of our planet to provide everything, including all our materials, fuels, food, clothes, and to clean our air, recycle our water and manage our waste. Our growing population and the way we live in this new human world are making us more demanding than ever of our planet's resources and processes. But, as we continue to change Earth, we reduce its ability to meet these needs and, as a result, we are facing crises in fresh-water availability, food production, climate change and 'ecosystem services', the immeasurable functions that the biosphere performs to enable our survival.

In the Anthropocene, we have already started to push global processes out of whack. In some cases, just tiny further changes could spell disaster for humans; for others, we have quite a bit of leeway before we face the consequences. Most of them have some sort of tipping point beyond which it will be almost impossible to return to Holocene-like conditions. For example, glacier melt at the poles could reach a tipping point at which sudden runaway melting occurs and sea level rises by metres. Fears of big changes like these have led some scientists to describe 'planetary boundaries' – biophysical limits for human safety, such as the extent of land-use change and biodiversity loss – some of which they say we've already exceeded.[10] In leaving the relative safety of stable Holocene conditions, it is clear that humans face unprecedented challenges.

The key here is how we will deal with the consequences. It may have been desirable to keep within the internationally agreed 'safe' global-warming limit of 2°C (above pre-industrial levels), for example, but we will almost certainly exceed that by the end of this century, and so the question becomes how we can live in the warmer Anthropocene environment.[11] We have always altered ecosystems to serve our needs and presumably will continue to do so. Our habitable range, for example, is not limited to the tropics, because we invented clothes and other ways to keep warm, just as air-conditioning technology keeps us cool. We have improved the planet for our survival in a number of ways, including by staving off the next ice age, but we have also made it worse. Some of those negative consequences we can overcome through technological advances or migration or other adaptations. Others we will need to reverse; some others we will need to learn to live with.

The good news is that some problems are already being brought under control. Pollution is being curbed in many countries through laws and technological improvements; radioactive pollution has been limited through the international nuclear test ban treaty. The growth in the ozone hole has also started to slow because of the Montreal treaty banning ozone-destroying chemicals. Crucially, the rate of population growth is also slowing, with many countries now in negative growth.[12] Other problems, however, remain increasing and significant threats. And, while science may be able to identify biophysical issues, it cannot tell us how to react – that is for society to decide. Humans are no longer just another animal, we have specifically *human* rights that are expected to be achieved through development, including access to sanitation and electricity – even the Internet.[13] Delivering social justice and protecting the environment are closely linked; how poor people get richer will strongly shape the Anthropocene.

The enormous impacts we're having on our living planet in the Anthropocene are a direct consequence of the immense social changes we're undergoing – changes to how we live as a species. We now support a massive global population, but we have not simply multiplied the number of small hunter-gatherer communities. More than half of the world's people now live in cities – artificial constructs of densely packed, purpose-built living spaces, which act as giant factories consuming the planet's plants, animals, water, rocks and mineral resources. Humanity operates on an industrial scale, and has needs – currently, eighteen terawatts of the energy at any time, 9 trillion cubic metres of fresh water per year and 40% of the global land area for food. It has become a super-organism, a creature of the Anthropocene, a product of industrialisation, population expansion, globalisation and the revolution in communications technology. The intelligence, creativity and sociability of this humanity super-organism is compiled from the linked-up accumulation of all the human brains, including those from the past who have left a cultural and intellectual legacy, and also the artificial minds of our technological inventions, such as computer programs and information libraries like Wikipedia. Humanity is a global network of civilisations with a stream of knowledge already being channelled for human protection. And, just as a cloud of starlings suddenly flips direction en masse, it is difficult to predict humanity's behaviour. Although humanity is an enormous planetary force, our super-organism can be steered by individuals and its behaviour can be shaped by the societies within it – and the solutions are often to be found at the local level. We are essentially a conglomeration of chemicals that recycle other chemicals and the biosphere is capable of supporting 10 billion of us. The difficulty is doing so within social and environmental constraints.

The self-awareness that comes with recognising our power as a planetary force also demands we question our new role. Are we just another part of nature, doing what nature does: reproducing to the limits of environmental capacity, after which we will suffer a population crash? Or are we the first species capable of self-determination, able to modulate our natural urges, our impacts and our environment, such that we can maintain habitability on this planet into the future? And what of our relationship to the rest of the biosphere? Should we treat it – as every other species does – as an exploitable resource to be plundered mercilessly for our pleasures and needs, or does our new global power imbue us with a sense of responsibility over the rest of the natural world? Our future will be defined by how we reconcile these two opposing, interwoven forces.

There comes a time in a child's life when they first realise that the food they enjoy – the meat they eat – comes from an animal. That the lovely fluffy mammal they pet is also food. Some children become vegetarian and refuse to eat meat again. Most do not. At this moment in our history, we are like children, realising that the things we enjoy in life, that we depend upon, from energy to water to consumables, all come with environmental and social consequences that affect us. How we struggle to resolve this issue will determine the trajectory of the Anthropocene for years to come.

We are pioneers in this era, but we have a superior understanding of science, excellent communication and connectivity that breeds collaborative thinking. In the post-natural era of the Anthropocene, we will have to either preserve nature or master its tricks artificially. I wanted to find out how, and that would mean leaving my desk in London.

Just as latitude and longitude tell you all and yet nothing about a place, so the abstract numbers and graphs produced by scientists seemed to be telling me nothing about the new world we are living in. What's more, there is no other field of science in which the academic findings are so contested by society. People often hold extreme views on solutions to the problems of the Anthropocene – many even question established scientific fact. I was intrigued and I knew that I wanted to explore our planet at such a significant time in its history. It seemed to me that the most important people I hadn't heard from were the human guinea pigs of this new epoch – those who are already experiencing this changed world – and I wanted to see how they were coping. I wanted to delve behind the headlines, the barrage of statistics, computer models, the tit-for-tat arguments between green campaigners and corporations, the shock-doctrines and tired slogans. I wanted to investigate the truth for myself by looking at the situation on the ground, talking in person to the human players of this epoch, seeing with my own eyes the reality of our Anthropocene.

I decided to leave my job in London and set off on a quest: to explore the globe at a crucial moment in its living history, at the beginning of this extraordinary new human age. I looked at how people are learning to undertake nature's tasks. I found people creating artificial glaciers to irrigate their crops, building artificial coral reefs to shore up their islands, and artificial trees to clean the air. I met people who are trying to preserve important remnants of the natural world in the Anthropocene, and those trying to recreate the old world in new places. And I met people who are looking at ways to resolve the conundrum: to find a way for 10 billion people to live in greater comfort, with enough food, water and energy; and yet at the same time, reduce our impact on the natural world and its ability to carry out the processes we rely on.

As I travelled through our changing planet, I looked at the world we are creating and wondered what sort of Anthropocene we want. Will we learn to love the new nature we make, or mourn the old? Will we embrace living efficiently or will we spread out over newly ice-free lands? Will we eat new foods, plant new crops, raise new animals? Will we make space for wildlife in this human world? I experienced the Anthropocene from different perspectives and met the pioneers who are negotiating a development path through the complexity of our shared biosphere. This book is a journey around our new world, a series of stories about remarkable people living in extraordinary times. It's the story of ingenious inventions, incredible landscapes and about how we have come to own Gaia for better or for worse.

As humanity faces its biggest challenge in 10,000 years, I set out to discover whether our species will survive, and how.

1

ATMOSPHERE

*E*arth's Great Aerial Ocean, the churning sky of gases that gravity hugs to the planet, is the breath of life that ignites this unique speck of the universe. Breathe in, breathe out: the atmosphere is vital to life on Earth. It is an organ of the living biosphere – a great pulsating body that recycles the breathable air, regulates the temperature and climate, and protects us from the hazardous meteors and deadly cosmic and ultraviolet rays of space.

The atmosphere extends for an indivisible one hundred kilometres, and is invisible except through its meteorological moods that reveal clouds of water vapour or falling snow, electric flashes of lightning or the blush of a sunset.

The swirling currents of Earth's aerial and terrestrial oceans interact to create our planet's many weathers and different climates, and these dictate the conditions for life. Perhaps the most significant of these global weathers is the Hadley Cell, a pattern of hot moist air that dumps reliable rains on the lush equatorial belt, generating the planet's highly biodiverse tropical rainforests and swamps, while leaving parched deserts to the immediate north and south. The impact of this system can be seen from space as a sharp delineation of green to brown.

But life on Earth also dictates the atmospheric condition and its weathers. The world's first atmosphere was hydrogen and water vapour – it took around 2 billion years for the gas of life, oxygen, to pervade the air, courtesy of the early photosynthesisers. Those ancient blue-green algae, which survive today as unremarkable-looking stromatolites, used energy from the sun to make sugars from carbon dioxide, in the process releasing oxygen as a waste product.

The continual breathing of Earth's living organisms, from tiny ants to massive trees, depletes the atmosphere's oxygen and replaces it with carbon dioxide and water vapour. During daylight hours, especially in the summer, this respiratory exchange is offset by the photosynthesis of the world's terrestrial and oceanic forests of trees and algae. The various feedbacks between biota and air have created an atmosphere of roughly 78% nitrogen and 21% oxygen with the remainder being a mix of noble gases, carbon dioxide and traces of others.

It is into this intricate relationship that humanity has stormed, adding enough warming gases to the atmosphere to shift the delicate equilibrium of the past millennia and change global climate for centuries to come.

The atmosphere acts as a blanket against the unimaginably cold temperatures of outer space, and the main gas responsible for these cosy conditions is carbon dioxide. Carbon dioxide is invisible because sunlight passes straight through the molecule. However, it is opaque to the infrared rays that heat travels in, so, like the glass in a greenhouse, it warms the air. Sunlight travels unhindered through the atmosphere until it hits the surface of the Earth. If that surface is very reflective – like a shiny white glacier – then most of the rays will bounce straight back as light. But if the surface is dark – like black rock, soil or ocean – then this energy is absorbed as heat, which radiates into the atmosphere as infrared rays that can't pass through the carbon dioxide. In this way, heat gets trapped bouncing between the atmosphere and the Earth, warming them both and sustaining life.

We know from fossil records that the planet's climate has swung between tropical prolificacy that saw metre-long insects, and ice ages that killed off the majority of life forms. These catastrophic big freezes were the result of massive events like meteor hits or supervolcano eruptions that filled the atmosphere with so much dust that sunlight couldn't penetrate to the planet and killed the animals that produce that all-important carbon dioxide. At such times, the concentration of carbon dioxide in the atmosphere dropped as low as 160 parts per million (ppm) molecules.

For the past half a million years – the world into which humans evolved – the carbon dioxide concentration has hovered between 200 ppm (during ice ages) and the comfortable 280 ppm of the Holocene. Historically, the main fuel humans used was wood, emitting the same amount of carbon dioxide that the tree absorbed during its growth. But in the Anthropocene, the vast majority of our energy comes from burning fossil fuels – emitting the huge stores of carbon dioxide from plants and creatures that died millions of years ago. As I write this, carbon dioxide concentrations in the atmosphere are 40% higher than pre-industrial levels – 400 ppm – the atmosphere is warmer, more energetic and holds more water, giving rise to more extreme weather. Scientists are saying that there is no longer such thing as 'normal climate', by which they mean what was normal for the Holocene.

We are also using the atmosphere as a repository for other gases released during combustion and for a range of other pollutants, including refrigerants that attack the ozone layer high in the stratosphere that protects us from UV rays.

And, in the Anthropocene, the atmosphere has also become humanity's global voice. Just as visible light can travel through the air, so can sound, radio waves, and microwaves, enabling instant communication by radio, telephone and Internet. The atmosphere is as transparent to the human-generated pulses in the satellites it hosts as it is to the sun's vital energy, and allows our species to traverse the globe virtually in seconds.

In 1932, King George V became the first monarch to deliver a Christmas Day message by radio to 20 million listeners from Britain to the outposts of the *empire. In a script written for him by Rudyard Kipling, he addressed 'men and women so cut off by the snows, the deserts, or the sea, that only the voices out of the air can reach them'. The atmosphere of the Anthropocene is now full of these 'voices out of the air'. Imagine if we could see the beams emitted by our radios, laptops, televisions, mobile phones and other devices. For almost all of the planet's 4.5 billion-year history, the atmosphere has been lit solely by extra-terrestrial flares, like suns or meteors, or by electrical storms. Now, the skies are infused with artificial lights of different wavelengths as our devices communicate with each other and with us. And that's just in the invisible spectrum. In the visible spectrum we have lit up our world so brightly that towns and cities can be seen from space at night and, for city dwellers, the stars fade into oblivion.*

Satellites enable us to look down from space at our home as no eye has done before. The same cameras show us in unprecedented detail just how much we are changing our world. Using the Internet, we can pool our shared knowledge and intellectual resources to solve new problems, to cooperate in different ways and to transcend the geography of our planet to inhabit a virtual room no matter where we are physically.

The atmosphere has also become a playground for our aerial adventures, a medium for rapid and direct long-distance travel around and beyond our planet into space. Humans can now journey from London to Sydney in less than a day. We can trade between communities within time frames that allow fresh blueberries to be picked by a human in South Africa and eaten by another hours later in London.

Our technological invasion of the skies has allowed us to communicate across our species in a way that no other life form has. The atmosphere is un-ownable, common to all Earth-dwellers – it gives life with the first breath and life is extinguished with the last. In this chapter I look at how our changes to the atmosphere will help decide how societies develop over the coming decades.

I meet Mahabir Pun outside the tiny airstrip in Pokhara, some 200 kilometres west of Nepal's capital Kathmandu. He is a shortish fellow in his mid-fifties with an inflatable ball of a stomach and thick black hair that emerges at extraordinary angles above his square face.

'Gaia, come. Come!' he says urgently, setting off ahead of me at a rapid pace and agitating his hair further, so that it stands wildly up on one side.

As I trot along behind him, people gather to watch the unusual spectacle of the pale, sweating foreign woman dressed for Arctic exploration with a bulging backpack following a local guy in light cottons and open sandals.

A political demonstration earlier in the week has led the Maoist government to impose a military-enforced curfew in the area, banning all vehicles including motorbikes, buses and taxis, so Mahabir has had to walk several kilometres to meet me. But here, as in every other place that lacks functional governance, people are resourceful. Casting a sly look around, Mahabir motions for me to get on one of two motorbike taxis, while he takes the other, and we speed off.

Pokhara is a lake town, shimmering within a halo of mountains. It is closest of anywhere in Nepal to achieving the new prime minister's promise of turning the country into the 'Switzerland of Asia'. Enticing cafés and shops crowd the lanes beside the lake. Brightly clad clusters of men, women and youngsters gather at a little jetty that delivers worshippers to the pretty Buddhist temple on an island a hundred metres away. Women wearing saris are knee-deep in the lake dousing rainbows of laundry and shampooing their long black hair. Fish leap clear of the surface and birds circle overhead looking for snacks.

Rising above the town is the oddly shaped peak of Fishtail Mountain, whose sheer granite sides point a geological finger into the blue sky. It is mid-December in the Himalayas, there should be ice on this lake and snow descending far down the mountainsides. But only the highest peaks are white; pink flowers bob at head height on green stalks that sway in the sun. We stop and I remove another fleece.

The picture-postcard prettiness includes some less appealing details, I begin to notice. A fetid slick of vibrant green run-off from the town's cafés and businesses is discharging raw sewage and some sort of oily pollutant directly into the lake. Dirty, poorly clad children are poring over discarded plastic and other solid waste littering the banks – while I watch, one boy walks a few metres away, pulls down his shorts and defecates at the lake's edge. Looking upwards, I see that the quaint country homes lining the street are in fact filthy dilapidated mud-floored shacks, offering little protection or comfort for their large families. We're a long way from Switzerland here. And this is one of the most improved parts of the country.

In trying to grasp the enormity of the development task facing the poor world at the beginning of Anthropocene, Nepal is a good place to start. Sandwiched politically, culturally and geographically between two of the world's fastest emerging economies, Nepal has avoided following either the Chinese or Indian model for national growth and slid further into decline. It is one of the ten poorest countries, with more than one-third of the population living below the poverty line on less than $0.40 per day and half of children under 5 malnourished. Around 90% of Nepalis live in rural areas, many depending for survival on subsistence plots too small to support them, with little or no access to electricity, clean water, sanitation, education or health care, and national shortages of everything from rice to kerosene. More than a decade of Maoist insurgency and civil unrest has wrecked the economy and crippled infrastructure. Nepal has been incapable of even basic governance over the past few decades, and relies on an army of aid charities to avoid mass starvation – the number of NGOs in the country soared from 220 in 1990 to more than 15,000, now contributing around 60% of GDP.

Desperate times? A century ago, most people in Switzerland lived in similar conditions to this, and were even less likely to reach their fiftieth birthday.

Around the world, 40% of people (2.8 billion) have no access even to a communal toilet, which is a major factor in the 2.4 million deaths every year from diarrhoea. About 80% of illnesses are caused by faecal matter (people living without sanitation can ingest as much as ten grams of faecal matter a day). If Nepal is to make the leap in development that Switzerland made, it will need to grow its economy to make similar social investments in health, education and infrastructure. Nepali women will be able to do laundry at the push of a button, freeing up time for education and income-generating activities. Nobody will be using the public lake as a toilet. By 2048, it's predicted that the average income earned by a person in Asia will be dollar for dollar equivalent to that earned by someone in the United States. The question is how they are going to get there in the changing conditions of the Anthropocene, and without exacerbating the environmental challenges humanity faces. I've sought out Mahabir to discover how humanity's recent exploitation of the atmosphere is being used to smooth that path.

It is a five-hour tortuous drive to the tiny town of Beni ('the place where two rivers meet'), in a Toyota that dates from 1973, as the driver tells me proudly, giving the chassis a fond slap that causes the side panel to reverberate and almost detach from the rest of the car. The threadbare re-treads skid and swerve in and out of potholes along a narrow road that disappears alarmingly into gorges on either edge. We're chasing sunset, but it wins, plunging us into darkness for the final, hair-raising hour of the trip.

We overnight in a spartan hotel – made of timber, like all of Beni's buildings – and set off again at first light. There is no road to Nangi. Reaching Mahabir's remote mountain village involves a full day's hike up near-vertical paths, and it's not long before my pack is straining at my shoulders and my legs complaining at the unaccustomed exercise. In an age where I'm more used to judging distances by the time it takes to travel them by car, plane or other oil-fuelled transport, it's quite an adjustment to talk in journey times of hours or days by foot.

My laced-up hiking shoes are stifling in the sun. Mahabir had warned me that we would be trekking at altitude where there was likely to be thick snow at this time of year. 'Tonight freezing, tomorrow night more freezing,' he tells me cheerfully, as I eye his open flip-flops. Until recently, everyone in his village went barefoot, he says. Even in the snow? 'Yes, of course. But now even the poorest person has sandals.'

The ascent is immediately steep and continues so for nine hours. Every time the path diverges and I hopefully query it, the answer comes emphatically from below: 'Up, up.' It is with a certain satisfaction that I notice Mahabir starting to look a little damp and taking rather longer than before to plod up this interminable stairway.

It is beautiful, though. Vultures spiral up from below us on thermals that take them high into the eye-watering blue. The mountains seem to grow vaster as we climb and I start to experience 'peak mirage' – each time we approach a peak, the path unfurls higher and the peak recedes further up. Children often draw sky as a stripe of blue high above the grounded green of domesticity. It feels as though every step is taking us closer to piercing that blue, penetrating that mysterious space where men have placed angels and gods.

The atmosphere is vast and unknowable, but as familiar to us as it was to our distant ancestors. Who has not lain under a tree and taken pleasure at how the phantasmal wind shivers its leaves, or delighted in the puffs of clouds cruising by, or peered at night through the breathable air to the stars beyond. Until recently, only winged creatures could transcend our planetary home and explore the three-dimensional Great Aerial Ocean of the atmosphere. The closest we earthbound humans got was through arduous climbs like this, ascending slowly and painfully through the clouds to taste the chilled, thin air beyond. It wasn't until the end of the eighteenth century that hot-air balloons carried men high above the sod, giving them a bird's-eye view of our home and enabling direct travel between destinations 'as the crow flies'. Now that we can dance through the atmosphere with our toys and technologies, we can achieve a truly global perspective on our natural and artificial worlds, and perhaps even reconcile the two. Satellites orbiting the planet can allow us to track tagged marine and land mammals, measure forest loss, and compare Arctic ice coverage over decades. We can measure the transition from Holocene to Anthropocene in real time as the planet changes.

Handily placed stone rest-stops are built into our steep path at twenty-minute intervals and we make use of them all – stopping to sit a while, offloading our backpacks and admiring the view. There's something noble about conquering a peak: this 3,500-metre hill is my Everest and I take the same pride in my pathetic achievement as Hillary.

We see no other foreigners, just local people commuting up and down between villages that are unlinked by roads, and traders carrying impossibly large baskets of firewood and oranges from the higher slopes to the markets below. 'Oranges are growing very well in the past few years as it's become warmer,' Mahabir tells me. 'Many villages higher up are growing oranges now.' We have an orange-pip-spitting competition and Mahabir giggles in delight when he doubles my distance.

'Normally this whole area is covered in snow from October,' Mahabir says, looking at the muddy earth. 'Recently we have been getting less and less snow. We used to get two metres in the winter, and it would stay for weeks. Last winter we only had two centimetres, and it comes later. It means the winter crops have no water and are dying. The price of wheat and barley will be high this spring,' he predicts.

As we warm the atmosphere, the vast, churning blanket of greenhouse gases that protect life on Earth from the freezing cold of space is changing, affecting the snow here in Nepal and the price of food around the world. As we release more and more stored carbon into the air by burning fossil fuels, we are heading for a 4°C post-Holocene warming this century alone. That's 2°C higher than the 'safe' level determined by scientists. The impacts of this atmospheric carbon are affecting every part of the planet.

No single person or community came up with the idea to put the greenhouse gas carbon dioxide into the atmosphere. The oil-based market-driven economy is a characteristic of civilisation that arose out of the human love of energy and its promise of power and wealth. One gallon of oil contains an amount of energy that would take a man eight days of labour to produce. What is wealth if not the key to freedom, a way of throwing off the shackles of labour and the circumscribed life – the liberty to have, be and do as you choose; the dream of no man having dominion over you? It's intoxicating stuff.

Around the world, scientists and governments now understand the relationship between oil and global warming, and are discussing how to steer us towards healthier ways of achieving this hit. But replacing the efficient package of energy found in fossil fuels with alternatives is far from easy. Poor countries like Nepal, where people still get much of their energy from pre-industrial renewable sources, are feeling the effects of global climate change even while they yearn for the benefits of reliable power that fossil fuels provide. It's a problem I will encounter throughout my travels.

A road is planned for Nangi, Mahabir tells me, but until then, the only way for people to communicate or trade beyond the distance a voice carries is to physically meet each other or send an emissary. For millennia, people have made journeys like this out of necessity, and yet, in my home life, making the journey to meet someone in person now is often so unnecessary that it carries its own message, that of deference or love, for example.

As we climb, we chat in panting bursts. Mahabir, who despite his grubby outfit and self-effacing manner is something of a celebrity in these parts, tells me about his quest to transform his tribe's villages through the unlikely medium of Wi-Fi connectivity. His plan leapfrogs the traditional model of connectivity – improved roads followed by landline connection – and exploits the atmosphere instead.

Nangi village, home to around 800 people of the Pun tribe, has no telephone line or cellphone reception and consists mainly of subsistence vegetable farmers, yak herders and those who leave to seek their fortune as Gurkha soldiers. Mahabir was taught in the valley by retired soldiers who had never been to school themselves. They used wooden boards blackened with charcoal, writing with soft limestone from the local cliff. He first used a pen and paper in seventh grade (age 13) and a textbook in eighth grade, but even these rudimentary lessons were expensive for his father, a retired Gurkha from the British army, who had to sell all his land to pay for them. So Mahabir left school at 14 and worked for twelve years as a teacher, supporting his family and helping his brothers through school.

It took two years of writing daily application letters to universities and institutes in America before Mahabir was finally accepted with full scholarship on a degree course at the University of Nebraska in Kearney. 'I knew I wanted to change things in our villages. I wanted to bring an income in and better education and medical facilities,' he says. Twenty-odd years after arriving in America he returned to Nangi with his dream and an equally important folder of contacts.

It is dusk and we have climbed 2,500 metres higher by the time we are greeted with an excited rush of village children who present us with garlands of sweet-scented marigolds and escort us the last few yards to Nangi. Mahabir shows me to my home for the night: a small round wattle-and-daub hut with a stone roof. I make my introductions by candlelight and share a tasty curry of home-grown vegetables cooked on a smoky dung-fuelled stove by the school's science teacher, before falling into exhausted sleep.

In the morning Mahabir leads me through the small village, past women grinding masala spices and kneading dough for chapatis on wood and stone, past a circle of community leaders and elders sitting cross-legged and deep in discussion on the cold ground, to the school. Our short walk is sprinkled with smiles and greetings – everyone is glad to see Mahabir. He points out a rather grand, newly finished hut. 'Girls' composting toilet,' he says, taking me inside. He smiles and pats the internal wall approvingly, as I stand awkwardly either side of the hole, trying not to notice the smell of, um, toilets. 'The compost works very well for growing vegetables,' he says.

As nations develop, societies work in increasingly technical, mechanised and complex ways, and entirely new jobs emerge to support these industries, most of which demand literacy and numeracy. Globalisation favours those who speak international languages, and the people who will shape our lives in the Anthropocene will be those whose understanding and experience goes far beyond small village life, and those who are able to negotiate the accumulated learning, wisdom and knowledge generated by millions of global citizens via the hive mind of the worldwide web. It starts with school – with reading and writing and the self-confidence and awareness that sprout from those uniquely human skills. Done effectively, education is the bridge out of poverty, and educating girls is now recognised as a trans-formative development goal. Educated women marry on average four years later, have at least two fewer children and provide better health care for their families, for example. And not only is the income generated by an educated person higher, the average income of the community is also raised. 'When you educate a girl, you are educating a nation,' a 6-year-old girl once solemnly recited to me in Uganda. So what stops girls being educated? I've heard everything from worries she'll be too clever for marriage, to worries that she'll no longer be 'pure' or that she'll get pregnant. But the biggest factor is poverty – girls are first to be pulled out of class to work when money is tight. And as they get older, it's a question of toilets. Schools that lack clean private toilets – and many have no facilities at all – lose girls once they reach puberty and begin menstruating, and they also struggle to keep female teachers. Development comes down to the importance of toilets, like the one Mahabir showed me.

Nearby, there is a fenced-off vegetable patch with plastic sheeting over half of it. 'We started experimenting with growing vegetables later in the year, so that we would have some fresh greens all year round,' he explains. 'At first we needed the plastic sheeting as a kind of greenhouse, but the past three years, the warmer weather means the plants have grown perfectly well without it.'

At the far side of a rectangular patch of mud that serves as the football pitch and general assembly area for the Pun tribe is a row of low, wooden school huts. We walk over to them and Mahabir pulls back the door.

I'm not sure what I was expecting, but this gleaming array of computers and monitors flanking both long walls is a startling sight. Girls and boys, many barefooted, sit studiously working away, the only sound the clatter of keypads. 'You want to check your email?' Mahabir asks me, grinning at my surprise. The computer and Internet facilities here would be unusual in a school in London – here, they are astonishing.

In the Anthropocene, the world no longer needs to end at the village perimeter. Just as social development goals now include a right to electricity, it is no longer acceptable for people to be denied access to Tim Berners-Lee's brilliant toy. Through it, we are no longer a few individuals collaborating with a few more. We are a bigger more beautiful creature: the organism of humanity, 'Homo omnis'. We can communicate not just with remotely located people, but with everybody simultaneously – we're even attempting to speak to aliens located elsewhere in the universe.

The atmosphere of Earth has been lit up in the Anthropocene by the billions of invisible beams of our communicating devices. And it has happened in a remarkably short time. The first transatlantic telegraph was sent in 1858 by Queen Victoria to the US president James Buchanan, and by 1902, cables encircled the world across the Pacific and Atlantic oceans, connecting even faraway Australia. A century later, the phone in my pocket, beaming signals through the atmosphere, allows me to check weather reports and traffic cams, chat to my grandmother in Sydney, broadcast live to a television studio and pay my bills. Smartphones are getting so clever and human-responsive that they will soon be our personal dashboard, telling us how much exercise we do, monitoring our calorie and vitamin intake, our sleep pattern, heart rate, stress levels, cholesterol and so on. Further into the Anthropocene, some researchers believe, we will increasingly think of our smartphone as a partner – even in emotional terms.[1]

In East Africa, I saw how mobile money services, such as M-Pesa, are enabling phone users to transfer cash and pay for goods with the speed and convenience of an SMS text message.[2] A customer pays cash to his local corner-shop agent, who then tops up his mobile money account using a special kind of secure SMS. He can then transfer money to another person or pay for something by sending a text to the recipient's mobile phone account, which transfers the money straightaway. Even people without mobile money accounts can receive payments in the form of a text code, which can be exchanged for cash by their local corner-shop agent. For the millions of Africans who don't meet the criteria for a bank account, or who live too far from a branch, mobile money presents an opportunity to save securely for the first time. Kenya's M-Pesa is now used by over two-thirds of the adult population (more than 17 million people) to pay for everything from school fees to grocery and utility bills, taxi fares to airline tickets. It allows people in remote, rural areas to trade their wares in markets thousands of kilometres away, urban migrants to send money rapidly to their families in their home village, and for the government and aid agencies to distribute timely emergency cash to starving people living in slums.

Mobile phones aren't just bringing access to money, though. A Nepali peasant with a smartphone on Google, now has more access to information than the president of the United States did fifteen years ago.[3] In the Philippines most communications between the government and citizens take place by SMS texting. In Malaysia, flood warnings are sent by text message, and across the world, from quake-struck Haiti to the famines of East Africa, evacuations and relief for natural disasters are coordinated by SMS. In India, tribal groups are using mobile phones for 'citizen journalism', spreading information and giving voice to disenfranchised groups.

During the Arab Spring of 2011, citizens organised themselves to fight oppressive regimes using mobile phones, even bypassing government Internet and network clampdowns by accessing social media sites such as

Twitter and Facebook through apps and proxy servers. Across Africa, voting in national elections with a smartphone can cut election fraud by 60%.[4] In Afghanistan, the police receive their government salaries through mobile phone banking because it cuts down on fraud. Into the Anthropocene, mobile phones could even start to democratise markets. Enterprising individuals using crowd-funding tools like Kickstarter have a way to access markets that have been the exclusive domain of big corporations since the days of the East India Company.

It's no wonder that the way our species communicates globally has become fundamentally different in the Anthropocene. In 2012, the UN telecoms agency predicted that by 2014 cellphones will outnumber people on this planet, with 70% of new phone subscriptions coming from the developing world; by 2017, there will be over 10 billion networked mobile devices around the world, carrying 130 exabytes of data a year. Up until 2003, humans had created 5 billion gigabytes of digital information. In 2010, the same amount of information was created every two days; by 2013, it was every ten minutes. By 2020, 5 billion people are expected to have access to the Internet via mobile devices, an extent of connectivity that governments and development organisations couldn't have dreamed of just twenty years ago.[5]

This is made up of users in the poor world becoming part of the collective human conversation, where they can have influence beyond the restrictions of wealth, geography, caste, gender or other ways people have traditionally been stifled. The human species in the Anthropocene is a changed, networked animal. With this technology we have exceeded the limitations not just of our human body but of our hive – we've become a global community. The secret of our enormous planetary influence is our cooperation as a species, and our technological exploitation of our atmosphere-based communications system takes this cooperation to a new level. It is an accelerator of humanity's impact, and as such can be used to increase our destructive traits, or it could prove our salvation: a tool that enables development and human progress, showing us in real time how we are affecting other humans and the rest of the biosphere.

Mahabir well understood the opportunities that communications technology could bring to remote villages when he set about transforming Nangi.

At the end of a line of regular-looking computer hardware, I spot something a little different – a couple of wooden boxes housing circuit boards. 'Ah, these are the first computers that I built with recycled parts donated from old computers, because we couldn't afford new ones,' Mahabir explains. In 1997, Australian students donated the four adjacent computers, and the rest were sent over subsequent years by people in the US and Europe. Without a telephone line, no way of funding a satellite phone link, and with the country in the grip of insurgency, Mahabir realised that to bring twenty-first-century communications facilities to his village, he would have to be imaginative. In 2001, he wrote to a BBC World Service radio show asking for help in using the recently developed home Wi-Fi technology to connect his village to the Internet. Intrigued listeners emailed with advice and offers of assistance.

Backpacking volunteers from around the world smuggled in wireless equipment from the US and Britain after the Nepalese government banned its import and use during the insurgency, and suspicious Maoist rebels tried to destroy it. By 2003, with all the parts in place, Mahabir had linked Nangi to its nearest neighbour, Ramche, installed a solar-powered relay station using television antennae fixed to a tall tree on a mountain peak – and from there sent the signal more than twenty kilometres away to Pokhara which had a cable-optic connection to the capital Kathmandu. Nangi was on the Internet.

'I used a home Wi-Fi kit from America that was recommended for use within a radius of four metres,' he says. 'I emailed the company and told them that I had done twenty-two kilometres with it – I was hoping they might donate some equipment, but they didn't believe what I told them.'

One of the advantages of Wi-Fi is that it doesn't require costly and resource-intensive infrastructure – mile upon mile of cables and copper wires do not have to be laid over complicated terrain. Development in the Anthropocene need not be as dirty and as invasive of the natural world as it has been to date. More than forty other remote mountain villages (60,000 people) have now been networked and connected to the Internet by Mahabir and his stream of enthusiastic volunteers, and many more are in the pipeline. 'The villagers are now able to communicate with people in other villages and even with their family members abroad by email and using VoIP [Voice over Internet Protocol] phones,' he says. 'And they can talk for free within the village network using the local VoIP system.' I realise that Mahabir and the kids have been using VoIP for longer than me. Having always had access to a landline phone, I've only started using VoIP – Skype – within the past couple of years to make cheaper overseas calls, whereas the village adopted the technology a decade ago.

Teachers are a rare commodity in this part of the world, but the children are no longer being taught by barely literate soldiers. The Wi-Fi network means that a teacher, based here or even in Kathmandu, can teach classes across many villages, face to face with students via the monitors, answer questions and receive and mark homework. Mahabir's 'tele-teaching' network also allows the few good teachers in the region to train others. He is also developing an e-library of educational resources in Nepali that will be free to use, and working with the One-Laptop-One-Child organisation, which he hopes will provide laptops to children in the region. Thanks to Mahabir's work, a generation of children that would otherwise miss out on an education until the country gets around to training some teachers, instead have unprecedented opportunities to learn and discover a world beyond the dreams of their parents – it's a good definition of development.

But how does he power such a system so far from the grid? 'We built a hydropower generator in the stream at the bottom of the village,' he says. He wants to install another, bigger turbine when they can afford it, so that the entire village has power – at the moment, the precious electricity is reserved for the computers and server.

As we embark on another full day's climb up to Relay No. 1 with spare parts to fix a broken component there, we come across another of Mahabir's networked villages. Here, towering incongruously among the simple stone-roofed huts is a huge white satellite dish. 'We tried for years to get some sort of phone system here,' Mahabir explains. 'Then a few months ago, we got sent this dish by an NGO for a satellite phone and television. By then of course we had the wireless network Internet phone so we didn't need it. Anyway, it would be far too expensive to make calls on that.' Still, the villagers have erected it on the roof of the school where it sits like a totem to the useless. Nobody in this village has a television set, let alone the electricity to power one.

Mahabir was quick to realise that the connectivity had numerous other important applications. In the past year, the village has built a telemedicine and dentistry clinic, in which village midwives and nurses can talk using the webcam directly to doctors in a teaching hospital in Kathmandu. And nurses have been trained in reproductive medicine, childcare, wound and accident management, and basic dentistry.

The Wi-Fi has also improved livelihoods here, allowing yak farmers to talk to their families and dealers several days' walk away, and enabling people to sell everything from buffalo to home-made paper, jams and honey. Finding a sustainable income stream is key to keeping the other social development projects going, and Mahabir is betting on tourism. Many of the villages are located on beautiful but little-visited trekking routes in the Annapurna mountain range, and they have started advertising campsite facilities and trekking guide services for tourists. The local teenagers and adults know the routes well, and Mahabir is organising training for them, including some rudimentary English. And with the help of Western volunteers, the villages have come together to build their first tourist hostel just below Relay No. 1, on a remote stretch of the mountains. 'We are setting up secure credit-card transaction facilities using the Internet so that more tourists will come, which will help finance the education and health projects,' he says.

Mahabir, the one-man revolutionary, has still more plans to transform the village, including a yak cross-breeding farm. The warming rate here in the Himalayas is five times higher than the global average and it's forcing yak farmers into ever more remote and dangerous locations, because the thickly coated animals can't live below 3,000 metres. Mahabir is trying to cross the yaks with cows to produce a useful pack animal that is hardy, can live at lower elevations and also produces good milk. 'The first sixteen cows we took up there for breeding got taken by snow leopards, so we've had to guard them more carefully,' he says.

Cattle are vital for the villagers because they produce the dung that is used to fertilise the poor mountain soils, enabling their crops to grow. But the cattle need to eat and, ideally, something other than the villagers' crops. In another of his inspired projects, while all the villages around have been destroying their sparse forests for firewood, timber and agriculture, Mahabir has fostered a substantial nursery from which he plants about 15,000 trees a year in Nangi, and more than 40,000 a year in the surrounding area. It provides the villagers with firewood and the cattle with fodder. While many people in Nepal's hill villages suffer food shortages, the people of Nangi look well nourished – some of the teachers are even a little plump, which is hard to believe considering the slope they have to conquer just to get from their homes to the school each day.

As Mahabir calls up instructions to a guy at the top of a swaying tree who is grappling with tools to fix the relay equipment, I realise that development in these remote villages need not be hostage to a failed government. For much of the Holocene, people like the residents of Nangi would have been limited socially and economically by the geography of their village. A true visionary with determination like Mahabir can effect change village by village, incrementally constructing a web through the atmosphere. But how much faster and effective Nepal's development would be if it were backed by national coordinated programmes, good governance and regulated private industry with access to markets, as is occurring elsewhere.

The democratisation of online information, education, communication and markets, means that the Anthropocene has the potential to lead to a more equal global society – a 'flatter Earth', in which the dominance of Europe, the United States and a few other rich nations is challenged by eastern and southern rivals, such as China, Brazil and India. At the beginning of the Anthropocene, there are already signs that humanity's well-being is improving – there are now fewer 'failed states', more countries practising a degree of democracy, and a global reduction in poverty compared to just a few decades ago. In 2008, for the first time, the number and proportion of people living on less than \$1.25 a day fell in every continent, and the trend has continued since.[6] Our exploitation of the atmosphere with mobile and Internet communication – and the entrepreneurship that follows – has played a significant role in this trend.

We humans may have brilliantly exploited the atmosphere to communicate as an interconnected super-species, but at the same time we have been horribly indiscriminate about what else we've put up there.

The ugly face of humanity's atmospheric interference – the many gases we have emitted – now threatens to overwhelm our civilised and natural world. In as much as they have put off the next ice age, perhaps indefinitely, carbon gases have been to our advantage, but the greenhouse effect of those emissions is impacting every part of our planet from farmlands to deserts to oceans. We are dumping so many different pollutants into our aerial ocean that we are not only changing our climate and weather systems, we are also poisoning ourselves.

Air pollution is not new – ancient Rome was notorious for its smoky streets from wood and coal fires, and in 1306, King Edward I of England actually banned coal burning on penalty of death. Needless to say, it was ineffectual. It wasn't until a London smog of 1952 killed an estimated 4,000 people in just four days (and a further 8,000 in the weeks and months after) that a Clean Air Act forced Londoners to burn smokeless coke rather than coal. Similar acts in the 1950s and 60s transformed the atmospheres of New York and other cities in the then-developing world – Westerners still breathe a cocktail of pollutants, but they are mainly invisible carcinogens, like ozone and oxides of nitrogen, rather than soot and sulphurous emissions. However, the citizens of the current developing world are now experiencing similar conditions to those mid-century pea-soupers, but on a much greater scale.

The 'dark satanic mills' of Britain's Industrial Revolution, and the coal stations that powered them, blackened the skies and caused countless deaths – air pollution from European coal power stations continues to kill more than 22,000 people every year, scientists calculate.[7] Greenhouse gas emissions from smokestacks and exhaust pipes continue to pump out. But the visibly filthy skies of the past few centuries have cleared because of more stringent pollution controls that have forced factories and power plants to install scrubbing technologies and other practices. And, because the dirtiest manufacturing has moved out of western Europe.

China, where the vast bulk of dirty industry is now based, has an atmosphere so filthy that only 1% of the nation's urban population is breathing air considered clean by European Union standards, a 2007 World Bank study found, although much of the report was redacted by the Chinese, who feared social unrest.[8] Visiting Beijing in springtime, I was struck by the eerie absence of the sun. The pollution, which stung my eyes and throat, shielded the sun so thoroughly that although the cloudless days were light, the source of this light was impossible to see. And that was the city after its 'clean-up'. China has followed Europe's example and outsourced its dirty industry from the wealthy cities, such as Beijing and Shanghai, inland to rural and less developed central and western areas, as well as to poorer countries, such as Indonesia. In doing so, China's air quality will improve, just as Europe's has, while the atmosphere of developing countries will worsen. Ultimately, the only way to halt the filthy process is to clean up industries such as construction and manufacturing, by using the latest plant designs in the poorest countries – and also to vastly improve efficiencies and recycling. Eventually, many dirty industries will become obsolete and be replaced by newer ones, providing an opportunity to design pollution avoidance from the outset.

The atmosphere of the Anthropocene is remarkable not because it is infused with a range of chemicals and particulates – natural events such as volcanic eruptions can produce that – nor because it is the first time humans have produced their own emissions, but because it is the first time that humans have done so on the global scale of the planet's biggest natural events.

Our vast population – already more than 7 billion – is part of the cause. An increasing proportion of humanity relies on the goods, services and energy produced through dirty industrial activity. And on top of this, people are producing their own home-made pollution, the combined effect of which is turning the invisible air brown.

Kathmandu, Nepal's only real city, is inscrutable through thick acrid smog. Pollution and dust generated in the bowl-shaped valley is not easily dissipated – there is limited rainfall to wash it away and little wind. The air is so laden with dirt that the shopfronts all wear a film of grime. Lacking customers, underemployed staff spend their hours pointlessly dusting and sweeping – shifting the miasma, which resettles in seconds. Visibility is so poor that flights are often cancelled or delayed, but perhaps not as often as they should be: there were five plane crashes here in 2012–13 alone, killing more than sixty people. Most of the haze, around two-thirds, comes from burning biomass for cooking, like the dung fires I saw in Nangi, with fossil fuels supplying the remainder. Plumes of smoke rise from the wood and dung fires burning in every household and merge with the emissions of factory chimneys and agricultural clearance fires. In the streets, motorbikes and cars with badly tuned engines crawl bumper to bumper, farting out sooty black puffs. The resultant fug hangs over the region for the entire season and well into the spring. It spreads in mile-long clouds of brown haze for thousands of kilometres from the Yellow Sea to the Arabian coast, and its sooty particulates have even been discovered in the Arctic ice at Svalbard.[9] The brown cloud can be seen as a stain over Asia in satellite images, but it has become a highly emotive issue. When the pollution layer was at first called the Asian Brown Cloud, India complained, and so the United Nations Environment Programme renamed the haze 'Atmospheric Brown Cloud' during its follow-up study.

The warming effect of the brown haze adds to that of the greenhouse gases, and is hastening the retreat of the Hindu Kush glaciers and snow packs, including the snow that melts to drive Mahabir's turbines, partly because rising air temperatures are more pronounced in elevated regions. And black unburnt carbon in the filthy clouds is being deposited on these white peaks, reducing their reflectivity and exacerbating ice melt. It's a reason for the five-times-higher warming rate seen here, oranges growing higher on the slopes towards Nangi, and Mahabir's overheating yaks.

The dense blanket of pollution hovering shroud-like over Asia is also affecting monsoons and agricultural production. It's a complex relationship: the soot particles, ozone and water vapour in the haze absorb sunlight, heating the atmosphere, enhancing warming by as much as 50%; while at the same time, the sulphate particulates cool Earth's surface by shading.[10] The shading aerosols alter the global hydrological cycle because less sunlight hits the sea, so there is less evaporation and therefore less rainfall. A decrease of about 40% in the monsoon rainfall over the northern half of India to Afghanistan, and a north–south shift in rainfall patterns in eastern China has already been observed, reducing crop yields.[11] The brown cloud also reduces the efficiency of precipitation because it makes it harder for large raindrops to form, leading to drought-like conditions. Its effects can be felt all the way to Australia.

Any changes to rainfall immediately affect plant growth, including agriculture, which is further impacted by particulates deposited on the plant leaves. These reduce the amount of light that gets through, limiting photosynthetic activity, and can also cause acid damage to the plant cells. And elevated levels of ground-level ozone reduce the yield of certain crops including wheat and legumes – one study estimates that the brown clouds have already reduced Indian rice yields by 25%.[12]

The haze is also a health hazard, linked to an increase in acute respiratory infections, particularly in children; lung cancer; adverse pregnancy outcomes; heart attacks and other conditions. In India alone, it is estimated that nearly 2 million people die each year from conditions related to the brown cloud. Household solid fuel used for cooking, a major source of brown haze, kills more people each year than malaria – wood smoke alone kills more than 1.5 million a year, mostly women and children. By 2050, urban air pollution is set to be a bigger killer than dirty water and poor sanitation, according to the OECD, with 3.6 million premature deaths a year predicted mainly in China and India. In many places where we live, we have turned the planet's vital fresh air into a poisonous dangerous vapour.

But the atmosphere of the Anthropocene may not be permanently stained. The good news is that dealing with brown haze presents a much easier and faster solution to regional – and global – warming than acting on carbon emissions. The rewards of decreasing soot emissions from biomass combustion could be sizeable and rapid – because unlike carbon dioxide that persists for a hundred years, the brown cloud pollutants only hang in the atmosphere for a matter of days. Slashing black carbon alone could bring an astonishing 40–50% reduction in global warming, and that's in addition to all the health benefits.[13]

This would mean more stringent vehicle emissions standards, which many developing countries from China to India are already starting to enforce. And it means fundamental changes to how people cook and warm their homes in places like Kathmandu. Supplying a $30 clean-cook stove to the 500 million households who cook with open fires could be done for just $15 billion. Distributing clean-cook stoves would not only reduce the pollution burden, but also free girls and women from gathering and carrying firewood, a task that endangers their health, puts them at risk of rape and prevents them from going to school. A big win all round.

The pretty village of Phakding, on the busy trekking route to Everest base camp, is immediately different to others I have visited, but it takes me a few hours to realise what is missing. I twig when I spot Ani, the owner of my guest house, cooking on an electric stove: the smoke, which has been a continual presence of every street and house since I entered the country, is strangely absent. I breathe in deeply through my nose to check. There is the faintest whiff of a fire burning on a far-off ridge but otherwise nothing smoky. I can smell the garlic and chillied onions cooking in Ani's kitchen, the sweet dank smell of drying vegetation mingling with fragrant marijuana and flowers, and the earthy sourness of buffalo manure.

I chow down on vegetable stew with chapatis, and press Ani for details of this smoke-free nirvana. It's an NGO initiative that has seen the whole village embrace micro-hydropower, using a flowing stream to drive a turbine that produces electricity for the village. 'We were using more and more kerosene and diesel because of all the tourists that were coming, and it was getting so expensive,' she says. Initially, when the micro-hydro was suggested, there was scepticism from some of the villagers, who thought that converting to the new energy source would cost too much. After all, those who couldn't afford the increasingly costly diesel simply chopped wood in the forest and supplemented their fires with dung and other waste. But as more and more tourists came, and the village's energy requirements rose, the sparse, high-altitude forest was becoming frighteningly depleted. Arguments were brewing between those who wanted to keep the forest for buffalo feed, and those who needed it for fuel.

Micro-hydro has changed all that, Ani says. 'It powers everything for free – all the lights, the music, cooking. And the kitchen is so much cleaner without the black soot,' she smiles. The forest has even started growing back. The other villagers I speak to agree that the new energy supply has been an improvement, even if one important controversy remains: whether chapati can ever be properly tasty when cooked on anything other than an open fire.

The glacier that feeds the village's micro-hydro sits high in the mountains, a dirty apostrophe surrounded by snowy peaks on Nepal's northern border with Tibet. 'We used to play on the glacier as children, and it came right down to the monastery,' says Ringin Laama, a local yak herder. 'But now it's about two kilometres further back.' The mountainside beneath the glacier is heavily scarred, clearly marking its original extent on the rocks. Every year, the glacier retreats much further, according to Laama. 'I think it will be completely gone in ten years' time,' he says. 'Strange to think.'

Curved walls made of large boulders have been constructed along the mountain face above each small cluster of houses. The ground here used to be permanently iced all year, and the melting has exposed fissures in the rocks. The Himalayan mountain range – the 'Abode of Snows' in Sanskrit – is one of the youngest on Earth and it's still highly energetic with active seismic and tectonic processes. Landslides, already common here, are becoming more frequent in the warming climate, with deadly consequences. Laama traces a large scratch in the mountain opposite with his finger. It extends down through the rubble of former houses and terminates in a huge boulder.

Elsewhere, landslides are damming up rivers, resulting in a build-up of water. The risk then is of a catastrophic release: a flash flood that occurs rapidly, with little warning and transports vast amounts of water and debris at high velocity. Every year, hundreds of people die – in 2002, flash floods and landslides killed 427 people in Nepal and caused $2.7 million-worth of damage.

Glacier-fed rivers are set to swell over the next decade as melting accelerates. But it will be a short-lived sufficiency – once the glaciers have gone, there will be no more meltwater. It means that the turbines that power Ani's kitchen will lie still. Perhaps anticipating this, some of the villagers have invested in a rival cooking source, supported by another NGO. I visit the home of Alok Shrestha, a man so self-sufficient he makes methane cooking gas from his household's sewage, animal manure and other waste, which are shovelled into a biodigester under his house. The biodigester is a large feeding box for bacteria, which metabolise the nutritious waste and produce useful methane in the process. Shrestha taps this for his cooking stove and has plenty left over to power lighting and a small generator that recharges batteries.

I have visited several homes around the world where people are making their own biogas from a range of wastes, including one in Peru that was powered by guinea-pig poo, and all fuel efficient cooking stoves with smoke-less flames. Firing up a stove for brief, if regular, intervals is one thing; powering equipment that requires a continual source of electricity is another. For example, the hydroturbine on which the hopes and dreams of Mahabir's Pun tribe rest is powered by a stream fed by snowmelt higher up. No snow means no meltwater and no power, and levels have been diminishing in the once-deep stream. When the water fails, the development of that entire region, which outstrips its neighbours on every measurable scale from literacy to health, will be at the mercy of the government for grid provision.

The government has just a few years to build enough reservoirs to trap the melting waters and power the country towards twenty-first-century development, before lights go out. It is hard to see how that will happen. Drought conditions are already causing blackouts as the nation's hydropower struggles with lack of water and poor energy infrastructure. As throughout the developing world, most of the blackouts are planned load-shedding intervals, where the government tries to manage the poor supply by rationing electricity among different regions in turn. The load-shedding is seriously impacting Nepal's fledgling electric car industry, its 700 Safa Tempos (three-wheeled electric passenger vans), which cleanly ply Kathmandu's streets, charging at thirty-two stations and transporting around 100,000 people a day. More than $8 million is invested in the industry, and its five manufacturing companies employ thousands. But power outages have pushed the industry to the verge of collapse – people (90% of whom are self-employed single women) who purchased a Safa on loan schemes backed by NGOs can't afford to pay the instalments because they can't use their cars and have to take taxis instead. If the industry does go under, it means more filthy cars on the streets – which means more carbon in the atmosphere and more brown haze. Which means more warming.

The atmosphere of the Anthropocene is quite unlike any atmosphere the planet has known, and the effects of humanity's impacts on our aerial ocean will leave their mark on the world for millennia to come. The chemicals we are introducing into the air will find their way into the oceans, rocks and living parts of our world. Corals and trees are ingesting a different ratio of isotopes (forms) of carbon from the one they took in during the Holocene, because they are now absorbing carbon dioxide emitted from fossil fuels. But despite this, our changes to the atmosphere itself are as transient or permanent as we make them. If, tomorrow, we stopped releasing gases into the atmosphere, switched off our millions of signalling devices, ceased all aerial transport, within a matter of years most of our atmosphere would return to Holocene-like conditions. Within a few centuries, even the carbon dioxide levels would drop down to pre-industrial norms.

We are of course not going to stop releasing chemicals tomorrow, though. The amount of almost every pollutant humans emit is increasing, and will continue to change the climate. Despite being told repeatedly that our climate is changing, by numerous scientists, agencies and media, it is nevertheless hard to fully appreciate how significant this is – we may intellectually believe the change, but to emotionally understand and realise what it means is a different matter.

Our climate is one of humankind's most powerful reference points. It fundamentally describes where and how we live, our culture, environment and even our place in time. Climate is what defines the Holocene geological epoch. The climate is what determines biodiversity regionally and globally, it decides the ecology, the hydrology (how much water there is) and weather. It determines, for example, whether malaria is more likely and whether wheat can grow.

Living in a changed climate is like living in a different world – or rather, our world in a different geological time. Instead of the climatically stable Holocene, we are entering the uncharted territory of anthropogenic, or human-caused, climate change. We will feel its effects even as we try to insulate ourselves from the changes and adapt to them. Climate change will increasingly affect our food production, the integrity of our cities, energy production, global politics, and the way we interact with other people and other species.

Human behaviour and the way nations develop will decide the atmospheric conditions as the Anthropocene unfurls. And the atmosphere of the Anthropocene will also play a deciding role in how humans develop. Poverty-stricken, backward Nepal is teetering on the edge of a bright new future: it has the promise of a functioning democracy, and the benefits of a decade of NGO experimentation in projects from micro-hydro to clean-cook stoves, even while it battles the legacy of atmospheric warming from industrialisation elsewhere. Whichever way it teeters, the children of Nangi have in many ways escaped the destiny of most of their contemporaries. Because they are already a part of the great human conversation, theirs will be a more assured Anthropocene, with opportunities to overcome the limitations imposed by geography.

The effects of our transformation of the atmosphere will crop up frequently in this book – I'll show how they are intertwined with other changes we're making to our planet. Many of humanity's solutions rest on our innovative technological invasion of the skies, innovations just like Mahabir Pun's in Nangi. As Google's Larry Brilliant said about world-changing: 'It starts with ordinary people. Ordinary people do extraordinary things, and then we lionise them. We make heroes out of them. And that's a problem, because it makes other ordinary people look at these heroes and think that they can't achieve the same things. But that path is open to everybody. Anybody at any time.'

2

MOUNTAINS

When early Earth's shifting sludge of molten rock hardened into a crust, some 4.3 billion years ago, our planet got its first land cover. As this crust cooled atop the churning lava, like the skin on a pan of custard, it contracted and twisted, leaving some parts higher and thicker than others: the first mountains were created.

But the earth is never still. This solid ground, this apparently permanent land is imperceptibly shifting. Over the billions of years, the planet's bubbling-custard turbulence has shattered the crust into scattered fragments, or sent its many islands crashing together to form huge continents. Several of these vast, coalesced supercontinents have merged and separated, each time creating an entirely reconfigured planet – the most recent and best known of these is Pangaea ('all-earth'), which formed 300 million years ago before splitting up. Each time the drifting plates of crust have crashed into each other, the collision sends one edge above the other, creating a mountain.

When the plates have drifted apart, the wrinkles are pulled smoother and the mountains sink. So some mountains on the planet are on the up, like the still-growing Himalayas, and others are dropping.

Mountains can also appear suddenly, through the process that gave birth to the original land masses, volcanism. Every so often, a bubble of molten rock spews out of a crack between the plates, and piles up on the surface of the land or seabed, creating a new mountain. Kilimanjaro in Tanzania and Kinabalu in Borneo appeared in this way.

When mountains first arise, they are sharp and jagged like the Himalayas, but over time, they round down as their surfaces erode, crumbling gradually away through glacial or river flows, or in the sudden slips of a landslide. Exposure to the air, wind, sun, munching microorganisms, and rain, also wear away mountain rocks in a process called 'weathering', which locks away carbon dioxide from the air as it reacts with dissolved minerals in the rocks.

Mountains are unusual because they have multiple climates. Usually, in order to experience a different temperature or weather system, you need to travel hundreds of kilometres north or south, but heading just a hundred metres up or down a mountain can have a similar effect. That's because the air molecules in our atmosphere are not evenly spread out – they are far denser in a blanket near the ground. The higher you rise, the fewer molecules there are in the air to radiate back the sun's heat, so it's colder. That's why mountains – even those on the equator, like Mount Kenya – have snow and ice on the top. When the altitude is combined with latitude, such as in Antarctica, the entire range can be hidden under deep snow and ice.

This change in climate produces interesting island-like ecosystems, with some species found only on specific altitude-defined spots on certain mountains where they have been isolated from their cousins for thousands of years.

The relative coldness up a mountain also generates the world's largest source of fresh water, as moisture-laden air condenses against the peaks, relinquishing its load as rain or snow. And much of this remains where it accumulates, in mountain caches. For perspective, consider this: 97.5% of the

Earth's water is ocean or salty groundwater; of the remainder, just 0.01% is held in the clouds and rain, 0.08% is in all the world's lakes, rivers and wetlands, 0.75% is in groundwater, while 1.66% is in glaciers and snow-packs. That means well over half of the world's fresh water is stored in glaciers.

That's how it was in the Holocene. But, as humans heat the planet in the Anthropocene, mountains are changing dramatically. Species seeking their usual living temperatures are climbing up the slopes at an average rate of twelve metres a decade, driven by global warming – they will need to migrate an estimated one hundred metres upslope for every 0.5°C of temperature rise.[1] Clearly this is easier for animals than plants, but they too are moving – European vascular plants have shifted an average 2.7 metres uphill in the past seven years, for example.[2] Other species have been marooned on peaks because their previous ranges have been occupied by human settlements or become farmland. Once at the top, though, there's nowhere further to go, and thousands of species risk extinction, particularly those living on tropical mountains. Conservationists are now carrying out 'managed relocation' of species to places with a more a suitable climate, in the hope of saving them. In some cases, climate migrations have been beneficial, allowing people to cultivate fruit and vegetables higher up a mountain. But elsewhere, disease-carrying mosquitoes are now living at higher elevations where people with no immunity are being infected, with deadly consequences.

Of most concern to humans is the loss of ice. On average, glaciers have become fourteen metres thinner since 1970.[3] Almost every single glacier looked at by the World Glacier Monitoring Service since 2000 has retreated, including all of the European ones, most of the tropical ones from the Himalayas through Africa to the Andes, and south to the mountains of New Zealand.

Humans have always worshipped mountains as gods or the home of gods, they have built temples high on their slopes and made pilgrimages to their peaks. In the Anthropocene, we are subjugating these geological marvels, turning them darker, drier and more homogeneous, stripping them of their unique flora and fauna, even decapitating them for the minerals inside. We *are changing the shapes of Earth's mountains – when they lose their protective snow, the exposed parts crumble away. Mountains including the Matterhorn in Switzerland are disintegrating.*

Humans are still enthralled by the highest peaks, but the trails they make now to these heavenly reaches are more likely to be marked by litter than prayer stones. Nevertheless, even as we desecrate them, we depend more than ever on mountains for our fresh water.

In this chapter, I look at how the changes we are making to our mountains are affecting the people who live on them, and how humans are trying to recreate Holocene mountain conditions in the Anthropocene.

Temples rise above mudbrick houses and castles. Prayer flags flutter from every roof and woollen cloaked men and women with brightly coloured waist-sashes stand in the street gossiping. I'm in northern India's remote Trans-Himalaya, in the the ancient kingdom of Ladakh. Consisting entirely of mountains, this, the highest inhabited region on Earth, is home to an 80% Tantric Buddhist population, settled by pilgrims and traders travelling the ancient Silk Route between Tibet and India or Iran.

In Stakmo village, farmers are preparing for harvest. Two men sit, chatting in sing-song tones against a dry-stone wall, sharpening their scythes with a blade pressed between their knees. An old woman with long, ribbon-woven plaits leads a donkey and calf over to her whitewashed mud-brick house. In the field behind, a yak munches alfalfa and swishes its horse-like tail. Bright marigolds nod crazily around a single apricot tree, and there is a barely perceptible tinkle of wind chimes. The scene feels timeless.

And yet, much has changed, the villagers tell me. 'By mid-September, we would wake up with completely frozen moustaches,' says Tashi, a 76-year-old farmer, who wears a woollen hat and large, pink-tinted sunglasses. Buddhist prayer beads hang around his neck and his dark sun-crinkled face is cleanish-shaven. I'm above 4,000 metres (13,000 feet), but nevertheless, it is not cold enough to freeze moustaches – from the clear, cloudless sky, the sun beams down intensely, as it does for more than 300 days of the year, and it's burning my European face. The roof of the world is heating up.

Wedged between Pakistan, Afghanistan and China (or, more accurately, Tibet), Ladakh was a latecomer to the Indian state of Jammu and Kashmir and remains a contested territory. At night the Indian and Pakistani border patrols take potshots at each other; the Chinese come and paint Indian rocks red and the Indians respond by painting Chinese boulders green. But Stakmo feels very far from such nationalistic posturing. The villagers are more concerned with the ancient and essential task of coaxing food from the mountains' mustard-coloured desert soils. Global warming is at play here, disrupting Ladakhi lives more effectively than any international land squabbles. Humanity is heating the region so fast that the mountains are changing colour before people's eyes: from white to tobacco as the glaciers disappear. And with them, Ladakh's only reliable water source.

In his lifetime, Tashi has seen two big glaciers vanish in this valley alone – he points out their locations to me and I see only the same dry, sandy and pink rocks that fill the eye between valley and sky. Just the very top peaks are white, and the only glaciers I spot are at least 5,500 metres up. The warmer climate is not the villagers' biggest concern, though. In fact, they rather like not having to be confined to their houses so early in the year. The most painful change is the new unpredictability in precipitation. A catastrophic pattern is developing for moisture at the wrong time of year.

This part of the Trans-Himalaya, after the Rohtang Pass, is in a precipitation shadow. It's drier than the Sahara, with no rain for months on end. The westerly winds don't reach here and the monsoon from the east doesn't surmount the high pass. The snows used to arrive after October and build during the winter. Then, in March, the snowpack would begin melting, providing vital and timely irrigation for the sowing of the area's barley crop. But the past decade has seen a gradual reduction in snowfall – the winters of 2012–13 were particularly dry, with serious consequences. Harvests are failing, drinking water is trucked in by government tanker, traditional self-sustained communities are breaking up as young people migrate to the cities or plains for work. Worse, when the precipitation does come, it arrives as rain during the harvest season, ruining what few crops the villagers have in the fields, before disappearing to lower elevations.

The changes have also reduced the sparse natural vegetation. Thupstan, another Stakmo farmer, tells me that he used to let his livestock roam the mountains eating wild grasses. Now, though, he has to give over some of his valuable planting fields to grow alfalfa for the yaks and goats. And wild creatures too are feeling the pinch. Last week, Thupstan found fifty ibex in his field eating his vegetables. The ibex attract wolves, which eat his goats. And the ibex and wild yaks are destroying his stone walls, knocking them over and blocking the irrigation channels with boulders.

Nearby, in Ladakh's principal town of Leh, the rainfall is also causing problems. This is a region that had never experienced rain before the past decade. Houses are built from unfired mud bricks, and the roofs are sticks bound with mud and yak dung, with a hole to let the fire smoke escape. These homes are built for snow, which covers and insulates them in winter. The new rain is literally washing them away. The more wealthy people are starting to concrete their houses.

A little rain at the end of summer is no substitute for snowfall during winter. It is quickly drained away in the rivers and there is little replenishment of the groundwater. The springs in Leh have been dry for months now, as more and more people pump out the groundwater. Wells sit dry and unused. Some of this is the result of a booming tourism industry. New hotels and guest houses are fitted with flushing toilets, twenty-four-hour showers and washing machines. It is completely unsustainable. The guest house I stay in has a traditional Ladakhi composting toilet, but few others still do. My landlady nevertheless gets all the water we use from a hundred feet below ground with a generator-driven electric pump.

The explosion of tourism, and a ruinous government policy on subsidies, are partly to blame for the new water shortages, but much of the problem is down to climate change, owing to rising global greenhouse gas emissions and the regionally produced brown haze. Data are pretty much impossible to obtain – the military jealously guards all such information – but the locals are unanimous in their conclusion: the glaciers here are disappearing – and fast.

People here are especially vulnerable, because they have such a brief summer. If farmers don't sow their single crop of barley, peas or wheat in March, it won't have time to mature for harvesting in September before the harsh winter sets in, with temperatures that drop below -30°C. The problem is, the glaciers that remain are too high at above 5,000 metres, and don't fill the irrigation channels until June – too late for the sowing season.

Meanwhile, demands on the region's water sources are growing. In the Anthropocene humans are infiltrating the farthest reaches of the planet in large numbers. Even those places that once only supported a few families are now host to regular flights from rich cities, bringing swarms of temporary migrants and their lifestyle expectations. As in so many places in the developing world, tourism has brought new wealth and possibilities to the people of Leh, but without water, this fertile patch in a mountain desert will return to dust.

The Himalayas is the largest area covered by glaciers and permafrost outside of the polar regions, containing 35,000 square kilometres of glaciers and an ice reserve of 3,700 cubic kilometres. Glacial melting is accelerating every year, with current annual retreat rates of seventy metres for some glaciers. Mountains are changing dramatically and so fast that we can use recently produced Google Earth images to watch the white bits shrink. Melting rates have already exceeded those predicted by the international community of climate scientists (IPCC) – they expect 70% of the region's glaciers to disappear the same way as Stakmo's by the end of the century. Meltwater from small mountain glaciers alone already accounts for 40% of current global sea-level rise, and is predicted to add at least 12 centimetres to sea levels by 2100.[4]

As mountain glaciers shrink, lakes are created from the meltwater, hemmed in by the moraine of rocks and debris that are left by the retreating ice. As with dams caused by landslides, when glacial lake dams breach, the outburst of millions of tonnes of water can cause devastating flash floods. Satellite images have revealed around 9,000 glacial lakes in the region, of which more than 200 have been identified as potentially dangerous, capable of breaching in a so-called glacial lake outburst flood (GLOF) at any time. Many only appeared during the past half-century, and have been growing steadily since. People have always lived in danger zones, such as the slopes of a volcano or the banks of a flooding river – often to exploit the richer soils – but the risk has been from an 'act of God', a natural event. In the Anthropocene, we are increasingly producing our own danger zones and imposing them on communities that have no traditional preparedness for such events. The Imja glacial lake in Nepal, for example, is now two kilometres long and nearly a hundred metres deep. When it bursts, the deluge of water could reach sixty kilometres away, swamping homes and fields with rubble up to fifteen metres thick, leading to the loss of the land for a generation. Hydrologists in Peru are now building tunnels to drain away an Andean glacial lake, after another ruptured its banks killing 10,000 people – tapping controlled outflows from these lakes could provide much-needed irrigation and hydropower for local populations.

Over the Holocene, glaciers around the world fluctuated with changes in temperature or precipitation, but in recent decades, glacier melt has increased rapidly and become global. In the Anthropocene, humans are steering natural processes and cycles. We have pushed the planet beyond its natural state and crippled its capacity to self-regulate or to bring back the glaciers we've melted. And as glaciers melt into lakes, overall loss of humanity's precious water increases because water evaporates faster than ice. Snow and glacial melt in the Himalayas are the source of up to 50% of the water in ten of Asia's major rivers, including the Ganges, Brahmaputra, Indus, Yellow, Mekong and Irrawaddy. These are the most populous river basins on Earth, with more than 1.3 billion people depending on them for everything from agriculture to fishing. In the Anthropocene, we will either have to discover ways to live without the fresh water that mountain glaciers store for us, or replace Earth's largest fresh-water reserve with massive concrete reservoirs. The former option would certainly jeopardise the lives of millions of people, not to mention eroding wetlands and other ecosystems. The latter is urgent – globally glaciers have, on average, lost almost a quarter of their mass in the past sixty years. Around the world, reservoir-building is already under way by governments, albeit on a woefully inadequate scale. China is constructing fifty-nine reservoirs to catch and store meltwater from its declining glaciers in Xinjiang province, a high-altitude desert, but it is incredibly expensive and logistically impossible to replicate everywhere the vast acres of ice with concrete tubs of water. Ideally, reservoirs would be built underground to reduce evaporative loss, but that only makes them more expensive. Nevertheless, the Anthropocene will surely experience a vast programme of reservoir-building.

However, there is another option. I've come to Ladakh to meet a remarkable man, who is taking on the global-warming challenge – and winning. The person they call The Glacierman dresses not unlike Clark Kent: beige sweater, sensible lace-up shoes. But, unlike comic-book superheroes, he's 74 years old. He invites me into his beautiful family home in the small village of Skarra, near Leh, where, in a bid to peddle the 'regular guy' image, he presents his charming wife and daughter and we drink the peculiar local butter-chai and snack on almonds and apricots.

Chewang Norphel is no ordinary villager. He makes glaciers.

Norphel takes a barren, high-altitude desert and turns it into a field of ice that supplies perfectly timed irrigation juice to some of the world's poorest farmers. So far, he has built ten artificial glaciers since he retired as a government engineer in 1995, and their waters sustain some 10,000 people. It's hard to describe what an extraordinary feat this is. In one of the most climate-change-ravaged regions, Norphel, a one-man geoengineer, has effectively conjured up water, doubling agriculture yields as assuredly as if he'd swooped in wearing a cape and stopped global warming in its tracks.

In a display of energy and enthusiasm that is exhausting to witness, Norphel skips across the boulder-strewn landscape above Tashi's village. He wants to show me his latest artificial glacier design, but I'm finding it tricky simply to breathe the thin air, 4,000 metres up. He carries a small backpack: tonight he will sleep in a tent 1,000 metres higher up, at temperatures that dip to -10°C, so as to continue his work in the morning. 'When it is very cold and very difficult work, I have to remain focused. All I can think about is making the most successful glacier,' he says.

Engineer, hydrologist, glaciologist, backyard enthusiast, Norphel has created his own field of expertise using scientific principles and training but the tools of an uneducated peasant. 'What he has achieved in such circumstances, in remote parts of this mountainous desert, is remarkable,' says Pankaj Chandon, coordinator of the WWF's Indian High Altitude Wetlands Conservation Programme, based in Leh, who has followed Norphel's progress over the past decade. 'It is testament to his sheer force of character. But also, he has come up with a unique, innovative idea that provides water when it is needed. It is a fantastic adaptation technology for the climate changes that we are experiencing in this region.'

Norphel has always been focused. As a child, born into a farming family in Leh, he would take every opportunity while out minding the herd to scratch times tables and algebraic equations into the dirt with a stick. 'I begged my father to let me go to school, and he agreed as long as I also kept up my farming duties. So I would rise at 4 a.m. and take the cows and goats for grazing before school. After school, I would rush home to help in the fields.'

In the 1940s, when Norphel was growing up, there was just one school in Leh, which taught in Urdu (not Ladakhi), and only up to primary standard. As the youngest of three brothers, Norphel would ordinarily have been sent to live in a Buddhist monastery, in part to reduce the family costs as his father would not have been able to afford secondary school. So, at 10 years old, Norphel simply ran away, travelling more than 400 kilometres to go to school in Srinagar, Kashmir. The only poor boy at his school, he paid for his education by cooking and cleaning for his teachers.

Graduating in science at the college in Srinagar, Norphel knew two things: he loved mathematics and science, and he wanted to help the farmers he'd seen struggling so hard during his early childhood. One of his heroes at this time was his father's cousin, who had been to London and returned to Leh as Ladakh's first engineer, built the town's airport and the Leh to Srinigar road.

There was no university in the state at that time, so Norphel travelled south to Lucknow for a civil engineering degree, this time being taught in Hindi. He loved the rigour of the subject and the practical application of physics and material science. 'You can really make a difference with engineering. You can solve people's problems quickly and in a way that they can see,' he says. 'Simple projects, such as a well-placed bridge of good design, can make things so very much easier for people who have to otherwise walk a day or more out of their way.'

For Norphel, the whole point of his training has been towards using his knowledge in the service of his fellow Ladakhis. Engineering is a vocation for him in the same way that medicine might be to a doctor. Like Mahabir, his determination and effectiveness is transforming lives.

As soon as he qualified, Norphel returned to Leh to join his father's cousin in the government of Ladakh's rural development department as a civil engineer. It was an exciting time to be working, but also extremely challenging. There was hardly a road or bridge when he started in 1960, and everything had to be built by hand. 'We had no funds even for pick-axes and shovels – people were using animal horns to dig in some places – but roads were the most urgent requirement,' he says. 'People had to travel everywhere by pony, and where the tracks were very poor, all the ponies would have to be unloaded so that the animals could cross the broken part, and then reloaded after. Journeys that took weeks can now be made in hours.'

Over the next thirty-five years, the enthusiastic, raven-haired engineer became a familiar sight in Ladakh's villages. Unlike other government experts on secondment from elsewhere in India, Norphel became known for his genuine engagement in the villagers' problems, and they grew to trust him. More than 90% of the population were subsistence farmers, living and working in tightly knit communities. There was no money around – everything was done through trade and cooperation – and when Norphel needed labour for his projects, people willingly came forward. 'There is scarcely a village in Ladakh where I have not made a road, a culvert, a bridge, a school building, an irrigation system, or a zing [small water-storage tank fed by glacial meltwater],' he says.

He approached each problem scientifically, experimenting by altering the variables until he arrived at a satisfactory solution – and always remembering that his designs had to be sustainable, using locally available materials. For example, he built a number of canals where instead of using an expensive cement lining that cracked during winter, he allowed weeds to grow and thicken, their roots naturally sealing the canal lining.

By the time of his retirement in 1995, priorities were shifting. Road-building was still important, but Ladakhis were becoming aware of a far more serious problem – one that threatened their livelihoods. 'Every village

I visited it would be the same thing: water scarcity. Glaciers were vanishing and streams were disappearing,' Norphel says. 'People would ask me to bring them water. Their irrigation systems were drying up and their harvests were failing. The government was starting to bring in grain rations.' Norphel was determined to do something. 'Water is the most precious commodity here. People are fighting each other for it: in the irrigation season, even brother and sister or father and son are fighting over water. It is against our tradition and our Buddhist teachings, but people are desperate. Peace depends on water.'

Inspiration came within a hundred metres of his house, one bitingly cold winter morning. 'I saw water gushing from a pipe and was thinking what a shame it is that so much abundant water is wasted during winter-time – the taps are left open to stop the water freezing in the pipes and bursting them,' he says. 'Then I noticed that on its route to the stream, the water crossed a small wooded field, where it was collecting in pools. Where the trees provided shade, it was freezing into ice patches. By early March, the ice patches melted.'

Norphel realised that if he could somehow copy this on a much larger scale, he would have a way of storing up this winter water in an artificial glacier that would melt at just the right time for crop sowing and irrigation. It was a beautifully simple concept but achieving it would be fraught with difficulties. 'People laughed when I first presented the idea and asked for funds,' he says. 'Officials and villagers were sceptical, "What crazy man are you? How can anyone make a glacier?" I was told.' But Norphel soldiered on. He held meetings with village elders, and explained the concept. Gradually, his relentless enthusiasm caught on.

He had no equipment; no altimeter or GPS reader, not even a bulldozer. Perhaps just as challenging was the societal change that had occurred over the past decade. As water scarcity increased and the roads brought in trucks with government-subsidised grains, many villagers had left their fields to find work in the new tourist industry in Leh or elsewhere in India.

The old trade and cooperation system was abandoned in favour of a new money-based economy. 'The attitude completely changed: If I wanted any of the villagers to repair a canal or help build a new glacier, I had to pay them. No one does anything for free any more,' Norphel says.

Norphel's ingenious idea was to divert the winter 'waste' water from its course down the mountain, along regularly placed stone embankments that would slow it down and allow it to spread and trickle across a large surface depression a few hundred metres from the village. Here, the slowed water would freeze and pack into a glacier. He shows me the glacier site, pointing out the path he sends the water on until the rocky valley starts to take shape in my mind and I see how the glacier forms. Siting is everything. The glacial area is shaded by a mountain face during the winter months, when the sun is weak and low. By March, when the sun rises high enough, the thick ice sheet begins melting, pours into a water tanker and through a sluice gate to the farmers' irrigation canals. The meltwater also helps recharge the groundwater aquifer. This water is so precious that during the irrigation season a man has to sleep by the sluice gate to guard against water theft.

The rocks beneath the ice sheet channel mountain breezes, cooling the sheet further. And Norphel points out second and third artificial glacier sites at successively higher elevations. 'By the time this lowest one has melted, the middle one will start to melt,' he says. 'Then the highest one and, finally, the natural glacier at the top of the mountain.' He is grinning now, and I can't help joining him: it's such a great invention.

He built his first artificial glacier with very little help, above the village of Phuktse. It was an immediate success, supplying an extra thirty days' water to irrigation channels. 'When people saw the benefits of the artificial glacier, they started helping me and we stretched the length of the glacier to two kilometres,' Norphel says.

'It was like a miracle, people quickly started to cultivate more land and started planting willow and poplar trees between their fields,' says Phuktse farmer Skarma Dawa. 'This technology is very good because it works and it is simple and there's very little maintenance required.' They are built using local labour and materials at a fraction of the cost of a cement water reservoir.

Norphel has built nine glaciers since. They average 250 metres long by a hundred metres wide, which he believes provide some 6 million gallons (23,000 cubic metres) of water each, although there has been no accurate analysis to date, and the undulating ground makes it difficult to guess the volume of ice in each glacier.

His work has earned him recognition from those he has helped – 'I have a shelf-full of home-brewed beers and a trunk of *khatag* [ceremonial silk scarves given by Buddhists]', he says – but there has been little interest from the scientific world. 'I am trying to collect data on how and where the glacier forms best, and which parts precipitate first and why, so that I can improve on them and people can use the technique elsewhere. I lack scientific equipment. I have only my own observations.'

Norphel says he has already had some interest in his glaciers from NGOs working in Afghanistan and Turkmenistan. 'In some areas, reservoirs are a much more practical solution.' 'But in terms of water storage and release at the irrigation season, you can't beat artificial glaciers.'

Creating glaciers from scratch, while pretty awesome, is not entirely new. People may have been doing it as far back as the twelfth century. Legend has it that when Genghis Khan and his Mongol warriors set their sights on what is now northern Pakistan, the local villagers thwarted their advance by growing glaciers that blocked the mountain passes. The practice of 'glacier grafting' is known to go back centuries in Baltistan, an ethnic-Tibetan region of the Pakistani Karakoram mountains, where people rely entirely on glacier meltwater for irrigation. The technique, which has an important ceremonial component, involves dragging ice from a so-called 'female' glacier (a fast-moving, surging glacier), and from a 'male' glacier (a slow-moving, rock-strewn glacier) and planting them at a specific site – usually on a northern side of the mountain, above 4,500 metres. The ice is planted on top of boulders and interspersed with gourds, which burst and freeze, after which the 'mated' ice is insulated with a covering of cloth and sawdust. Similar techniques, whereby ice is planted above air-channelling boulders, are practised in Argentina, often in shaded areas like caves. So-called 'rock glaciers', in which the ice sheet forms from frozen snowfall, produces a purer meltwater that is often preferred to 'true glaciers', which contain gathered material and melt into a milky run-off.

Recreating the glaciers lost to human-produced global warming is an imaginative solution to the very real problems that alpine villagers face and, perhaps because its impact is local and grounded, this particular geoengineering technique seems uncontroversial. In richer countries, such as Switzerland, ski-resort managers already spend thousands of dollars on artificial snow and ice, and on preserving the cold stuff where it still exists, using giant reflective blankets. In 2008, a German professor constructed fifteen-metre-high, three-metre-wide wind-catching screens to channel and trap the cool winds flowing down the mountain on to the Rhône glacier in Switzerland. If it works over time, he intends to repeat the process on other glaciers.

Norphel doesn't have access to technological blankets or screens – he can't even analyse the efficacy of his glaciers with real accuracy. But, while I am with him, he receives his first scientific visitor, Adina Racoviteanu, a geography graduate at INSTAAR, University of Colorado at Boulder, who is passing through en route to her glacier field stations further east. She offers to make him a topography map of the artificial glacier site using her handheld GPS monitor. Norphel's eyes light up in boyish excitement. 'That would be wonderful,' he says, and the pair spend a happy couple of hours taking readings across the site, achieving in that time what would take Norphel weeks to do with his tape measure and plumbline. The device they are using, on loan from Racoviteanu's institute, costs $3,000, but before she leaves for her 'real' glacier, she tells Norphel that models are available for as little as $300. 'If I could get one of those, how much easier this would be,' he sighs.

Norphel reckons that more than seventy-five other Ladakhi villages are in suitable locations for his artificial glaciers, each of which provides an estimated 6 million gallons of water a year, but lack of funding is holding him back.

We make our way down the valley to Stakmo, stopping by Tashi's house. 'This man is a hero,' Tashi tells me. 'The artificial glacier he has given us allows me to grow potatoes, which need to be planted earlier in the season, and my harvest is so much bigger. I grow tomatoes and other vegetables as well now. I make three times as much income.' The new irrigation has allowed him to take advantage of the warmer conditions. Climate change is ushering in novel farming opportunities across the region (where water is available), and a whole range of vegetables – aubergines, apples, sweet peppers, watermelon – are now growing at high altitudes where previously farmers struggled to sow barley among the ice and desert.

Tashi's new fortune may be short-lived, though. Climate change is also altering the precipitation patterns here, bringing less snowfall during wintertime when it is needed to contribute to the artificial glaciers. 'These glaciers are not magic formations,' Norphel says. 'They need to build over winter.'

The artificial glaciers are not a long-term solution to the climate-change problems people are facing here, but they do provide a breathing space for some of the poorest people to adapt. Further into the Anthropocene, this entire region is likely to become uninhabitable for the majority of farmers currently living here. Norphel is giving these Buddhist people a few more precious years in the homes, landscapes and communities that their ancestors have prepared them for, where their traditional songs and tales are set, and their language is understood.

*   *   *

Norphel is not the only independent agent defying humanity's onslaught by geoengineering glaciers on Earth. In the Peruvian Andes, people are attempting to literally paint a mountain back to whiteness.

Licapa, at 4,200 metres up, is a village of people whose livelihoods are based around farming alpaca, the domesticated camelid of South America. This part of Peru, a hundred kilometres west of the town of Ayacucho, is one of the poorest in the country and was hit particularly hard in the 1980s and 90s during a decade of terrorism led by the Shining Path, a violent Maoist guerrilla group based here.

When I arrive at the village, below the Chalon Sombrero mountain peak, women are doing laundry in a small, grubby-looking pond, while a group of men repair one of the stone houses. These highlanders, who speak Quechua, the ancient language of the Incas, have spent the past twenty years trying to rebuild their broken communities, homes and lives, helped by various government schemes. But climate change is against them.

Salamon Parco, a young father, is fighting a personal battle against global warming. When he was the same age as his 5-year-old son Wilmer, he tells me, a river ran through the valley, watering the alpaca pastures. Women never used to wash in the pond, he says. But the glacier at Chalon Sombrero, 5,000 metres above sea level, disappeared completely twenty years ago, and with it the water. All that is left is a black rocky summit above a rocky channel where a river once ran.

Like Stakmo, it doesn't often rain in Licapa, and what rain does fall is confined to January and February. The rest of the year, the high-alpine grasslands rely on glacial meltwater and, in its absence, turn yellow and die. More than 1,000 people have already left the village because they couldn't feed their families, migrating to shanty settlements around Peru's capital, Lima. Parco, with a wife and three young children, has considered the same. 'But my home is here. What would I do in the city? I need to try and make it work here first,' he says. Instead, Parco and his friend Geronimo Torres are spending every morning painting the black mountain white, hoping to bring back the glacier on which 900 people depend.

They began painting the mountain in May and by my visit in September, they have turned three hectares of black rock white. The remarkable experiment, backed by $200,000 in prize money from a 2009 World Bank climate-change adaptation competition, was conceived of by the rather eccentric and unlikeable Peruvian entrepreneur Eduardo Gold. The money, which Gold tells me he has yet to receive, is going to be used to build a factory in Licapa to produce lime paint for whitening the mountain.

The experiment is based on the principle that a black body absorbs more heat than a white one. By increasing the reflectivity of the black rocks using white paint, the mountain should be cold enough to retain the ice that forms on it – and eventually a glacier will be made. That is the hope, anyway.

There are plenty of sceptics, including Peru's environment minister, who has said the money could be better spent on other climate-mitigation projects. And Gold, who has no scientific qualifications, has also been judged with some suspicion by agencies and public bodies.

Nevertheless, Parco tells me he is already seeing results. 'In the daytime, the painted surface is 5°C, whereas the black rock is 20°C. And at night, the white surface falls to -5°C,' he says. Ice has begun forming on the painted rocks overnight, although it has melted by 10.30 a.m.

The plan is to dig a small reservoir of water above the painted section and pump water up to it using a wind turbine, which would then be released during the night in a slow trickle over the paint, where they hope it would freeze. In time, the ice would build up and the process would be self-sustaining, because glacial conditions would be there: 'Cold generates cold,' Gold says. Parco and Torres have seventy hectares to paint in total, a job they had thought would take two years. They started the job with two other men, but fifteen days later, this other pair dropped out because there was no money to pay them.

'We are still painting the mountain because it works, and because we have no choice,' Parco says. 'If there is no glacier, then there is no water for us and we will have to move away.'

I ask Lonny Thompson, an Ohio University glaciologist, who has been studying Peru's glaciers for the past forty years, what he thinks of the idea. Painting the mountain may have some success in the short term in a local area, he tells me, but it is not feasible over greater regions. 'Nobody is going to paint the entire Andean chain white,' he says. What is needed now that the glaciers are disappearing, is man-made water storage to replace them. 'This means a big programme of building dams and reservoirs, which is tricky in such an earthquake-prone zone, but necessary.'

It is unlikely that Parco's remote village of Licapa will be prioritised for new waterworks in the next few years. Painting a mountain white may, however, produce enough ice in the next couple of years to buy the villagers time to adapt to a different livelihood.

Attempts to whiten the Earth's surface to increase its reflectivity are also being considered on a far greater scale elsewhere. Methods to reduce the amount of the sun's energy that heats the planet – called 'solar radiation management' – have the potential to rapidly counteract regional or even global warming. With global temperatures almost certain to exceed the 2°C of warming this century that scientists consider 'safe' for humanity, quick-cooling options look increasingly attractive. Deflecting the sun's energy back into space would do nothing to counteract the ocean acidification effect of atmospheric carbon dioxide – which I'll come to later – but it is a valuable way of buying time while societies decarbonise, adapt to warmer conditions and new climates, and figure out an effective and efficient way of removing the carbon dioxide we've put into the atmosphere.

Some engineers are proposing erecting Earth-orbiting space mirrors that would bounce sunlight back out before it even enters our atmosphere. Terrestrial proposals include whitewashing roofs of houses and public buildings, planting lighter and more reflective crops (perhaps using genetically modified varieties), and covering deserts or ocean in reflective materials. With enough paint and willpower, strategic mountaintops could be sprayed white from the air, perhaps.

Since the 1980s, Almería in southern Spain has developed the greatest concentration of greenhouses in the world, covering 26,000 hectares. Dubbed the 'sea of plastic', this Anthropocene landscape is remarkable not only because Europe's driest desert now produces millions of tonnes of fruit and vegetables, but also because the greenhouses reflect so much sunlight back into the atmosphere that they are actually cooling the province. While temperatures in the rest of Spain have climbed faster than the world average, meteorological observatories located in the plastic expanse have shown a decline of 0.3°C per decade.[5] It turns out that the plastic acts like a mirror, reflecting sunlight back into the atmosphere before it can reach and heat up the ground. At a local level, the plastic greenhouses offset the global greenhouse effect.

More controversially, filling the atmosphere with airborne particulates would also cool the Earth by shading it from sunlight. This happens naturally after a volcano erupts, such as Pinatubo in 1991, which lowered global temperatures by more than half a degree for two years after the event.[6] In the deep past, supervolcano eruptions threw the planet into ice ages, causing mass extinctions. The same effect, albeit on a far lesser scale, can also be seen on shipping lanes because ships typically burn heavy fuels that issue smoky sulphurous emissions that seed measurably colder airstreams across the oceans. Sulphur particles – like the ones found in the Asian brown haze pollution – have a shading effect that reduces the amount of sunlight reaching the Earth's surface by as much as 15%, and are masking humanity's warming by as much as 80%.

No one would suggest increasing industrial air pollution as a solution to global warming, of course; instead, engineers are looking at other atmospheric ways of reflecting the sun's energy. Injecting salt particles into low-level, stratocumulus clouds could make them brighter and more reflective, providing a local cooling effect. Clouds of perfect reflectivity and altitude for this occur naturally in semi-permanent sheets in three places: off Chile–Peru, Namibia–Angola and North America. Jim Haywood, an expert on brown haze at the UK Met Office Hadley Centre for Climate Research, has carried out modelling studies on the Chile–Peru stratocumulus clouds, which indicate that modifying the cloud would produce a significant cooling effect. However, modelling also reveals other potential consequences: spraying the West African clouds appears to reduce rainfall over the Amazon, which would be bad; spraying salt into the clouds off Chile seems to increase rainfall over arid Australia, which could be useful. British engineer and inventor Stephen Salter, who pioneered wave-energy technology, thinks that rather than focusing on the three main clouds, it would be better to monitor oceanic warming zones in strategic locations around the world and spray salty nuclei into the air above them to create cooling reflective clouds and help modify dangerous weather. 'Typhoons like Haiyan [which devastated the Philippines in November 2013] could have been significantly dampened before it made landfall by spraying the cloud,' Salter says. He has designed a fleet of floating towers that could pump seaspray into clouds to whiten them, and he calculates the total cost of deployment on a scale that reduced global temperatures by half a degree per year would be less than that of a single international climate conference.

Meanwhile, other scientists are looking at what effect sulphur particles would have if they were pumped tens of kilometres into the stratosphere, mimicking a volcanic eruption but on a lesser, though more long-term, scale. Their experiments, restricted so far to laboratories, are looking at how reflective of the sun's heat different particles would be, and whether the cooling particles would have any unwanted side effects, such as destroying the ozone layer.

The changes humans have made to mountains in the Anthropocene have largely been driven by temperature or precipitation – both of which we still have the power to alter, whether by reducing our greenhouse gas emissions or by reducing the sun's power to heat us. The Anthropocene could become a time of more nuanced climate change, where temperature and precipitation are modulated to humanity's needs, where weather is planned. It's an extraordinary idea.

Humans have always modified their environment – it is only because of our exquisitely adapted brain that we thrive worldwide, essentially by insulating ourselves against the natural environment. Whether we raise average global temperatures by two, four or even six degrees, enterprising members of our species will no doubt adapt successfully. Given centuries, the entire human population would likely manage to live comfortably under such conditions. The problem is that the rate at which we are warming the atmosphere is too rapid for humans to adapt. Nevertheless, the concept of artificially cooling the atmosphere is highly controversial, given the atmosphere is a global commons. Perhaps it is because the intent is so explicit; although humans are artificially warming the atmosphere with greenhouse gas emissions, the intent behind burning fossil fuels has always been to produce energy, not to warm the planet. Some argue that even research in this area should be banned because it implies intent to carry out the practice; others say that it draws effort away from climate-change mitigation – from decarbonising our energy production. But surely freedom of inquiry should be preserved – carrying out scientific research into whether something would work and what its consequences might be does not make a scientist an advocate for deployment, and there are scientific questions that need to be answered, such as the impact on rainfall, and whether or not it would even be technologically possible, before society can start to decide whether or not to deploy such techniques.

This new, Anthropocene field of geoengineering is a fascinating area of research and the scientists working in it are some of the most remarkable and thoughtful people I've encountered. It is eerily reminiscent of the atomic research carried out in the 1940s – today's geoengineers are working at the cutting edge in an exciting, entirely new science, spending their lives making discoveries and designing amazingly powerful technologies that they fervently hope will never be deployed. Each speaks sincerely about the risks involved with deployment, and reiterates that slashing emissions of warming gases is the best way of dealing with the problem. There may be very real and serious consequences of using reflectors. Models indicate that cooling the northern hemisphere (to slow catastrophic ice melt in the Arctic) would slash rainfall in poor countries in the southern tropics; one solution would be to simultaneously deploy reflective coolants above the southern hemisphere. Another problem is the so-called 'termination' issue. In order to keep global temperatures down and counter-act future warming, these reflectors would need to be sprayed continuously and perhaps in greater amounts. If the spraying programme was terminated, global temperatures would rise very suddenly by whole degrees – this would be much more dangerous for humanity than gradual global warming caused by our rise in emissions.

However, the predicament that humans currently find themselves in – facing catastrophic climate change and yet increasingly reliant on the fuels that exacerbate the problem – means that planetary cooling tech-niques are likely to be seriously considered. It is, after all, what humans have always done when presented with a challenge – engineer a way through it. Proponents of solar radiation management, such as Paul Crutzen, point out that they are simply mimicking volcanic activity and have the potential to quickly and cheaply reverse the warming effect of a doubled carbon dioxide concentration. And, since we already know what happens when a volcano erupts, it is arguably one of the safest methods – safer than the impacts of global warming, for example. The technique could be used continually to avoid catastrophic climate change, or in times of severe drought or heatwaves, hopefully under the auspices of an internationally agreed treaty. And, carefully deployed, such techniques could potentially maintain or bring back glaciers over entire mountain ranges. But, while the world ponders the feasibility, ethics and wisdom of global-scale cooling, Norphel and other architects of the mountains are devising practical and effective local solutions to a warming planet.

3

RIVERS

*W*ater that falls from the sky, leaks from a lake, rises from a spring or melts out of a glacier will make its way from land to ocean in a river. These freshwater conduits are life-giving. They irrigate forests and meadows, create wetlands and deltas, transport nutrients and sediments, and nurture entire ecosystems – self-contained watery worlds of animals, plants and microorganisms.

And yet, rivers owe their very existence to life forms. For billions of years, terrestrial fresh water flowed to the oceans in vastly broad, shallow sheets – like floods – across the planet's hard barren surface. It took the arrival on land of root-based plants, around 420 million years ago, for rivers to evolve. Plant roots weakened the surface of rocks, making them crumble, producing mud that eroded channels through which water then coursed. The plants' strong root systems then further channelled the water, strengthening the muddy banks and creating a deeper, meandering path that we would recognise as a *river. As these proto-rivers flooded and receded, sediments were periodically dumped, creating deeper, richer soils where huge woody plants took root. Forests diversified and enhanced the channels, helping produce the vital network of wetlands that exists today.*

The world's rivers drain nearly 75% of Earth's land surface, from the icy polar regions to the steamy tropics. Although they hold only about 0.0001% of the world's water (and less than one-third of all fresh water), rivers are a key part of the global hydrological cycle, describing the geography of accessible fresh water for plants and animals. Hundreds of thousands of species have evolved to rely for all or part of their lifecycles on freshwater bodies, from the trickling source of a mountain stream, the torrential violence of waterfalls and rapids, the calm deep waters that flow between riparian forests, to the wide-open, sediment-flooded wetlands and deltas.

The rejuvenating flow of Earth's powerful arteries appears timeless. Dinosaurs lived and died on riverbanks that exist today. They fed on fish, some of which – such as sturgeon and gars, and the arowana and arapaima of the Amazon – still swim the rivers. These ancient creatures are joined by a vast array of newer fish, reptiles, mammals, birds and insects that contribute to making freshwater ecosystems some of the world's most diverse.

Humans have also been a part of this ecosystem, relying almost exclusively on rivers and lakes for drinking, bathing, food, waste disposal and transport. Fresh water is so essential to humans that you can map society by it. River deltas have proved fertile culturally as well as agriculturally – the great religions have rivers as gods, such as the Ganges, or as important parts of their narrative, such as Moses' journey on the Nile, or Jesus' baptism in the Jordan. Historically, cities were built in fertile river valleys and at river mouths. Agricultural run-off of sediment, water and nutrients created rich coastal deltas that could support greater food production. This and the good maritime and river connections for trade and transport made deltas ideal places to live. Human civilisation was born on a riverbank. The Tigris, Euphrates, Indus and Nile spawned humankind's first great experiments in urban living, shifting our species firmly on to the trajectory we have followed ever since.

In the Anthropocene, humanity is draining the world's rivers and other sources of fresh water. Climate change has altered the global water cycle from Holocene norms, intensifying evaporation and precipitation. There is now greater flooding, worsening droughts and a general loss of predictability that makes planning more difficult for people trying to adapt. Greater water extractions by humans for agriculture, industry and energy mean that many rivers have dried up, while others are now too polluted to use.

Of all the ways we've engineered Earth in the Anthropocene, little rivals our audacious planetary-wide replumbing of the world's waterways. We have straightened and diverted them, buried them, dammed them and drained them for irrigation, filled or emptied them of fish, dug their beds for construction materials, used their flows to drive turbines for hydropower, and even created our own canals to bridge cities and divide continents. Humans now control more than two-thirds of the world's fresh water. We've captured so much water that we've redistributed its weight around the world and the globe now spins a fraction slower.

In the past century, we have drained half of the world's wetlands, built 48,000 large dams and diverted most of the world's large rivers – only 12% still run freely now from source to sea.[1] Major rivers, such as Mexico's Rio Grande, China's Yellow River and Australia's Murray River, frequently no longer reach the sea. Inland seas, such as the Aral or Lake Chad, have dried up with the use for agriculture of their feeder rivers. Dams, diversions and extractions are preventing river sediments from flowing downstream to maintain deltas against erosion. That, combined with groundwater extraction in coastal cities, is causing two-thirds of major deltas to sink.[2] Around a quarter of people rely on groundwater that is being extracted faster than it is being replenished, more than 800 million have no safe drinking water at all, while four in five of us live in a place where the water supply is at risk.[3] It's not just humans that get thirsty, all species need water, and ecosystems around the world are suffering from a decline in supply – 30% of freshwater species are now endangered, the highest proportion of any ecosystem.[4]

And our demands on the planet's rivers are growing ever greater. Despite all the ways we've cosseted ourselves in the Anthropocene against the hazards and unpredictability of the natural world, we remain desperately dependent on rivers for drinking, for agriculture, for fisheries and, increasingly, for our energy.

In many ways, the Anthropocene will be shaped by how we manage our rivers – it's already proving emotive and political territory in different parts of the world.

The southernmost habitable region of our planet is an untamed wilderness of glaciers and mountain peaks, subantarctic forest and scrub desert, volcanoes and turquoise lakes. Home to condors, puma and blue whales, Patagonia is the tail end of the Americas, one of the last accessible nowhere lands on Earth and the jumping-off point for Antarctica. It contains the Southern Ice Field, the world's most important reserve of fresh water after Antarctica and Greenland. Forests of Antarctic beech bear testimony to a time when these lands were part of the warm Gondwana supercontinent, while frequent earthquakes and fiery volcanoes are evidence of continuing geological movement.

This extraordinary landscape is the focus of a bitter international battle over plans to build a cluster of hydroelectric dams on three of Chile's mightiest rivers. It is an issue so divisive, it is tearing apart some of the country's biggest corporations and risks unseating the president. I went there in the hope of learning whether in the Anthropocene, people will choose the promise of cheap electricity and associated economic development, or the preservation of a natural wilderness that few will ever visit.

Deep in the heart of Patagonia, I find the churning glacial blue of Chile's most voluminous river, the Baker. The river cuts fast and furious through the mountains here, a tumultuous pulse that roars in defiance of any checks or dams. It is wild, wet and loud. Rainbows flash in the spray and the rocky banks glisten in the wash. I fancy I hear bird call drowned out in the background, but I cannot be sure. Beyond the surging river, all is still and silent. Two large hydroelectric dams are planned for the Baker – a cacophony of concrete, steel and asphalt to tap the river's immense natural power for city-dwelling humans thousands of kilometres away. I try to picture my surroundings submerged beneath a reservoir, imagine an access road, and the bustle of a large human workforce in this remote location . . . and fail.

Patagonia is desolate. Its very emptiness is part of its charm and it has always drawn those escaping society. Butch Cassidy and his gang sought refuge in these wastelands, as did a variety of Soviet defectors, Welsh Christians and English fortune-seekers. It's a rainless, inhospitable place, incessantly windy and freezing cold. But the skies are vast, the austral light is incredible and the rocky desert is a palette of extraordinary colours.

Above me soar enormous black condors seeking death in the dusty grasses. In the absence of trees, other birds of prey sit on the road, leaving it until the last moment before taking off in front of my wheels. I see guanacos (a type of llama) and a black and white skunk as I drive for hour upon hour. The desert rolls on in valleys created by glaciers, and the expanse is strewn with large incongruous boulders, called 'erratics' because they don't belong to this rocky ground. They were dragged down from the mountains and dumped here by ancient glaciers.

After almost a day of driving through the vast bleakness, I am relieved to see a sign of human habitation. Poplar trees struggle to protect a lone house, standing small-windowed against the chill. It is one of the region's estancias – ranches set up by those who pioneered these lands nearly a century ago, clearing the rocks and natives to profit from a booming wool industry. Further on, I come across a bubbling carpet of sheep, herded by gauchos on horseback and their dogs. The scene looks timeless and utterly natural, and yet it's an illusion – sheep were only introduced to the country in the late nineteenth century.

Eventually, I reach Coyhaique, capital of the Aysén region of Patagonia. In rolling hills at the foot of a basalt massif, it is a compact, ordered town whose folk live mainly by fishing and cattle ranching. For many, life is not dramatically different from that experienced by the pioneers; but graffiti around town reveals a new disquiet. *'Patagonia Sin Represas!'* ('Patagonia Without Dams!') is perhaps the politest of the slogans, reflecting anger over plans for the Baker River, and for several further dams on the untamed Cuervo and Pascua rivers.

Like most hydropower, the energy would be produced by building up a head of stored water behind a dam that can be released in powerful bursts past turbines to generate electricity. To convert the relatively shallow river flows of the Baker and Pascua into deep energy stores means creating reservoirs, and together the dams would flood 6,000 hectares of land. But the biggest opposition is reserved for the accompanying electrical transmission line. Some 6,000 pylons, towering as much as eighty-five metres high, will transport the direct current 2,450 kilometres north to Santiago, Chile's capital, and on to the energy-hungry mines in the desert beyond. The electricity line alone will require one of the world's biggest clear-cuttings, a 120-metre-wide corridor through ancient forests, fragmenting the ecosystems en route.

Critics say the dams, pylons and transmission line would destroy forever a true wilderness for short-term energy gains; proponents argue that hydro-electricity is a clean source of energy, that Chile needs the 3,500 megawatts per year of power to meet its development goals of becoming South America's first developed nation by 2018 and, lacking oil or coal reserves, has no viable alternative.

The country needs to triple its installed capacity by 2025 to meet its energy requirements – currently half the nation's electricity comes from hydropower and the other half from imported fossil fuels. The Patagonian dams alone could generate one-third of Chile's electricity, which surely makes the sacrifice of a few remote rivers a small price to pay for such bountiful energy? However, those opposing the dams argue that Chile has one of the longest coastlines in the world, 7,000 kilometres, which is ideal for wind, wave or tidal energy projects; 10% of the world's volcanoes, so plenty of geothermal potential; one of the world's strongest solar-energy zones in the Atacama Desert, all of which could provide energy more locally to its point of use without disturbing pristine Patagonia. Confused, I consult an energy expert.

Claudio Zaror, a chemical engineer and energy advisor to the government, is a quietly spoken, slight man who endured the worst of Pinochet's brutal dictatorship. Kidnapped by the secret police in his twenties, tortured and held for years in a cell sixty centimetres square, he was a lucky 'disappeared' – he survived. After decades of deprivation, he wants the lives of ordinary Chileans to improve and has little patience for what he considers unnecessary sentimentality over some remote rivers. For Claudio, the issue is clear-cut: 'We are a developing nation with nearly 20% of the population living in extreme poverty – I want that number to reduce and we need energy for that.'

Every year the country needs an extra 500 megawatt installed capacity – another 8% annually – because of an increase in population, consumption and industrial growth, Claudio says. 'If it doesn't come from the Patagonian dams, it will have to come from a fossil-fuel source with all the carbon emissions that entails, because other renewable energy sources are prohibitively expensive. Environmentally and economically, hydropower is our only feasible option,' he says.

Climate change is bringing new urgency to the situation, Claudio adds, because droughts are becoming more frequent and severe across the central region, where most of the nation's hydropower comes from. 'During the 2008–9 drought, less than 15% of the base-load was met by hydro and we had to import diesel for the power plants at $118 per barrel.' Meanwhile, with 92% of the country's glaciers retreating owing to climate change, the glacier melt means there would be strong river flow in Patagonia for the short to medium term.

However, the dams issue remains divisive, not just in Aysén, but across the nation. Surveys show more than half the population is against the proposed dams, but the gap is small enough to cause the government problems whichever decision it takes. The controversy has spilled internationally, as people from around the world claim a stake in this globally unique wilderness. Even the venerable *New York Times* has waded in with an editorial calling for the dam proposals to be scrapped.

Over the last century, humans have built the equivalent of a dam a day – the vast majority since 1950. Two-thirds of the world's major rivers have now been disrupted with more than 50,000 large dams – there are more than 85,000 dams in the US alone, stoppering large and small rivers and in most cases utterly transforming natural flow. The most famous of these, the Hoover Dam, constructed in the 1930s, is largely responsible for killing the mighty Colorado River before it reaches the ocean. With a 40% increase in global hydropower predicted by 2050, humanity in the Anthropocene has designs on most major rivers, and controversy over how to use these planetary arteries is only set to increase. In Europe and North America, most of the hydropower potential has now been exploited – indeed some dams are being removed and rivers 'renaturalised'. In Africa, Asia and South America, though, hundreds of hydrodams are being planned to provide essential electricity for some of the world's poorest people, and in some of the most ecologically important environments from Patagonia to the Amazon to the Congo. However, the people receiving the new electricity are usually not the same people faced with losing their environment, livelihoods and homes.

Globally, hydropower is an attractive low-carbon source of energy, which unlike solar or wind can produce a continual supply of electricity no matter the weather. Around 20% of electricity worldwide already comes from hydropower. The infrastructure can be relatively inexpensive, is 80–90% efficient and comes with its own battery: the reservoir. This is such a good device that solar- and wind-power generators are increasingly looking to use 'pumped hydro' to store their surplus electricity, using it to pump water high up to a reservoir for release when the sun doesn't shine or there's no wind. Dammed reservoirs are, of course, also a great way of storing water for drought and modulating damaging floods.

Yet dams, for all their attractive benefits, are also saddled with a lot of negative impacts. Creating the reservoir often involves flooding fertile land, sometimes displacing thousands of people. Communities may lose their land, houses and culturally important sites such as ancestral burial grounds or a landscape that carries strong meaning for them. If the area to be flooded is not adequately cleared of vegetation, methane – a greenhouse gas with twenty-five times the warming potential (over a century) of carbon dioxide – will be released from rotting material. Nearly a quarter of humanity's methane emissions come from big dams. Stalling a river in a reservoir allows some vegetation to build up and rot anyway, which can poison the water for fish.

The weight of so much water can also cause earthquakes, leading to dam breaches and catastrophic loss of life. In other instances, heavy rains can leave dam managers with the dilemma of whether to try to hold the waters back but risk bursting the expensive dam walls, or releasing the flow, risking flooding people downstream. In many cases, the flow has been released with devastating consequences for lives and livelihoods. In this way, dams that are intended to mitigate flooding can actually result in more serious sudden deluges.

Downstream of a dam, natural seasonal floods that revitalise wetlands and fertilise paddy fields cease. The flow may be so reduced that farmers cannot irrigate their fields and streams are no longer navigable. Migratory fish are often prevented from reaching their spawning areas, other fish have reduced vegetation and may be split from their breeding populations, affecting ecosystems and fisheries. And dams are a barrier to sediment flows. Instead of being flushed downriver, sediments get backed up against the dam walls, which damages the turbines and causes the reservoir level to increase over time. Downstream, though, the effects of losing nutrient-rich sediments is far more problematic. The fertility of the entire system can be impacted, with soils lost during seasonal rains not being replaced. The upstream–downstream demands often straddle national borders leading to conflict over precious water.

However, the economic benefits can be huge, and the new reservoir can be a haven for wildlife, such as birds, or provide new fisheries and much-needed irrigation security. The Aswan Dam on Egypt's Nile, for example, was highly controversial when it was built in the 1960s. Yet for all the environmental damage it wreaked on the downstream river system, you'd be hard pushed to find an Egyptian that advocates its removal – the dam has been an outstanding economic success, bringing improved harvests from better irrigation despite drought conditions, as well as hydropower and flood protection worth billions of dollars. Even there though, the river is contested. In 2013, Ethiopia voted to strip Egypt of its right to the majority of the Nile, the source of nearly all of Egypt's water, paving the way for construction of a massive hydropower dam on the Sudanese border.

As with many development opportunities, hydrodams can be constructed in a way that is minimally socially and environmentally invasive, or in the cheapest way to make the fastest possible return on investment.

In August 2008, HidroAysén, the company behind five of the Patagonian dams, submitted its environmental-impact assessment to the Chilean environmental agency for regulatory approval. The thirty-two government departments charged with assessing the report found it so wanting that the company was instructed to address more than 3,000 comments and given a nine-month extension to do so. In October 2009, HidroAysén submitted its response in a 5,000-page addenda document that once again fell short of public-agency requirements, more than half of the departments making highly critical comments. These included criticisms that the environmental-impact assessment contained a lack of data on seismic risks in an area known for earthquakes and volcanoes, total unaccountability of glacial lake outburst floods, lack of data on impacts to key natural habitats in and outside of national parks, local communities, biosphere reserves of global importance, wetlands and aquifers.

However, with the two powerful companies behind the projects – HidroAysén and XSTRATA – enjoying the backing of Chile's right-wing president, Sebastian Piñera, plans for the dams have rolled on undaunted. Over the past few years, the plans have been approved, appealed, thrown out, reappealed, reapproved, and so on.

I'm in Coyhaique to track down architect and keen mountaineer Peter Hartmann, who is the regional head of CODEFF (Chilean Friends of the Earth), one of the main groups opposing the dams. We meet in a busy café, where Peter picks me out immediately, unfurling his long thin body to lollop over, arms outstretched in greeting. In the booming flat tones of the partially deaf, he invites me to stay with him. 'We'll chat there,' he says, waving away the waitress's offer of menus.

We set off along a dirt track that worsens as it rises up the mountains above the town until I am thrown crazily from side to side as the truck negotiates increasingly deep ruts. But it's worth the journey. Peter's home turns out to be a beautifully crafted wooden house with a grass roof and windows that glow in the sun, reflecting the city below and an incredible rock colossus above. Over a shared maté – the South American herbal infusion sipped hot through a metal straw from a small gourd – Peter describes his many objections to the dam projects, from his concerns about ecosystem destruction to the visual disturbance of having intrusive power lines running through the unspoiled mountains and valleys he cherishes. 'You are used to seeing electricity pylons and cables everywhere where you live, so you don't realise how ugly they are and how they ruin a landscape,' he says. 'But here, we don't have big artificial structures interrupting the natural view. It's one of the last places on Earth like this and I want to keep it that way.'

Peter, who is helping lead the offensive against the dams, is a charming, generous and endlessly fascinating host. He provides me with an enter-taining history of the area while he prepares a stew for our lunch from indigenous vegetables including tasty lilac-coloured potatoes, known locally as '*meca de gato*' ('cat shit') for their undeniably similar shape. Peter is one of Chile's very few vegetarians.

His passion for the area comes from decades of intimate knowledge. He has climbed its mountains and rock faces, navigated its freezing rapids and defended it against polluting industry and unsightly infrastructure. Large areas of valley and slopes around Coyhaique still bear the scars of the first European inhabitants. Arriving just decades ago, fleeing conflict in Argentina or seeking grazing lands from elsewhere in Chile, these cattle herders caused unimaginable destruction in their quest for arable soils beneath the jungles of Aysén. Lacking the resources or the will to clear the forests by axe, they simply set fire to it. Some 4 million hectares – half of Patagonia's forests – were destroyed in the 1940s and 50s in the world's biggest fires, which raged uncontrollably, fuelled by the dry timbers and the tinder-like flowers of the native bamboo plants. The devastation is still evident: graveyards of uncleared, un-decomposed trees lie where they fell. The thin soils, no longer secured with tree roots, and made weaker with the hooves of non-native sheep and cattle, simply pour off the moun-tains, silting up the rivers, reducing the limited arable land further. The once mighty port of Aysén, now silted to less than a metre deep in places, is no longer usable and a new port had to be built at Chacabuco. People in Coyhaique whisper confessions: 'I personally burned several acres.'

Where cattle don't graze, the forests have recovered. 'We must learn from these mistakes we made in the past and not add to our destruction with megadam projects,' Peter says. 'Our proposal is to keep this unique, unspoilt region as a living reserve instead of destroying it like other parts of the world.'

Persuasive as his argument is, I want to understand what has spurred so many first-time protesters to take to the streets over these dams. We head off in Peter's ageing Chevy, passing incredible vistas of high mountains and gushing streams. Deciduous trees in every shade of yellow and red cover the higher slopes, while evergreens occupy the lower. We hunt out rock paintings made by the few indigenous nomads that passed through this region and search in vain for the *huemal*, an endangered native deer and the Chilean national symbol.

We stop at a straw-bale house owned by Francisco Vio, a tourism entrepreneur. His home heating and electricity is powered entirely by solar panels with propane back-up for the winter months, when he gets just four hours of sun a day because he is in the shadow of a large mountain. Inside, the house is cosy and well insulated against the cold in a region where most people live in corrugated iron or timber shacks with barely a barrier against the freezing conditions. Wood for burning is cheap – a truckload, which lasts a month, is just $80, and although this represents a third of the minimum-wage salary, it is still cheaper than the initial outlay for insulation. Francisco is campaigning against the dam project alongside Peter. He first came to the region in 1986 as a hitchhiker from Santiago, fell in love with it and resolved to return and live here with his family. 'What does development really mean?' he asks me as he bounces his toddler on his knees. 'Does it have to be a lifestyle where you consume more, create more trash, destroy the natural areas that give you a sense of well-being and make living worthwhile? We don't need so much more electricity to develop as a nation. There is another way.'

Opposition to the dams is based on an aesthetic, the idea of wilderness that cannot be replicated – an idea of untouched nature. Like everywhere on Earth, the influence of humans is already here in the sheep, cattle and burned forests. But in the Anthropocene, when so many wild places have been so dramatically altered, the idea of Patagonia is of increasing value to many. And while environmental activists like Peter may not hold many cards in government circles, wealthy landowners do.

Peter drives the battered truck into the grounds of a handsome estancia.

Sergio de Amesti, an agricultural engineer turned cattle farmer, manages some 3,000 hectares in the Simpson Valley, the most valuable and productive arable land in a region where 85% of the ground is rocky mountain or glaciers. The planned transmission line would run right through his land. Amesti was the regional secretary for the ministry of agriculture under Pinochet's regime – just the sort of private-enterprise-minded character that the government might have counted on to be in favour of the dam project. But Amesti is not. 'The main selling point of my meat is that the cattle are reared in a pure, idyllic region – uncontaminated, unpolluted, noise-free and visually pure. Huge pylons would destroy that image and lower the value of my meat and land,' he says.

I meet other locals in a worse predicament, including a visibly angry beekeeper, Gabriella Loshner, one of around 200 people whose homes will be flooded by the new reservoirs. Compared to other megadam projects around the world, it is a tiny population – the relatively low social cost of the Patagonian dams is something that the project's supporters, including government ministers, repeatedly emphasise. ('There's no one and nothing there,' more than one bemused minister has declared about Patagonia.) By comparison, the Three Gorges Dam in China displaced 1.2 million people and flooded thirteen cities, 140 towns and 1,350 villages. Brazil's proposed dam at Belo Monte in the Amazon would displace 20,000 people, many of them from indigenous tribes. And hydroelectric dams planned on the Mekong in Laos would affect millions of people in the river basin and delta.

Other Aysén residents are campaigning on behalf of non-human creatures threatened by the proposed dams, including *huemal* deer, native fish otters and unique cold-water corals at the river's outlet. During the last glacial period, 10,000–20,000 years ago, the river reversed its direction of flow from the Atlantic to the Pacific, an historical quirk that infused the Baker River system with unique biodiversity. Unlike rivers to its north and south, the Baker contains an endemic population of fish, such as members of a primitive catfish genus and *Odentethes hatcheri*, a type of silverside.

Dams on the Baker will prevent fish migration, and even the subtle alteration of nutrients can have far-reaching effects, says Brian Reid, an energetic American limnologist (someone that studies fresh waters), who we visit at the Patagonia Ecosystem Research Centre in Coyhaique. 'Damming a river turns it into a lake and completely alters its function,' he explains. Brian is interested in the river's silica levels, which are an important component of diatoms, a major group of planktonic algae that support a large ecosystem. If the levels of silica go down compared to the levels of nitrates, another group of algae called flagellates are favoured, which are responsible for toxic oceanic 'red tides'. Diatoms are larger than flagellates, so animals can feed on them more efficiently, making the whole system more productive.

Brian fears that dams on the Baker River could significantly alter the silica levels downstream. 'Damming of catchments across Europe has resulted in so much particulate trapping that it has reduced silica levels in the Baltic and Black seas. Production and efficiency of marine organisms there has gone down and it has affected fisheries.' The head of the Baker River is a pro-glacial lake, producing significant levels of silica. 'I row the river in a raft when I'm sampling, and you can hear the turbidity, the tiny velocities that keep everything in suspension – it sounds like a bowl of Rice Krispies,' Brian says. If it is dammed, it will result in a warmer reservoir that would be more productive, but a loss of suspended sediments flowing to the ocean.

No system on Earth is ever truly isolated from another, which is why the human changes we make to even small parts of the planet can have such enormous consequences. Building a hydroelectric dam hundreds of kilometres inland can affect cod numbers far out at sea. In the Anthropocene, our Earth-changing capabilities are more sophisticated than ever, but we have barely begun to comprehend the complexity of our impact. Until now, this has meant that we address each eventuality as it occurs, in a cascade of reactions to each action. But, as scientists get better at modelling the outcomes of our various interventions, we should be able to fine-tune our geoengineering to benefit people and ecosystems.

The way that many hydrodams operate, for example, has an unnecessary impact on wetland ecosystems. So-called 'hydropeaking' – flooding and draining a reservoir in artificial daily pulses – can be devastating for fish. The dramatic rise and fall of water levels during dam releases – sometimes of several metres – is too extreme for plants and animals to cope with, resulting in dead zones around the shores of reservoirs. Fish that lay their eggs in the shallows among submerged tree roots, for example, may find a few hours later that those sites are high and dry with the eggs desiccated, sometimes with the loss of an entire species. Most dams use hydropeaking because it's most profitable, releasing most energy at the peak of demand. A less damaging option is a run-of-river design which, instead of relying on a head of water to build up in a big reservoir behind a dam wall, simply allows natural river flow to drive the turbines. Run-of-river dams don't disturb the upstream ecosystem because no reservoir is created, they don't get silted up, and they don't result in the abrupt upstream–downstream temperature difference you can get when a reservoir is drained from its lowest (coldest) layers. They are only suitable where a river has significant drop, which the Cuervo has, prompting campaigners to call for the dam company XSTRATA to change its dam plans there.

But the Cuervo plan has bigger problems. The proposed dam lies directly above the Liquiñe–Ofqui fault line, on a triple point where the Nazca, South American and Antarctic tectonic plates meet. It means that there is a likelihood of a volcano or earthquake at the site, and yet there has been no study to investigate this, Peter says. In 2007, one month after XSTRATA submitted its report declaring the siting to be on a seismically inactive zone, the area experienced a massive earthquake that dislodged boulders into the fjord below, triggering a tidal wave that killed people on the opposite bank. 'The government threw out their report,' Peter laughs. Earthquakes have wrought considerable damage at dam sites around the world, including in April 2010, when a quake at Yushu in China's Qinghai province killed tens of thousands of people in minutes. The Yushu reservoir, sited on a seismic zone, may actually have triggered the quake due to the weight of the water on the underlying geology. In the Anthropocene, humanity's dam-building is shaking the Earth.

There are other dangers too. Patagonia is one of the fastest-melting glacial regions in the world, which has already resulted in catastrophic outburst floods, debris-laden torrents carrying away entire forests. At times, these glacial floods have caused the Baker to rise by four metres and even turn around and run upstream for days at a time. 'They are preparing to construct dams on what is probably the most unstable river system on the planet,' Peter says, flinging his arms incredulously.

Back in town, I visit the offices of dam company HidroAysén. Veronika, an earnest and sweet-natured woman, is adamant that the dams would rescue local people from poverty by providing much-needed employment and helping development in the region. I press her on this vague 'development' term, and she describes how she was one of the fortunate few who escaped from her isolated village for a year's education in Puerto Montt, a small city further north. 'Most people here have no choices. There are not many restaurants or shopping malls or decent education opportunities,' she says. 'It is very difficult even to get to the next village because the roads can be very bad and impassable, especially in winter.' Dam-building requires improvements to the roads and surrounding infrastructure, and shops, restaurants and other services will be built to serve the workers, she reasons.

Her colleague, Rodrigo, is also in favour of the dams because he sees them as the only viable energy option for Chile. In 2009, reliance on Argentinian gas led to disaster when a domestic crisis there created a fuel shortage and the country cut Chile off. 'We can't rely on another country to provide our energy,' Rodrigo says. The energy-security argument is one the dam company is focusing on in an orchestrated PR campaign that includes television advertisements threatening catastrophe if the project is blocked. One shows the lights going out mid-surgery in an operating theatre.

After repeated back-and-forth between HidroAysén, XSTRATA and the government, and seven court appeals on environmental grounds, Chile's supreme court ruled in favour of the dams in April 2012. Only the transmission line awaited approval.

But astonishingly, one by one, the project's backers have begun pulling out, bowing to public pressure. President Piñera's own ratings plummeted after he publicly supported the dams following massive – occasionally violent – protests against them. Chile's second biggest bank, BBVA, announced it would not be assisting HidroAysén with loans for the project, citing environmental and social concerns. And then, in December 2012, Colbun, the big Chilean energy company, announced it was selling its 49% stake in HidroAysén. In December 2013, Chile elected the left-wing president Michelle Bachelet, who has spoken out against the dams. For the first time, the $10 billion megaproject looked on shaky ground.

The battle is far from over, though. Electricity rates on the central grid have risen by 75% in six years, straining pockets and the economy, especially in energy-intensive mining operations north of the desert. With Chile relying on copper for as much as one-third of the national income, the spectre of cheaper hydropower from the south will not vanish soon.

Unlike other megadam projects around the world, the Patagonian proposals are not primarily of humanitarian concern. But they do raise fundamental questions about what we really mean by sustainable development in the poor world, and clean energy in the Anthropocene, and what price we are willing to pay as a global community to preserve unique areas of our planet.

If the transmission line gets the go-ahead in Patagonia, electricity production could begin as early as 2015. If not, Chile will be one of the few developing countries to choose to protect its natural environment over short-term financial gain.

It's hard to imagine such a thing happening in poorer, one-party Laos.

In the steamy hills of South East Asia's Golden Triangle, on the Thai–Burmese border, I take a slow boat into tropical Laos, beginning a journey along the Mekong River that will end 2,600 kilometres downstream in the South China Sea. The Mekong is the planet's twelfth longest river and one of its richest biodiversity sites, supporting over 1,300 varieties of fish and the world's biggest inland fishing industry. The river basin is home to some 60 million people in six nations, who depend on it for food, water and transport. It couldn't be more different to Patagonia's Baker River, yet, here too are highly contentious plans to dam the river for hydropower. As we enter the Anthropocene, the Mekong has become the most visible focus of international debate over the future of the world's great rivers.

The boat is wooden, flat-bottomed and broad enough for two rows of hard benches to run its length. We cut a winding route through hills and mountains draped in lush vegetation. Vines and creepers extend down from the canopy, cloaking the forest in a continuous verdant blanket from bank to peak. The river is interrupted by oily granite and limestone karsts that protrude from the banks and out of the water in peaks that mirror the larger green ones. From some of these rocky outcrops, bamboo canes point over the water. Fishermen from indiscernible villages haul in nets or wade through the water in swimming pants to catch something silver and fast that flashes beneath.

The mighty Mekong is narrow and shallow here, sucked dry on its journey from its source in the Tibetan Plateau, 4,500 metres higher in the Himalayan snows, through China's Yunnan province – until recently a region of such unspoilt beauty that Shangri-La was said to have been located there. The upstream extractions water China's wheat baskets and the country is growing thirstier every year. China has constructed several hydrodams on its portion of the Mekong, both for power and water storage. The remote Xiaowan Dam, completed in 2008, which at almost 300 metres high is the world's tallest, holds a reservoir some 170 kilometres long, and sends power all the way to Shanghai.

In some places the water appears to be boiling, bubbling up with unseen rocks. China intends to quell these rapids, blasting them with dynamite and streamlining the Mekong so that it is navigable all the way to Ho Chi Minh City in Vietnam. It has busted a stretch north of Thailand, but protests there put off their projects south – for now.

We pull up at banks occasionally, where a village is hinted at by far-off stilted houses growing out of the forest. Local people come aboard carrying baskets of fish and vegetables or live chickens – the boat transport is the only link many of these rural people have to life and markets beyond their village. Almost all of the country's 6.5 million inhabitants live on the Mekong or its tributaries. At one spot, a woman tries to board with two large monitor lizards and a big dead rodent on a string. The boat becomes uncomfortably crowded.

We glide though Luang Prabang, once the shimmering capital of the kingdom of Laos. French colonial villas, with painted shutters and verandas that hang with bougainvillea are interspersed with gold stupas and richly decorated Buddhist wats. Orange-robed monks with tranquil faces weave through the sleepy town. Officially it is a city, but with just 100,000 inhabitants, it scarcely merits the insult.

Above me, slash-and-burn scars denude the mountain slopes. It is a traditional but very destructive farming technique, in which forests are chopped and burned to make agricultural land. It leads to ruinous soil erosion and high carbon emissions, and it is one of the reasons for the success of opium in these parts – the poppies survive poorer soil compared to other crops. Opium use, traditionally practised by several tribes here, was manipulated by every colonial power across Asia. Now, the global opium industry is almost entirely in Afghanistan; nevertheless, I am offered it on several occasions.

Laos is on the list of least-developed nations. One reason for this is the cluster bombs littering the country that make it tricky to plough a field or build a road without painstaking and expensive mines clearance. One-third of the 2 million tonnes of bombs dropped on Laos during nearly 600,000 US missions during the Secret War in the late 1960s failed to detonate. More than forty years later, the bombs continue to kill and maim people especially during the sowing season when they are triggered by farmers turning the soil. The number of casualties is increasing, particularly among children who are seeking out the orange-sized bombs, because of the lucrative new market for scrap metal in China. Another hindrance to development is the Communist government, which after 1975 saw 10% of the population flee the country or be interned in re-education camps. Those who left were primarily the educated, middle-class intellectuals, a migration that cost the country a generation in development time, while its neighbours Thailand, Vietnam and China spent those years playing rapid catch-up to the established Asian Tigers (Hong Kong, Singapore, South Korea and Taiwan). The vast majority of Lao live at subsistence level, fishing the rivers of the landlocked country or rice farming across its spine and plains. What forest remains is an essential source of food and other materials for the largely rural people, half of whom don't have electricity. Their livelihoods are further threatened by unfair compulsory land purchase, in which the government pays little compensation for taking land from families for infrastructure projects or in corrupt deals with wealthy people or businesses. It has become such a contentious issue that even the state-controlled media reports it.

Laos has long nurtured aspirations of becoming the battery of South East Asia. With the majority of the Mekong's waters gushing down its length, it has unrivalled hydropower potential. It was the French who first dreamed of taming the Mekong – at the turn of the twentieth century it took longer to travel from Saigon to Luang Prabang than to Paris. But their attempts at railways, rapid-busting and canals came to nothing. A few decades later, international companies flocked to Laos with new plans for hydrodams down the length of the Mekong that would transform the country and hasten its development. War and political conflict put paid to that idea.

Until now. Eleven hydroelectric dams are at advanced planning stage on the untamed Mekong River. The Communist government promises electricity and the other riches of development for the people of this poor nation, but the environmental and social consequences of stifling this uniquely biodiverse river are potentially enormous. While the government sells off the country's natural resources from river to forest, reaping quick profits from electricity sales, rubber plantations, timber and mining, the people of Laos are in danger of losing everything.

A few days further downstream, past the charmingly petite capital of Vientiane, I stop in the small town of Thakek, surrounded by striking limestone karsts and protected forests. The animals here are shy, mainly because they are still hunted and eaten, so I see few. Forest-harvesting villagers pass me with assorted edibles in their baskets, including mushrooms, snails, insects, squirrels and a Mekong River weed that they dry. They munch on red ants and I try one – it pops in a citrussy burst in my mouth. I set out to rent a good off-road bike, ending up with a 100 cc scooter with a broken petrol gauge and speedo. I get a new inner tube for the rear tyre, get pumped and head out of Thakek on a three-hour voyage into the jungle, following a tributary river back from its Mekong entry in the hills. I pass fields being prepared for rice planting, fat black and pink buffalo wallowing happily in ponds and muddy pools, children firing stones from slingshots, and women in sarongs breastfeeding. Road-building, that ubiquitous Anthropocene signature, is under way even here. In a few years, the boat trip I made will be obsolete or maintained only for tourists. A 1,500-kilometre road from the Burmese port of Mawlamyaing, via Thailand

and Laos to Danang in Vietnam, is almost complete. Already, journeys which used to take two weeks by sea between Bangkok and Hanoi now take only three days overland. The new 'east–west economic corridor' will change Laos rapidly, spurring industrial and economic growth faster than any hydroelectric dam.

After about eighty kilometres, the road further disintegrates into mud and cavernous potholes, and the land becomes suddenly bare and brown. Engineering on a massive scale is under way here at the site of the Nam Theun II hydroelectric dam, the Lao government's pride and joy. I pass the powerhouse and then the road twists up and up through cooler damper air to Nakai village. Here I dismount, my backside numb from the ride and my feet and hands tingling from the vibrations.

I have arranged to meet American naturalist Bill Robichaud here, a man so crazily eccentric that 'he looks for animals in the jungle, but not to eat them!', as the Lao waitress explains to me in amazement. We meet in a surprisingly plush French restaurant, with prices to match – dam construction means foreigners, which means money. A couple of hundred metres from our table is the old river, already flooded into a large lake, drowning seventeen villages, the ancestral homes of 6,600 people, who have been relocated to nice-looking, newly built, traditional stilted houses above the village. The hydroelectric dam was only made possible with World Bank assurances to the dam's international partners that it would underwrite the project should things become 'problematic' politically in Laos. And the bank also loaned Laos a third of the $1.5 billion funds. But the money came with caveats, including that people displaced by the waters be compensated (hence the natty housing) and that the forest be properly protected.

This is widely regarded as a successful dam project, both socially and environmentally. Even anti-dam campaigners concede that care has been taken with the design, siting and efforts to mitigate its impacts. Unfortunately, even with what appears to be the best will, things aren't so simple. When asked where they would like to be relocated to, the villagers unsurprisingly said they wanted to stay near to their village, their friends and their river. The problem is that prime land in the village was not available – people were already living there – and so all that was left was infertile clay soils. The people were subsistence fishermen, but without the river they needed alternatives, so they were provided with crops. Everything died in the poor soils. Buffalo were given to them, which also died having nothing to graze on.

However, at the edge of the artificial lake, people are unravelling fishing nets to capture food to eat and sell – an enterprise that was not nearly as lucrative before the reservoir was created. Talking to them, most are cautiously positive about the project – they are already getting electricity for the first time, and the road is enabling communication with the outside world, as well as trading opportunities, and giving them options they never had previously. 'I don't miss my village,' one old man tells me. 'Now my life is much easier.'

The Nam Theun River didn't drop with enough of a gradient before entering the Mekong for sufficient profits to be generated by the hydro-power company. So, in an ingenious piece of engineering, the river has been dammed into a large basin, at the bottom of which a tunnel 250 metres long and nine metres wide has been bored down to the Xe Bang Fai River, which runs parallel to the Nam Theun but at lower altitude. This gives the water the rapid drop and strong flow needed for hydropower generation. While this will produce more than 1,000 megawatts of electricity, most of which is to be sold to Thailand for tens of millions of dollars a year, it also dramatically alters two rivers (affecting tens of thousands of people who depend on them) and impacts the Mekong, into which they both flow.

The dam is already having an effect on the ecosystem here. This forest is very special. Scientists rank it second only to Madagascar in terms of small-mammal diversity, and its 3,500 square kilometres have hardly been studied.[5] It is an important refuge for nine species of primate, tigers, leopards and elephants as well as newly found species, including some thought previously to have been extinct and known only from the fossil records. Among its oddities is the saola, an antelope discovered in the 1990s, known as the 'Asian unicorn' despite its two horns, which is a new genus and possibly a new sub-family.[6] It lives high in the Annamite mountains. And there is the *kha-nyou*, which is a totally new mammal family. It is a type of rodent, related to the porcupine, which looks a little like a big squirrel and lives among the limestone karsts.

It's Bill's job to oversee protection of this vast wilderness, and he's been given an impressive $1 million per year for the next twenty-five years to do so. But the dam has created plenty of problems. The raised water means that parts of the Nakai forests are now accessible to hunters on boats. The site was also flooded before all the vegetation was cleared, meaning that plant matter is rotting within the lake, poisoning the waters, killing fish and producing methane.

We meet a day after Bill has returned from a two-week visit to some of the remote communities that live within the protected area, and he has some tragic stories. The forest is home to ancient hunter-gatherer communities as well as subsistence rice and vegetable farmers. But in recent years, the Lao government has been systematically seeking out and expelling hunter-gatherers and housing them in villages, where they are given some land to farm. Government officials find the presence of hunter-gatherers a national embarrassment that doesn't fit well with its desired image of a modern developed nation. But once removed from the forest, most of these people become sick and die. Entire tribes have been lost in this way. They believe that their protective spirits stay on in the forest, too far to provide protection. It could be that they are exposed to new viruses or simply become depressed and unable to settle into agronomy. The remaining few want to be allowed to return, and Bill is negotiating on behalf of fifteen from one tribe.

I return to Thakek and follow the Mekong down the country until it shatters into a thousand rivulets, rapids and waterfalls at Si Phan Don (Four Thousand Islands). I stay in a sleepy village on Don Khone island, where there is no electricity save what the generator provides for a few random hours between 6 and 10 p.m. My fan hovers immobile above me through the sweaty night, its three blades paralysed above the mosquito net. Even the geckos, languid and fat on flying protein, seem to swelter in the heat, their calls dying mid-tone.

But all this will change, my excited host Mr Pan assures me. Soon will come electricity when the new dam is built. With soaring fuel costs, he can't wait to abandon his generator. 'And we will have Internet on the island with the dam's electricity,' he says.

Here, as everywhere else in Laos, the river is the heart of daily activity. I pass children playing in the waters, swimming and attempting to net small fry, a man bathes further down near to a woman who is washing up. Domestic ducks gabble further round the bend and I pass a floating vegetable garden. I take a twisty pathway down to Dolphin Beach, so-called because it hosts regular sightings of the rare Irrawaddy dolphin. At the beach I get chatting to local fishermen and end up joining them on one of the six daily sorties they make to gather their catches.

It's early on in the rainy season, which means that the trap of choice is a slatted bamboo contraption that catches small fish swept up by the rapids. It's one of ten specialist fish-catching devices that I count on the beach, each used in different circumstances, seasons or times of day. I hop aboard a fisherman's boat and motor out the few yards to the trap, which the fishermen spend three months creating. We gather handfuls of fish for the grilling rack on shore, and for our lunch. It's a semi-cooperative – a few fishermen who make and maintain the traps get to share the catch with their families. Few people catch enough to sell, but they have enough to eat well. I ask one of them what he thinks of the government's plan to allow a Malaysian power company to build a dam spitting distance from our boat

at Don Sahong, across the Hou Sahong channel, which is the most important fish-migration route in the region because it is the only one open
throughout the dry season. 'But we eat fish, not electricity,' he smiles. But
you would be compensated, I say. 'The problem is, we need fish to eat, not
money,' he repeats.

The Mekong is second only to the Amazon in terms of fish diversity.
And uniquely, more than 70% of its fish are migratory – some species
migrating annually from the South China Sea in Vietnam as far as Tibet.[7]
These include the world's largest freshwater fish, the incredible Mekong
giant catfish, measuring more than three metres long and weighing as much
as 300 kg – whose numbers have already declined by an estimated 90% in
the last twenty years through overfishing. If the Don Sahong hydrodam
gets built it will mean certain loss of a number of commercially important
fish, but perhaps more importantly, it will risk the livelihoods of tens of
thousands of people who depend on the fisheries for their daily food, such
as the sardine-sized carp, known as the *trey riel*.

In the afternoon I visit Mr Vong who runs a restaurant on Don Det
island. His business is closed while he carries out renovations. Over the
past few years, he has experienced bizarre flooding patterns where overnight his restaurant and the path beside it get inundated for weeks or
months at a time, and then the waters suddenly recede. Recently, he
discovered the cause: hydrodams on the Mekong in China, more than
1,000 kilometres upstream, that release and stop water regularly. 'There
have been so many fewer fish here, too,' he says. 'If they build the Don
Sahong Dam, I will have to close my business and move up into the
Bolaven Plateau. Here, I will surely be flooded.

'The Chinese dams are already causing problems, and the government
officers came and said to us: "When the waters come, take your chickens
and your buffalo and move to higher ground." We cannot live like this.'

I ask him what the people will do if there are not enough fish in the
river to eat, and he says he doesn't know. 'I need to eat fish. It's the way it

has always been. Maybe people will work in tourism,' he suggests. A major tourist attraction in the area is the Khone Phaphene Falls, Asia's biggest waterfall. But this stands to be another victim of the proposed dam, which will considerably reduce its flow.

Exploiting its rich natural resources will spur the impoverished nation into rapid development, the government claims. It will also push people from subsistence farming into more profitable enterprise – subsistence farmers pay no taxes . . . But 'poverty' is a very subjective concept. Laos has poor infrastructure (although it is improving thanks to Chinese funds), it has next to no medical provision, poor education and no social-service protection. But people do not starve here. Many are truly self-sufficient because they exist in a low population in a naturally resource-rich environment, even if the country is being rapidly deforested. People here gather fruit, vegetables, insects and other animals from the jungle and rivers – 90% live in rural villages – and they keep chickens and buffalo, plant rice and vegetables. In terms of cash, they are poorer by far than an Indian beggar; in terms of quality of life, they are perhaps rich.

For now. As the population grows (it has doubled since the 1970s) and the environment becomes degraded, polluted and lifeless, the Laos people will become as dependent as I am on food that must be traded for something else, on rice that must be shipped from another part of the country or world and exchanged for money that must be earned in an urban environment in manufacturing or the service industry. Lao people have something extraordinarily precious at the moment, and very rare indeed in this world: the ability to live a self-sufficient relaxed life in their home environment.

There is no glamour in poverty. And certainly many of the Lao I speak to are as eager as Patagonians for access to electricity and other important trappings of development. However, in the Anthropocene we have a choice over how development is attained and distributed among the poorest people.

In a country like Laos or Chile, which has few alternatives, exploiting rivers to generate energy makes a lot of sense, despite the social and environmental penalties. So, if we accept that many controversial dams are going to be built, how can their construction be made least damaging? Jamie Skinner, who was senior advisor to the World Commission on Dams and now heads the International Institute for Environment and Development's water division, suggests the answer might be to issue dam-builders with limited-length licences. 'In America, the licences are only for thirty or fifty years, after which there is a review. The reason many dams are being removed now, is that their licences have expired and the dams would no longer pass the more stringent environmental planning regulations,' he says.

Removing the permanency of dams would make them more palatable to environmentalists, especially if licences were only granted with the proviso that the firm could afford to remove it in thirty years. The problem is that in many countries, poor governance and corruption mean that such agreements could be worthless. Even in the US, where companies are legally obliged to put aside funds for environmental clean-ups, it is often the state that ends up paying. Limited lifespans are sensible for another reason too – climate change is altering rainfall patterns around the world, leaving many dams economically worthless. The international Hydropower Sustainability Assessment Protocol, released in 2011, is a method to rank dams in all phases from development to operation, and should help managers design better dams to an internationally agreed environmental and social standard and reduce conflict.

Essentially, Skinner says, dam planning needs to be a participatory process. The scientists can analyse the different engineering options and their power and environmental outcomes, but it is down to society to decide what constitutes an acceptable impact. Putting in gated spillways makes for a more regular flow that is less damaging to ecosystems, for example, but reduces power output, so is less profitable.

If local people feel adequately compensated, not just for land and livelihoods, but with a culturally sensitive approach to relocation, and if they get a share of the dam's benefits through electricity provision, for example, then dams can become far less traumatic and even be embraced by local communities. Equally, if we as an international community decide that some environments are simply too precious to dam, then we must offer compensation to those countries for their loss of potential power generation, and provide realistic alternatives for economic development.

Laos could use its powerful geographical position, and the fact that it's home to some of the most sought-after minerals and hydropower potential to its advantage, to ensure that its wealthy neighbours pay adequately for environmentally sustainable exploitation of the Mekong. In that way, Laos could afford to leave some important sites unexploited to the benefit of all.

The Don Sahong Dam would be built just two kilometres north of the Cambodian border, and I take an early boat across to Laos's sad, troubled neighbour. Cambodia is reeling from decades of brutality, famine, torture and genocide – struggling for an identity, trying to emerge from the past, but failing. It is a place where the ATMs dispense only American dollars, where the roadsigns are in French and where international charities perform basic government functions from health to education. In Siem Reap, a pleasant city thronging with tourists for the nearby Angkor Wat monument, women speak in clichés from American movies produced years before they were born. 'Love you long time,' they murmur at disinterested men. T-shirts with slogans like 'No money, no honey' hang from a market stall and are bought by young American women in tour groups, with pink cheeks and blonde hair and shiny painted lips, whose plump bodies spill richly from candy-coloured minidresses.

If the Mekong is the lifeblood of South East Asia, then the Tonle Sap, here in Cambodia, is its heart. It is the biggest lake in the region and its waters pulse through the seasons. For most of the year, the Tonle Sap is a round, shallow body of water covering less than thirty square kilometres. But as I journey along its length from Siem Reap to the capital, Phnom Penh, the new rains are beginning their swell. From June until November, the Mekong ushers in gallons upon gallons of flow, filling the Tonle Sap to more than fifty times its normal size: 16,000 square kilometres. The floodwaters refresh ponds and inundate forests, proving vital breeding and nesting grounds for the fish that migrate up the Mekong. Two-thirds of the fish in the Mekong begin their life in the Tonle Sap – it is the most productive inland fishery in the world, supporting up to 4 million people and providing three-quarters of the country's fish catch. It means that even though Cambodians are among the world's poorest people, they are some of the best fed.[8]

By November, the end of the rainy season, the brimming lake actually causes the river to flow backwards. This annual flow reversal is cause for great celebration in Cambodia and occurs in the stretch outside of the king's palace, to much festivity. The upstream dams would halve this annual pulse (perhaps spelling the end of the famous reversal), and dramatically impact the ecosystems it supports.[9] Around half of the river's water would be held in storage over the border in Laos, and instead of an annual flood, the waters would be released at the whim of dam managers, who can override the natural cycle at the push of a button.

In the early morning, I slip down a rain-slicked muddy bank of the Mekong to board a rusty ferry downriver to Vietnam. There are few other passengers – most people are using the new highway. How different the river is now from the liquid crease folded into the mountains of northern Thailand. Here, the Mekong is fat and deep and hurtles along to the South China Sea like a migrating fish on its urgent journey south.

At the border the river splits into islands and channels – the Vietnamese call the Mekong the dragon with nine tails. People here live closer than ever to the Mekong, in stilted houses that reach the bank via wobbly boards tied together with ropes. Next to the floating café where I slurp my noodles is one such house, with a young family that is bathing in the river: mother, father and then two toddlers. Next the laundry is washed, sloshed in the river and slapped on their bamboo deck. Then it is the bowls and plates that must be rinsed. The river also supplies their food. A net suspended under the house holds a fish farm, and they feed the fish through a hole in the living-room floor.

Several new bridges are mid-construction on this stretch, but for now people cross from one side to the other on little canoes, rowed by women in pyjamas and conical straw hats against the sun. Further downstream I reach Can Tho, the capital of the Mekong Delta, where this once-glacial water spills into the South China Sea. The region is home to 17 million people, who depend largely on farming rice and fishing. Here, humanity has already changed the river: fish numbers are declining because of overfishing, pollution and sediment build-up (as a result of river extractions that reduce the Mekong's ability to flush sediments into the sea). And the river is also turning salty as the sea level rises (owing to global warming), producing a measurable salinity of four parts per thousand as far as fifty-six kilometres inland.[10] Many farmers are having to switch from rice farming to shrimp farming, which involves higher upfront costs and aquaculture skills that they do not have. The government is responding by filling in the smaller channels and moving ancient fishing communities into factory work.

I rise early to boat to the famous floating markets. I was last in this region in 1995, and I remember the markets as an enormous medley of hundreds of boats, spilling over with vibrant produce. This time they are smaller with few boats. Buying and selling is still under way, but in this region the bridges are now built, and floating markets will soon disappear completely, replaced by large land-based ones for the new motorbike-owning community.

The direct relationship people have with their rivers is disappearing as these water sources are diverted and dry up. But our dependence on them is as vital as ever – access to water is already a leading cause of conflict around the world and, although potential water wars have been avoided through river-sharing agreements and treaties, 60% of the world's 276 international river basins lack any type of cooperative management. However, in many places conflict is being averted by water-rich countries trading with their less-endowed neighbours, and this will become increasingly important into the Anthropocene. Nearly the entire Paraguayan economy depends on selling hydroelectricity to Brazil. The two neighbours share the Paraná River, which is dammed at Itaipu. Other thirsty countries are considering buying water from their neighbours. The US, for example, plans to buy water from Mexico (piped from two planned desalination plants at Playas de Rosarito) and Canada. Analysts predict that by 2020 the world water market will be worth $1 trillion, mainly through a growth in demand in Asia and South America.[11]

Humans are also distributing water around the world through 'virtual water' trade – the trade in goods and services that are produced using water. Most of this – 92% of fresh-water consumption – is embodied in agriculture.[12] So, for example, more than three-quarters of the water used by Japanese people originates outside the country in, say, beef bought from Australia. The water footprint of an average US citizen – which at 2,840 cubic metres per year is more than double that of the Japanese – is 80% home-grown, with the bulk of the remaining 20% coming from China's Yangtze River basin. As nations become more water-stressed, however, a new strategy is developing. Governments are buying up foreign land that is rich in water, or angling for a controlling stake in how that water is used. All of these acquisitions are occurring in developing nations and most in places where there is poor governance, corruption, little regulation and where local people have few rights to their own land and resources. Saudi Arabia, South Korea, China and India have all bought up huge tracts of land across Africa, usually in the poorest countries' most fertile zones and river basins, to grow crops for export back home. The Nile River, for example, is now under pressure not just to feed its resident human population, but also those in other continents who have purchased parts of Sudan.

As rivers dry up, water shortages could prove our species' undoing in the Anthropocene. Consultants McKinsey predict that by 2030 global water requirements will have grown from 4.5 trillion metric tonnes today to 6.9 trillion tonnes, exceeding our current reliable and accessible supplies by 40%. Despite our ingenuity, we cannot as yet manufacture the scale of fresh water we need, so artificial storage is urgent across the globe. Since ancient times householders have trapped rain and river water. Elaborate water canals and tunnels have been created from Arabia to China to channel precious drops where they've been needed. Recently, humans have become such an ambitious hydrological force that we have overcome natural limitations, even building modern cities in the desert, including Las Vegas and Phoenix in the US, with brimming fountains and expansive golf courses. The water-thirsty mines and coal production of the Gobi Desert, and the inland cities, industry and farmland of northern China are also far from adequate water supply. Satellite images of Arabia reveal bizarre green circles painted over the desert sands – water mined from elsewhere. China, though, has embarked on the most ambitious water-diversionary project the world has ever seen. By 2050, it aims to pipe almost 45 billion cubic metres of water annually, along hundreds of kilometres, from the Yangtze in the south to replenish the highly polluted Yellow River in the more populous north. The project is besieged with technical difficulties, vastly costly and involves relocating thousands of people along the route – many observers do not think it will succeed. Nevertheless, as humanity's water demands continue to outgrow the world's natural freshwater sources, people will increasingly attempt to move rivers and lakes across continents.

Even though our towns and cities are no longer restricted to the geography of meandering rivers in the Anthropocene, it is worth noting that the pipes, reservoirs and canals that carry out a natural river's function come at considerable cost. The abundance of water in rich countries costs an estimated $750 billion a year in infrastructure and treatment, which is too expensive for poor countries to replicate.[13] Instead, it is often cheaper to maintain and restore natural ecosystems. New York City gets its water from the Catskill Mountains and Hudson Valley, sources so pure that historically it needed no filtering. However, in the 1990s that threatened to change, because of agricultural pollution and other contamination. The city considered the cost of building treatment works (around $10 billion) and calculated it would be cheaper to invest in a programme of land protection and conservation ($1.5 billion). The gambit paid off and water quality has been maintained at a fraction of new infrastructure costs and with the added benefit of maintaining a biodiverse ecosystem.

Saudia Arabia has recently gone a step further, artificially creating a natural system to carry out an infrastructure task. At the world's largest aluminium (bauxite) mine, at Ma'aden, in the Arabian Desert, engineers are creating an artificial wetland system to act as a biofilter for the mine's industrial and sanitary waste water. It will replace chemical filtering and treatment tanks and mechanical processes with vegetation and microorganisms that will use natural processes to remove toxins, metals and other impurities from the water. And it will save 7.5 million litres of water a day in the process, through an innovative below-ground closed-loop system that reduces evaporation of the precious desalinated water. The wetland will act as an oasis, providing a habitat for birds, and can be created in a fraction of the time and cost of a traditional treatment plant.

Into the Anthropocene, we will certainly see more infrastructure replicating the services naturally performed by rivers, lakes and waterfalls. But increasingly, we will also start to take a more nuanced approach to creating artificial wetlands; in which the multiple and interdependent roles of the natural system are taken into account when geoengineering a solution to meet our human needs. So, the design will not stop at the physical structure, but will include the biological by maintaining or creating a functioning, biodiverse ecosystem in which the drinking water, irrigation and energy needs can all be met to some degree. At the same time, foods and other products will increasingly be priced to reflect the water cost involved in their production – there may even be an international water market similar to the new carbon markets.

I am momentarily sad when I realise that the wonderful journey I made down the iconic Mekong River is unlikely to be navigable in a few years' time. But nostalgia for the untamed rivers of the Holocene is a pointless sentiment that cannot and should not preserve them in the Anthropocene. Instead, we need to look dispassionately at the planet's great water courses and understand their interactions with ecosystems, chemical and physical cycles, and with the human populations living on their banks. Only then can decisions be made, by everyone concerned, about how we use our rivers – and how to supplement their capacity to meet our needs.

4

FARMLANDS

F lying above the rumpled sheets of land that cloak our planet, over the broccoli of forests and the mirrors and ribbons of lakes and rivers, there is one type of covering that immediately gives the viewer pause – in the way the eyes are drawn to any patterns that emerge in nature. It might be regular green necklaces looping the contours of a hill; or a patchwork of yellows and greens, each neatly edged; or large, crisp, rectangular verdant stripes; or perfect circles of emerald and gold amid the grey sands of a desert. These patterns haven't simply evolved, they have clearly been designed by Earth's master gardener: humankind.

Agriculture was invented around 10,000 years ago, in the Levant, and was triggered by climate change. After the last ice age, the Holocene ushered in warmer conditions with long dry seasons that favoured cereals. These annual grasses could mature rapidly within a season, dying off to leave a dormant embryonic stage – a seed – that could survive the dry periods. Cereals invest their energies in large seeds – sugar and protein stores – rather than indigestible woody stems, making them perfect for human harvest and collection. The first farmers gathered wild grains to sow around settlements, where they could be harvested, stored during times of plenty, dehusked and milled. These people selected the tastiest, plumpest seeds, those that were most easily dehusked, from plants that grew best in poor soils. In time, not only were humans deliberately planting grains, they were domesticating them too, overriding evolutionary pressures to create crops where the seeds were released for the benefit of humans rather than the plant.

These early farmers began to radically change Earth's landscapes. Humans were burning naturally occurring vegetation, such as forests, to plant their artificially generated varieties in easily harvestable zones – the first crop fields were born. At the same time as changing local ecology, humans were changing other biodiversity by domesticating animals such as chickens, pigs and cows, which used the new grasslands to graze safely and provided meat, milk and manure.

Agriculture was reinvented independently at least three times across the world, and was so transformative that it spread globally as the only way of supporting larger populations. Hunting and gathering, which continued alongside farming, is too inefficient to feed settled villages or cities. It was agriculture that enabled civilisations and allowed them to flourish.

Shifting cultivation, the most ancient type of farming, had changed landscapes only temporarily – after the slash-and-burn, crop planting and harvest, the nomadic farmers moved on, allowing the land to become wild again. But when people began to settle, using the banks of rivers and deltas to irrigate and enrich the soil during annual floods, many farmlands became permanent. Originally, humans limited their agricultural endeavours to these naturally fertile regions, but as our technological prowess strengthened we began cultivating even extremely hostile zones, such as the deserts of Arabia and China, using complex irrigation systems.

Even with the most ambitious watering network, crops were still limited by *the minerals in the soil, particularly nitrogen, which is an essential part of proteins and DNA. Each centimetre of soil takes centuries to form, although it's a little faster in the tropics. First, rocks must be weathered by wind and rain, with the assistance of microorganisms, insects and lichen, whose decay feeds more organisms, including, in time, plants. It is the accumulation of this organic matter, living organisms and minerals that we call soil. Nitrogen is abundant in the atmosphere, but in a form that very few organisms can metabolise. For thousands of years, farmers relied mainly on recycling biological matter to make up for the shortfall in nitrogen and other essential nutrients in the soil, leaving the stalks and silage in the fields to rot and adding whatever other organic material they could, including animal and human manure. But as populations grew, more crops had to be provided from the same amount of nitrogen.*

With famines always only a poor harvest away, nineteenth-century scientists looked further afield for precious useable nitrogen. They found it in South America, where the dry climate of the Atacama Desert and nearby islands had enabled vast quantities of guano – bird poo – to accumulate. Indigenous farmers had been using it for centuries, but the discovery generated enormous interest in Europe and the United States. A massive export industry was set up, and the guano-rich nations of Peru, Bolivia and Chile went to war over the lucrative excrement. Chile won but victory was short-lived. In 1909, a German chemist called Fritz Haber invented a way of turning the nitrogen in air into soluble ammonia, a form that plants can absorb; his colleague Carl Bosch scaled it into an industrial process. The era of artificial fertilisers was born, a geoengineering innovation that would deliberately and profoundly change the planet's nitrogen cycle.

The effect of artificial fertilisers on crop production, and hence on population growth, was immediate. The number of humans that could be fed from a hectare of land more than doubled. Half of the protein in our bodies now comes from ammonia made in the Haber–Bosch process. Billions of people owe their daily bread, rice and potatoes to artificial fertilisers.

In the past century, the industrialisation and intensification of agriculture has taken over the world. Some 40% of the ice-free surface is used for crops or livestock, making agriculture the biggest artificial land surface and one of the most obvious signs of our influence on the planet in the Anthropocene. Agriculture has already cleared or cultivated 70% of grassland, 50% of savannah, 45% of temperate deciduous forest and 27% of the tropical forest biome.[1] 80% of new tropical croplands are replacing forests.[2] And 70% of our fresh water is used for agriculture.[3] Among the great diversity of plants and animals on the planet, most agriculture consists of just a handful of plant species planted in monocultures, and the domestication of a similarly few species of mammal and bird. As a result, the Anthropocene is a time in which certain species such as wheat, rice and chickens appear disproportionately successful on every continent. Indeed, the weight of all the humans on earth is less than half of the weight of cows, and yet most of the cows would not even exist naturally – they've been selectively bred by humans from a wild species that has been pushed to extinction.

In 1974, at the first World Food Conference, Henry Kissinger promised that 'in ten years' time no child will go to bed hungry'. There was reason for optimism. For years, global food production increased faster than the rate of population growth. Widespread use of high-yielding varieties of seeds, as well as massive investments in irrigation and fertilisers, brought tangible returns. For about two decades, the Green Revolution boosted food security and livelihoods for hundreds of millions of people, primarily in Asia. But, in the Anthropocene, humanity is hungrier than ever – nearly 1 billion people don't have enough to eat, while 1.5 billion people are overweight or obese.

How will humanity feed itself in the Anthropocene? More precisely, how in the next four decades will we produce more food than we have done during the last 10,000 years? Food production needs to double by 2050, in the face of uncertain water supplies, climate change and degraded soils, and when almost all the best agricultural lands are already being used.[4] I headed to the farmlands to find out.

*   *   *

I drive through Gujarat deep into the far west of India, from the choking city of Ahmedabad across parched flat lands fenced by prickly-pear cacti. Tankers at the side of the road are delivering water to women who wait in line with steel pitchers that they will hoist to their heads, some in double- or even triple-height stacks, for the laborious trek back to their homes. Piped water has yet to reach many villages here in this, one of the best-developed states in India, and their wells have dried and won't be replenished until the scant monsoon rains arrive in another seven months' time.

The village of Raj-Samadhiyala is bang in the centre of the state, about fifteen kilometres outside the dirty industrial town of Rajkot. Like most villages, people here are utterly reliant on their crops for food, and what-ever income they generate comes from selling any surplus. But this is quite unlike any other village I have seen in India, with clean, tree-lined avenues devoid of litter and sewage, and a glistening lake of water shining like an oasis in the dry dust. Pink, violet, yellow and red flowers crowd the verges, where they are watered by strategically placed hoses. In 2007, Raj-Samadhiyala hit the headlines for being the cleanest village in the country. A year later, it won the national 'Tirth Gram' award for being the most crime-free village. In this land of 1.3 billion people, 85% of whom live in rural villages, something extraordinary is going on in this village. The usual gaggles of children that line village lanes are strangely absent here – I discover them later in the school, sitting diligently in neat rows beneath a colourful depiction of Saraswati, the Hindu goddess of learning.

I am greeted by Raj-Samadhiyala's village chief (or *sarpanch*), Hardevsinh Jadeja, a sprightly man in his sixties who, despite the heat and rural loca-tion, is wearing a pressed business suit. His shirt collar valiantly maintains its stiffness, but starch is no match for relentless perspiration and before long the cotton slumps limply about Jadeja's neck, as he shows me around the supremely managed village. With obvious pride, he explains the intri-cacies of what he terms 'non-Gandhian guided democracy', whereby the village is divided into units of twenty-five householders who vote (voting is compulsory) for a representative. The representatives then decide the rules by which all the villagers must live 'or face strict punishment and even be kicked out from the village – this is extremely important'. The system works on the basis of social equity, in which all castes get equal vote, and the village community helps those who need it, in every way from physical labour to financial assistance. But idleness is not tolerated. This social structure, which has produced such an undeniably clean and functioning society, owes its success to the unique way the village manages and shares its most valuable resource: water. And that is due in no small part to Jadeja's ingenuity and foresight – he decided some thirty years ago that clean water and sanitation were essential to underpin development, and that his village should have them.

After each monsoon, the village wells would fill, indicating that the underlying aquifer here is rain-fed. The problem was that most of the monsoon rains were wasted, streaming off and disappearing before reaching the aquifer. With the monsoons in the region becoming briefer, the deluges stronger and the evaporation of surface water faster, due to climate change, the problem of vanishing water became more severe. Most of the boreholes were dry and the water table had fallen to a depth of 250 metres. Harvests were suffering, people were becoming poorer, neighbouring villagers were unwilling to marry their daughters to grooms from Raj-Samadhiyala, and young men – and then entire families – began migrating away to the cities. The village was so dry, it was declared by the government to be a 'desert area'.

But the village had a remarkable asset: Jadeja. He is unusually well educated, with a postgraduate degree in English literature, and was pursuing a successful career in the police force – and had just been offered the job of deputy commandant in the Central Reserve Police Force – when he was asked by the Raj-Samadhiyala village council to be their *sarpanch*. The force's loss was the villagers' gain, as Jadeja opted for the less secure job, determined to transform the fortunes of his village.

As long ago as 1985, Jadeja approached scientists at the India Space Research Centre in Ahmedabad, who used satellite mapping to reveal the underlying geology of the village, which is located in a region that saw volcanic activity millennia ago. The maps showed up lineaments through the subterranean rocks, allowing him to chart the probable flow of rainwater through to the aquifer beneath. Jadeja mobilised his villagers: where a lineament was seen on the satellite maps, they dug down until these subsurface dykes on the route to the aquifer were exposed. At the highest ground, relying on gravity, and where the lineaments ran, he created a catchment lake for the monsoon rains – nobody is permitted to draw from this precious reservoir.

Using remote sensing and geographic information systems, Jadeja was able to work out the best locations for above- and below-ground water storage. Under his directions, over the following fifteen years the villagers constructed more than fifty check-dams to slow the monsoon deluges and give the water time to percolate to the aquifer below. Many of these subsurface channels used to hold water but had dried over time, but when excavated and cleared, rainwater once again flowed through them.

The villagers then planted more than 40,000 trees whose shade helps reduce soil evaporation, and whose leaves can be used as mulch fertiliser. The fields were further protected with terracing and raised earthen walls to slow down rain that might erode the soils. And it works. The rainwater trickles down and fills the aquifer so that village wells and boreholes are full throughout the year. Despite a severe drought, there is even surface water in the reservoir. Since Jadeja devised the scheme, not a single water tanker has needed to stop at the village, he tells me with justifiable pride.

The precious water that the villagers have worked so hard to collect is used prudently. Black plastic pipes run alongside the plantings, delivering sufficient but measured drips of water to the base of each plant, where it is immediately absorbed by the roots. Drip irrigation increases yields, while saving labour, conserving water and preventing salination of the soil. In

Raj-Samadhiyala, the resultant bountiful harvests – from cereals and vegetables to cash crops like cotton – have produced social and material wealth. Despite the ongoing drought, farmers here earn ten to a hundred times more than their counterparts in neighbouring villages, and reap three harvests a year. Every house conserves its own rainwater and has reliable piped water and a toilet 'that must be used'. Littering or spitting are met with a fifty rupees fine, and it is mandatory to educate girls. There is zero crime here, Jadeja says, 'and migration happens in our direction', with a waiting list of families wanting to live in the village.

I visit Gulab Givi's beautiful two-storey house, with bougainvillea blossoming in his garden and over his sun terrace. Givi, a member of India's 'beggar caste', used to spend his days as his caste dictated until the village council intervened. He now owns a shop and land (with loans that he proudly tells me 'have been repaid in full'). Still more impressive is the recently improved situation enjoyed by the poorest villager, of the 'untouchable' caste, previously the village's toilet cleaner, who now has a good house and earns an acceptable wage as a sweeper.

Wandering around some of the drip-fed fields of beans – the third planting of the season – Jadeja shows me how he is deepening the catchment lake in order to reduce the surface area and so cut losses due to evaporation. 'We need to improve drainage for the fields – I am experimenting with different types of soil – because over the past five years the rains have been coming too heavy, too early and rotting the crops before harvest,' he explains. It's a challenge that I am sure Jadeja is equal to.

Like Mahabir Pun and Chewang Norphel, Jadeja is a visionary and charismatic leader who has been able to dramatically improve the fortunes of his community. While the Gujarat state government plans the ridiculously inappropriate 'Indus Oasis' – a casino and entertainment complex modelled on Las Vegas, complete with an artificial lake with a wave machine and beach for watersports – in the heart of the desert, Jadeja is quietly working to replicate his success in sustainable water management with other dry and hungry villages in the region. Raj-Samadhiyala has become so successful, it is looking beyond agriculture and starting up the state's first bicycle factory.

On the drive back to the city, past brown and empty fields, I wonder why other villages are not reaping similar rewards. 'They are not intelligent enough,' Jadeja had told me. But I am struck more by his comment that since the project's success, not a single national government minister has visited. And yet, to feed 10 billion in the Anthropocene, schemes like this must be replicated in a coordinated way. The agricultural Green Revolution, which boosted crop production across Asia through irrigation and fertilisers, lifting millions out of hunger, relied on prodigious water use. And this has only increased, together with use by industry. Now, the challenge is repeating this success but in a sustainable way. In an increasingly water-scarce world, agriculture needs to store more water and use less, including through methods like drip irrigation.

Some 69% of India has now been classified as 'dry land' (25% is desert), a poor prospect for a country where the main industry is agriculture. And it's a pattern being repeated across the world from China through Africa and South America. In north China, for example, 24,000 villages have had to be abandoned because of sand-dune incursion, while Nigeria loses 350,000 hectares of cropland to desert each year.[5]

With lakes and rivers dry or too polluted to extract from, and rain too unreliable, farmers are using aquifers, which now provide two-thirds of irrigation water. Some are periodically replenished by underground rivers or monsoon soak-through, while others hold 'fossil water', trapped thousands or millions of years ago between rock layers. Fossil water is not replenished: once it's gone, it's gone. And India uses the largest amount of fossil water for irrigation in the world (sixty-eight cubic kilometres per year), and to such a critical level that international experts are warning that the country is facing agricultural collapse and starvation.

Much of the shortage is due to abysmal water management. There is a lack of reservoirs and other storage, with almost no recycling of used water. More than 700 million Indians lack sanitation (half of Indians defecate in the open, with the minister for rural development in 2012 describing his country as the world's largest 'open air' toilet), untreated sewage and other waste flows directly into rivers and canals, and what piped water there is is tapped by desperate people without legal access, and plagued by leaks. Although India has managed to send a rocket to the moon, it cannot pipe water to the middle-class residents of even its most modern city, Delhi – or not for more than a few hours a day, and even then the water is often contaminated. In Rajkot (population 1.6 million), the closest city to Raj-Samadhiyala, the authorities only manage twenty minutes a day. People spend hours waiting with buckets for private trucks to deliver water of exorbitant cost and dubious quality. In other places, they use tube wells, which are either shared by the community or sunk into private land, in which case there will be a charge for the water.

The development of cheap diesel or electric pumps, coupled with cheap energy, led to an explosion in groundwater pumping over the past twenty-five years, with over a million new tube wells sunk each year. One-sixth of the nation's electricity is dedicated to irrigation pumping, which leads to frequent blackouts, especially in rural areas. Farmers have needed to sink ever more powerful pumps ever deeper. As a result, in many places, particularly low-lying Gujarat, groundwater has rapidly dropped below sea level, allowing salt water to seep into the aquifer and pollute the wells. The state has even started a programme to recruit farmers into alternative industries with more certain prospects such as salt-working.

However, Gujarat, like other dry states in India, already has a tradition of harvesting and storing precious rainwater to use for drinking and watering food crops. I visited several of these underground 'tankas' – stone cellars clad in waterproofing lime to hold rainwater that falls on the roof – in Amhedabad, Rajasthan and Maharashtra. The ancient tanka system, which dates back thousands of years, also includes community storage reservoirs, such as the deep step wells in the Thar Desert, and temple bathing pools. In Gujarat, as elsewhere, the British colonial rulers banned the use of household tankas, ordering them to be filled in because they feared they were being used as hiding places for rebels. In most places, the ancient system has fallen into disrepair, channels have become blocked up, animals have been allowed to foul them and new buildings no longer incorporate water storage. But in recent years, people have started to revive the old practice. In some villages, ancient stone reservoirs are being cleaned and repaired. And a younger generation is taking on the challenge of finding a way to live sustainably in the twenty-first century.

In a suburb of Bangalore, the 'Silicon Valley of India', I visit Vishwanath. Entering his home is an instantly calming experience. The 'Zen Rainman', as he calls himself, is a civil engineer by training and a creative experimenter by vocation, who is on a quest to find sustainable living solutions that are practical and affordable for the masses. As a result, his house has become something of a laboratory where he tries out various technologies. He built it along ancient principles, by first excavating the tanka cellar to obtain the clay for his sun-dried, pressed bricks. The architecture was designed to enhance air throughflow, the building's orientation heeds the position of the sun, and climbing plants drape the external and internal walls. As a result, there are no fans or air-con – unusual in a middle-class home here. The family's energy needs are met through PV solar panels and thermal water heaters, and there is even a solar cooker on the roof.

Also on the roof is a Chinese eco toilet and four different experimental plots, bio-zones representing a wetland, a grassland, a paddy field and a vegetable patch. The crowd of wonderful greenery and devices makes quite a contrast to the bare concrete of his neighbours' flat, baking roofs. Aside from the aesthetics, the farm with a view has an important purpose, to show that it is possible for city-dwellers to produce some of their own food. In the next few decades, as climate change makes the small rural farmers' existence too precarious and they make the inexorable journey to the cities of opportunity, small farms and subsistence farmers will be squeezed out, as they already largely have been in industrialised countries. But many will continue to tend plots in the city. I have seen all manner of vegetables and fruits flourishing in the meanest corners of crowded slums, and they make a valuable contribution to poor people's diets. According to the Worldwatch Institute, around 800 million people grow food in urban spaces, producing as much as 20% of the world's food. The Zen Rainman is figuring out how to grow a rooftop farm, using rainwater alone.

Even his small roof plots require a substantial amount of water, but Vishwanath's water conservation is easily up to the task. Bangalore has probably the world's most expensive water supply. All its water is pumped up to the city at altitude from the Cauvery River, using eighty megawatts to lift more than 100 million litres of water per day. On top of this, the city and suburbs use a combined eighty megawatts per day in pumping water from the more than 200,000 boreholes that descend 1,250 feet down into the fossil-water supply. The system is subsidised by the government in such a way that the poorest end up paying the most – they are effectively paying for the richest to waste water. This is because piped water and borehole pumped water is subsidised: householders using more than 25,000 litres a month get a 450 rupees rebate. But the poorest cannot afford piped water or to dig and pump a borehole, so they rely on privately owned water tankers to drive up and supply water at around three rupees per litre.

Vishwanath's home directly funnels rainwater from the roof and several other traps around the walls and garden into his tanka, which meets all the household's annual needs, and then some. Since building his home, he has designed and built several different rainwater harvesting systems for similar homes in other middle-class suburbs, as well as ones for the rural poor, which have been installed in more than 1,000 villages. Vishwanath's and Jadeja's examples show that the most ancient unit of social living, the village, can be radically improved through relatively simple and inexpensive means. In the first age dominated by citizens rather than village-dwellers, such village transformations could allow those who stay rural to increase their opportunities nevertheless.

From India to Africa, where farmers rely almost exclusively on the rain to water their crops. Just 10% of African farmland is irrigated, mostly in Egypt or South Africa, compared to 26% in India and 44% in China. In sub-Saharan Africa, irrigation has actually decreased over the past thirty years, as schemes built during the colonial era have fallen into disrepair. In place of pipes, aqueducts and other infrastructure is the yellow plastic jerrycan, such a ubiquitous sight throughout Africa it's hard to imagine how water was transported before its arrival. In some places, it is the only concession to modernity (apart from mobile phones); in cities, it is a reminder that for all the fancy concrete and glass buildings, basic water provision is still woeful. In semi-arid and desert areas, it is commonplace to see women balancing the bright yellow containers on their heads, walking for forty kilometres or more, three times a week, to collect water for their village. Such places, where water for drinking is very hard to come by, have traditionally supported nomadic pastoralists or shifting agriculture – only 14% of Africa receives sufficient rainfall for arable crop production and even this land is subject to periodic droughts. But as populations rapidly expand and deserts spread across arable land, farmers are trying to cultivate crops in semi-arid conditions.

In Kyakamese village, in the north of Uganda, I come across Byarindaba Robinah sitting in her field, hard at work harvesting peanuts under the full glare of the equatorial midday sun. Her flowery sleeveless blouse is damp with perspiration and her skirt is rucked up under her on the dusty ground. Her eyes are bright and intelligent.

'The rains have been very poor this year,' she tells me, tilting up a full-cheeked smile, as her hands continue busily stripping peanuts from the dry weeds. 'In a good year, I can get twenty bags from this acre, but this year I think I will just get three or four bags.'

Each bag of nuts will net her around 100,000 Ugandan shillings ($53 or £33), although this year the price is down, so she'll just make around 80,000 per bag – losing more than 80% of her harvest is an expense Robinah can ill afford, and yet she smiles and shrugs her shoulders. 'The rains are getting worse and more unpredictable every year – we don't even have proper seasons any more – but I will persevere with farming for as long as I can. It is a good kind of work to make food for people,' she explains.

Robinah, 35, is new to farming. She and her husband used to work as teachers in Masindi, a town in the north-west of Uganda, but they found they couldn't make enough money to support their four children through school. 'Schoolteachers' salaries are very low,' she says.

The couple bought a three-acre plot of land, built a house and began farming. Nobody in Uganda is really 'new' to farming. In a country where more than 90% of the population is employed in agriculture, most people grow up tending a vegetable plot at least. Robinah's parents were tenant farmers, working other people's land, until they died – her mother of Aids before she was 40; her father of a bicycle crash a year later (average life expectancy in Uganda is 51) – leaving Robinah an impoverished orphan who had to quit school to work. After her marriage, Robinah's father-in-law supported her through teacher-training college and helped the family out with loans to buy their plot.

The fields that she sits in should be more than enough to provide for the family. 'But the climate is changing,' Robinah says. Rainy seasons that used to be predictable, allowing farmers to plant and harvest at well-defined times, are now so sporadic that no one knows whether the rains will last weeks or disappear in a couple of days – whether to plant at the first sign of rainfall or to wait and test its constancy, risking missing out on what little water there is. Some farmers are learning to make the most of the small amount of rain they do get, by planting on raised ridges or earthen mounds so that the rainwater soaks into the roots rather than immediately running off. Low-key initiatives like this are time-consuming, but are readily combined with manual watering, drip irrigation or flooding with hoses.

Robinah had been planting fields of bananas and maize, but this year she moved into more drought-tolerant crops including cassava, sorghum and peanuts. Crops like cassava – a South American tuber that's widely grown in Africa – can mean the difference between starvation and survival for poor farmers. Cassava not only tolerates drought and poor soils, it actually thrives in hotter temperatures, with yields expected to rise 10% by 2030 because of climate change.[6]

What about improving your yield with fertilisers, I ask. 'I've never used fertilisers, they are just too expensive,' she explains. Labour is another difficulty – she has no oxen and so the field must be ploughed and harvested using hand tools. 'It's hard to find any labourers to help, and they are very expensive,' she says. 'And because it takes me so long to harvest, animals come in and eat my crops or destroy them.'

Robinah buys her seeds from the local market. She knows that she could buy better seeds from specialist shops, but there is a high incidence of poor-quality 'fake' seeds on the market and anyway, good seeds cost four times as much. To improve her yields, Robinah would have to spend money she doesn't have on seeds, fertilisers and pest control, in a gamble that depends on the increasingly unpredictable rains returning to reliability.

Robinah is comparatively wealthy and even she can't afford it. The week before, Stanbic Bank Uganda, Alliance for Green Revolution and Kilimo Trust pledged a $25 million fund for loans for small-scale farmers. I ask Robinah whether she would apply for such a loan and she shrugs in response: 'It would be up to my husband.' Women are not allowed to own land or property in Uganda, even though they are responsible for 75% of the labour.

In the Anthropocene, women are key to improving yields across the developing world. They make up over half the global agricultural workforce and more than 80% of it in sub-Saharan Africa, but their potential impact is stunted by sexist policies and cultural practices – they own only 1% of all land, for example. If women farmers had equal access to finance, education and other resources, their yields could increase by as much as 30%, and malnourishment could fall by 17%, a 2011 analysis estimated.[7]

Robinah's plot, like so many others across the region, would benefit enormously from an irrigation system. Unlike some other, very dry areas, which need some sort of water catchment or rainwater harvesting, she has a bountiful water table within a few feet of her plot. But, until a few months ago, even acquiring drinking water was a big problem. She shows me a cloudy pond of fetid water, about two metres in diameter, which people had used for drinking and washing, and which they shared with animals. Earlier in the year, her village applied to the Busoga Trust, a British NGO, for assistance in building a hand-dug well with a pump. The village supplied the labour, the cost of the bricks, and hardcore; the Busoga Trust supplied the pump, cement and technical help. Within three weeks, their pump was operational. The skin rashes, diarrhoea and stomach pains the community suffered from using filthy water disappeared within two weeks. Malaria incidence has plummeted because nobody uses the open pond to gather water and wash any more. Unlike many other NGOs working in the country, the Busoga Trust is not duplicating the efforts of other agencies and charities – it's doing an impressive job that improves lives almost immediately. I visited five different villages with the trust's coordinator, Ned Morgan, an earnest 26-year-old from Boston, whose fair complexion and freckles ill prepare him for his long hours outdoors in rural Africa. In each village, the easy-to-blush Ned was welcomed as a kind of pump-delivering water-god.

The army of NGOs providing drinking water to villages in Africa offers a blueprint for how the irrigation problem might be solved, one village at a time. While large-scale canal building and water-diversion schemes have worked well in places like India, the costs are prohibitive for many African governments. Instead, small-scale water-harvesting and storage projects at the individual or village level, coupled with simple groundwater pumps, could bring immediate results. Shallow hand-dug wells can be used where the water table is high enough in villages with no access to borehole drilling machinery. But it's an arduous process because the pumps are not designed for agricultural use, so villagers must water by hand using a can. Of course, providing potable water is something the Ugandan government should be doing, not an NGO. And in some areas it is installing handpumps, but the Chinese-made ones the government provides are easily dismantled and stolen for scrap metal by gangs.

I cross Kampala city through driving rain to meet Ambrose Agona, director of Uganda's National Agricultural Research Organisation. He's a man so enormous his office furniture seems toylike, but his smile is as generous as his handspan. We get chatting about the challenges facing the Ugandan small-scale farmer, particularly in the country's north. At one point, I have to lean in close to hear him, owing to the clattering noise of heavy rains on the windows – 'It is so terribly dry up there,' he is saying.

Most of the country has two rains a year, the first from March to June and the second from August to November. This year, the first rains failed across the region, leading to starvation and hunger even in normally wet regions, such as Jinja, a source of the mighty Nile. The drought hit the northern and eastern parts – the borders of Sudan and Kenya – most severely, where they have not had good rains for the past four years and people and their animals have starved to death. 'The rains used to be so predictable that farmers would sow their seeds before the onset, they were so confident of water,' Agona says. 'There are parts of Uganda that have been in permanent famine for the past five years.' Food prices in Uganda, the 'bread basket' of East Africa, have soared to levels where even staples like beans and sugar are out of reach for many people.

The general trend towards drying of East Africa over the past fifteen years (increasing severity and frequency of droughts, and shorter, more erratic rains) fits leading climate-model predictions for the region. Africans are expected to be the worst hit by climate change, suffering worsening droughts that will directly reduce food availability. Whether you believe it is relevant or not – and most people I speak to think it is – Africans have contributed the least to climate change, belonging to the only continent bar Antarctica that has not yet significantly industrialised (indeed, many parts have been deindustrialising since the 1980s).

In the Anthropocene, deaths from climate change will not be caused by the weather per se (apart from a tiny percentage of cases), but rather by the fatal synergy between climate change and a catalogue of other misfortunes: natural disasters such as locust plagues; fake seeds; low productivity due to poor health; poor governance and corruption (that sees, for example, much of the agricultural budget vanish); social and gender inequalities; poor infrastructure; and trade laws and protectionist agreements that favour rich countries.

If the eighteenth century was about Westerners stealing the people of Africa through the slave trade, and the nineteenth and twentieth centuries were about Westerners stealing the resources of Africa through colonisation and unfair trade, then surely the twenty-first century is about Westerners stealing the water from Africa by climate-change-reduced rainfall and increased evaporation, and through direct land grabs of the most fertile, wet zones.

Foreign corporations and governments are buying vast tracts of prime land in poor countries at knock-down prices to grow food and energy crops for export. Investors are generally welcomed to the country with water access and tax and infrastructure incentives by governments that are either corrupt or hoping for help to develop their own agriculture. However, foreign investors in land are very rarely in it for the long haul, most selling on within five years. They have little interest in sustainability and, in a matter of months, their vast, mechanised operations can devastate a site's soil, biodiversity, water quality and forests, and threaten the livelihoods of billions of peasant farmers. In many cases, investors require that their newly acquired land be cleared of people, or they buy 'unoccupied', state-owned or common lands. But there is almost nowhere on the planet – certainly no fertile land – that really is unoccupied. These lands are vital farming, grazing, or hunting-and-gathering resources for the world's poorest people, who have no formal title to it, so governments feel free to dispose of it at will.

Agricultural commodities speculation has soared sharply since 2005 (increasing twentyfold to 2008 alone), with 98% of commodity contracts being traded as futures for profit rather than for the actual goods.[8] It has contributed to a sharp increase in food prices, exacerbated by weather-hit crop losses, which has been linked to food riots in multiple countries and the bloody civil disturbances of the Arab Spring. For the world's poor, who spend 70% of their available cash on food, even small price hikes can be life-threatening. More than 44 million people were driven into extreme poverty by rising food prices between 2010 and 2011 alone – every 1% rise in food prices pushes another 16 million people into hunger.[9]

A major reason for the ousting of Madagascan president Marc Ravalomanana in March 2009 was the ninety-nine-year lease he had agreed with South Korea's Daewoo corporation, handing over 1.3 million hectares of Madagascar's best agricultural land. The firm had planned to use a South African workforce to grow 5 million tonnes of corn a year for export to land-poor Korea to ensure its food security. With two-thirds of the Madagascan population in poverty and more than 500,000 relying on international food aid, the deal was quickly repealed by the incoming president, Andry Rajoelina.

With such limited land available, and so many mouths to feed, many believe the best way forward is for large corporations to use the world's land for technologically advanced mass production. I'd suggest not. The better solution for a hungry, crowded Anthropocene is in supporting local farmers to improve productivity, and to buy from them.

I travel to Ologara village in Kumi district, eastern Uganda, to meet a woman who has transformed her family's fortunes in exactly that way.

'Farming is a very noble profession. It's the backbone of our country.' Breaking off from weeding, Winifred Omoding stands up to emphasise her assertion and, at full height, she's barely as tall as her sunflowers that bloom healthily around her.

Diminutive in stature she may be, but there's nothing small about what Winifred has managed to achieve. While nearly 80% of the people in the region depend on food aid to survive and her neighbours' plots are pitifully brown with shrivelled maize and sorghum clinging to half-height stalks, Winifred's fields are an embarrassment of green. Her sunflower, sesame and cassava thrive amidst the cacti and dust that surround her small village of 500 people.

It's taken Winifred many years of hard labour to get this far. Just a few years ago, civil war had left her family traumatised, her crops were failing, and she was struggling to feed her family. Now, the 41-year-old farmer is not only producing enough food for her husband and nine children, but making a healthy profit by selling the surplus. Winifred still has a long way to go before her land attains its full yield potential but it's a challenge that she's more than ready for. 'I enjoy farming very much,' she smiles. 'Even when I just sit with someone who talks about agriculture, I am very happy.'

It was not her intended occupation, though. Like the vast majority of farmers in the developing world, Winifred fell into it out of desperation, being unqualified for anything else. Both her parents were schoolteachers and Winifred hoped to follow them into the profession. But years of violent conflict put paid to that ambition. It's a story all too commonly played out in this part of the world – every country in sub-Saharan Africa (with the exception of Botswana) has seen development seriously impacted by regional or national conflict over the past three decades. The Ugandan town of Soroti and its surroundings, including Ologara village, where Winifred's family continues to live, was badly hit by a series of insurgencies in the late 1980s, following the election of President Yuweri Museveni. Heavily armed rebel groups, including the brutal Lord's Resistance Army (LRA), reduced the prosperous region, with its educated populace, its thriving cotton industry and manufacturing base, to a war zone.

A large proportion of the population fled to hastily organised 'safety camps' for internally displaced people. 'The camps quickly filled up and there was no space to accommodate us there, so we stayed in our village,' she explains. Those left behind, including Winifred's family, were subjected to repeated violence by the LRA, armed cattle raids by pastoral tribal groups taking advantage of the lack of law and order, and human-rights abuses from government soldiers that were supposed to be protecting them.

'It was a very, very hard time. Whenever we heard shootings, we would have to leave everything and run into the bush and hide,' she recalls. 'The rebels killed my older sister, who left behind two small children. I was in school, but then they killed our dad and my mother couldn't afford for us all to go to school. Education was very important to my parents so it was very hard for her to see us all leave school like that, but the rebels had burnt our house down, taken our seven cows and goats and sheep, destroyed our crops and uprooted our cassava as well. We had nothing.'

It's difficult to fully appreciate the extent to which conflict has set back agriculture here. People are beginning to return to their villages – although the disarmament process is ongoing and tribal tensions remain a problem in other parts of the country – but some have spent decades away from their farms and have forgotten how to manage the land, unable to pass on farming skills to the next generation. Even those who remained in the area were often unable to use their land, which became choked with weeds. All industry moved away or disappeared during the conflict, including cotton manufature, which previously supported a large proportion of the population. And with all the animals gone, there are no oxen to plough the fields or provide manure to fertilise the crops.

Winifred applied to join the police force, but that was ruled out because she lacked school-leaving qualifications. Aged 18, she married Ephrem Omoding, 27, who had left school early to become a soldier and fight the rebels. Around that time, Ephrem's father died: 'He was bewitched and then poisoned to death,' Winifred says in a matter-of-fact tone. With her father-in-law's untimely death, the couple inherited eight acres of land.

Winifred and Ephrem built a one-room, thatched-roof, mud-and-stick house for their growing family on their new land, and got to work on the slow struggle to produce food.

The inheritance was a lucky break for Winifred – the majority of farmers in Uganda are smallholders with just one acre of cultivatable land. More than 90% of Ugandans supplement their diet – and sometimes their income – from these 'gardens', even if they have another primary profession. Around 85% of all farms, producing half the world's grains and more than half of all calories, are smaller than five hectares.[10] And these plots are shrinking because the population is increasing, and land is split between sons in ever smaller fractions down the generations. Smallholder farming can be basic, labour-intensive work, but it is how the majority of rural people feed themselves, and is part of their identity. Efficient, vast monocultures would not be appropriate across most of Africa, in part because of their poverty of biodiversity (including a lack of pollinators), but also because climate change makes it too risky to put all the meals into one crop. It would be far safer and more sustainable to diversify food production, and include local and indigenous vegetables, which might survive to feed families if the main crop is spoiled by disease, floods or drought.

Ologara village is in a semi-arid region, meaning that it gets less than 800 millimetres of rainfall per year. Cacti and small shrubs are the only naturally occurring vegetation. Deforestation has left the soils thin and loose; when the wind picks up, dust storms darken the sky. People here cannot afford to leave their fields fallow for even a season – they practise crop rotation, but it is not very effective and the soils are exhausted and infertile, leaving them vulnerable to problems like striga, a parasitic weed that destroys sorghum. In Africa, most smallholders cannot afford to replace the nutrients and minerals taken up by their crops each season, so three-quarters of farmland is degraded, with harvests in sub-Saharan Africa declining by 15–25% per year.[11] Already, grain yields in most of Africa are only around one tonne per hectare, compared with 2.5 in South Asia and 4.5 in East Asia. African farmers have therefore had to extend their cultivated land through slash-and-burn, causing degradation of ecosystems, loss of water-storage capacity and soil erosion.

Soil is running out around the world. Every year, 75 billion tonnes – more than 100,000 square kilometres of arable land – is lost.[12] Around 80% of global farmland is now moderately or severely degraded and, in the past forty years, one-third of cropland has been abandoned.[13] In Europe, soil is being lost seventeen times faster than it can be replenished naturally; in China, it's fifty-seven times. Every year, more land is sealed beneath buildings and roads. And because cities have usually grown from settlements on the most fertile land, vast urban centres are now squatting on some of our best soils. Soil erosion has brought down empires in the past, and is a major concern in the Anthropocene. Soils aren't essential to grow food – I've seen vegetables growing on floating lily beds in Bangladesh, on nutrient-rich membranes in Australian greenhouses, and in plastic bags in Spain – but even if we could produce enough food without soils, it would be impossible to replace the many ecosystem services that soils provide, from water management to pollution remediation to supporting trees and houses.

As in Ologara, deforestation is a major driver of global soil loss – once shrubs and grasses have been cleared, there are no roots to hold the soil, nutrients and water in place. A dry spell can turn the ground to dust that is easily blown off the land, turning fields into deserts. If heavy rains then fall, as they frequently do following a dry season in the tropics, then an even greater amount of soil will be washed away in floods, creating landslides and silting up watercourses downstream. Farmers are fighting back by planting trees and shrubs that help keep soils moist, buffer the winds and slow run-off of rainwater. West African tree-planting programmes have achieved remarkable improvements in soil fertility, and in Indonesia, I found farmers planting vetiver, a deep-rooted Indian grass, to protect their soils from erosion. As a bonus, vetiver reduces weed and pest incursion, produces a marketable oil and is also useful for animal feed.

Recycling biomatter is the easiest and cheapest way of replenishing the soil and, in many places in the world, livestock dung is the only fertiliser available. But it is becoming scarcer because, in trying to preserve the few trees and shrubs around cropland, farmers are burning dung for cooking. One solution, which I saw on a farm in India, is to use the dung to feed a biodigester – a tank in which bacteria break down the waste. The methane produced by the bacteria was used to fuel cooking stoves, and the decomposed manure was spread on the fields as fertiliser.

But Winifred didn't have any animals to fertilise the soil or provide labour. 'Our ground was so bad that at first we got some of the men in the village and tied their hands behind them to the harness of the plough and in that way we ploughed our field using manpower,' Winifred says. 'It was a terrible time. Many people went hungry and many children died during the 90s.'

Winifred planted fields of cassava and tried out others, from maize to peanuts. 'I bought the seeds from the market, but they are not good. They are mixed up from different farmer-producers and so you don't know what you're getting. And there are so many fake seeds around.'

The couple struggled on for a decade, barely making enough to survive, even though they had a plot that was three times the size of many of their neighbours'. 'Many days we only ate one meal a day. Cassava and beans mainly,' she recalls. Eventually, the couple's luck changed. Political stability returned in the early 2000s, meaning the men gave up soldiering and could help in the fields. And in 2003, aid groups helped Winifred and other Ologara women form an agricultural 'microloan' cooperative. In exchange for making small deposits into the cooperative, the women could get small loans for farm-improvement projects. Winifred used her first one to hire oxen to plough and weed her fields.

Then, realising that she was never going to get good yields while she used poor seeds, she made contact with scientists at the government's new National Semi-Arid Resources Research Institute (NaSARRI), based a few kilometres away in Soroti, and asked for advice. Florence Olmaikorit-Oumo, the social-outreach worker at NaSARRI who helps connect farmers to the relevant scientists, remembers Winifred's first visit clearly. 'She was in a terrible way with her harvest having just failed again. But she was so intelligent and hard-working,' Oumo says. 'We knew she would have some success.'

The timing was right. The institute was trying out new drought-tolerant and disease-resistant varieties of crops suitable for semi-arid conditions, and was looking for reliable farmers to help multiply their pure seeds in the field. Winifred was a prime candidate. 'You could see that she really listened to the advice you were giving her and wanted to learn,' Oumo says.

In 2006, Winifred was given training in how to prepare her land by, for example, ploughing in the silage and stalks from the previous harvest and allowing this to rot in the soil before ploughing again and sowing the seeds. She was advised on which crops to grow: maize was out (too thirsty), sorghums, cassava and millets were in. Then she was given access to the institute's latest seeds, which she had to buy using a loan.

Winifred saw immediate results – the next season's harvest was so successful that she had a surplus to sell on the market and had also produced a few kilos of good seed to sell to other farmers via NaSARRI's network. Excited, she started looking at growing other crops for cash.

This, she says, was the turning point in her family's fortunes and also in her relationship to agriculture. 'Before, I farmed to feed my children,' she explains. 'But now, I think of it as a way to make our lives better and to become more rich. It's exciting to try new things.'

One of the new things Winifred has tried is an improved variety of sunflower developed by NaSARRI scientists, which is drought-tolerant and produces seeds that contain high yields of oil when crushed. She sells these seeds to farmers through a cooperative, which she is a member of, that has been given sunflower-seed processing machinery by an NGO to make oil. In this way, Winifred is able to add value to her crop because the bottled oil sells at a much higher price than the seeds that went to make it. And, as a bonus, the 'cake' by-product of crushed seeds is a highly nutritious animal feed that she sells, as do the other members of the cooperative.

The profits have allowed the couple to buy another four acres, and she is now planting a new high-yielding, drought-tolerant variety of sesame that produces the consumer-preferred white seeds, rather than the browner natural variety. 'It is ready to harvest in just four months, rather than taking six months like the local variety of *sim sim* [sesame seeds],' she says, while wading through her bumper crop that's ripe for harvesting.

Evidence of the improved yields is visible beyond her fields. At the edge of her cassava plantation, on a clearing by the roadside is their small mud-and-stick family home, overshadowed by a much larger, half-finished brick house with several rooms. 'We built the walls from our sunflower money, and we are hoping to buy a roof from the next *sim sim* harvest,' Winifred says proudly.

So what difference has this new productivity made to Winifred? All her children are now in private boarding schools, except her eldest daughter who, at 21, is training to be a teacher. 'She will continue in the profession of my parents – the profession I wanted for myself. Education is the most important thing for our country,' Winifred says with conviction. 'It will help Ugandans to develop our nation and economy and it will stop our young people from becoming thugs or rebels – if you are educated, you always have an opportunity to make some money.'

What if her children wanted to become farmers, like her? 'Farming is very hard,' she says, thoughtfully. 'But they should have a garden to grow some cassava.'

Winifred is only at the beginning of her journey. An orange-juice production factory is under construction in Soroti and Winifred sees an opportunity. She is planning an orange grove, but water will be a big issue. The rainy season has failed completely for the last two years. In the future, farmers like Winifred might benefit from geoengineering solutions to drought, such as cloud seeding (already being used in China), in which silver iodide is blasted into the sky to create tiny nuclei for water droplets to form around. This can produce rain, but could also lead to regional conflict and accusations of 'stealing' another area's rain. Another kind of aerial dispersal, solar radiation management (sulphate droplets injected into the stratosphere to reflect sunlight into space, which I mentioned in Chapter 2), could solve the problem of heat stress lowering yields. Without the heat, the extra carbon dioxide in the atmosphere would enhance photosynthetic activity, and the technique could lead to a 20% increase in global crop yields, researchers calculate.[14]

Winifred is not waiting. She has already taken out a loan for a treadle pump, and is saving up to pay for a hand-dug shallow well. This, she hopes, will provide water for her oranges, although she is still trying to get advice on how to construct irrigation channels. Winifred would also like access to a tractor. 'If the rains come, and the oxen are sick or not available, then you can miss your opportunity to sow seeds,' she says. 'With a tractor, you can plough, weed, harvest and dig stony ground. It would be wonderful.' There are other areas, too, she hasn't yet begun exploring, including trying pesticides to reduce diseases like leaf wilt in peanuts. And, in common with nearly all farmers in Africa, she can't yet afford fertilisers.

The Asian 'Green' agricultural revolution, which was funded by the United States Agency for International Development and the World Bank, passed Africa by, and irrigation, fertiliser use and mechanisation are rare to non-existent, leaving the continent the world's least efficient food producer. Farmers in the US grow five times as much corn per acre as African smallholders, and synthetic fertilisers are key to this. Their invention altered the nitrogen cycle and changed our planet. More than 100 million tonnes of nitrogen are removed from the atmosphere each year by the fertiliser industry, helping feed half the world's people. The problem is, only a fraction of this nitrogen ends up in our food. The excess lingers in soils, lakes and oceans, fertilising massive blooms of algae that can be toxic and use up the dissolved oxygen, suffocating other species. This causes dead zones for kilometres, and is a particular issue in China, where intensive farming has polluted whole river systems. A solution is targeted application – or 'micro-dosing' – directly to each plant, which minimises waste, and catchment reed beds can filter out any run-off before it enters waterways.

Organic farming advocates prefer pre-industrial methods of soil enrichment, such as muck-spreading, crop rotation with legumes (which harbour bacteria in their roots that can 'fix' nitrogen directly from the air, enhancing the soil's fertility for other crops while requiring little or no fertiliser), and so-called 'conservation agriculture', in which the stalks and leaves that are not harvested are left in the fields to mulch down. These are all useful, environmentally sensitive and cheap, and will remain important even when artificial fertilisers are widely available. The latter cannot replace lost organic matter in soils, and mulch is a valuable way of restoring structural integrity and ensuring that nutrients remain in the soil rather than washing out immediately – mulching alone can double yields in some cases. But, as with all fertilisers, manure runoff can pollute waterways, and making manure releases its own potent greenhouse gas, methane. Legumes can be very labour intensive and take up space that farmers might prefer to dedicate to cash crops. And mulching, although effective, uses material that could provide valuable animal fodder, cooking fuel or biogas feed.

Environmental considerations, together with a certain cultural mistrust of solutions concocted in a laboratory and then applied to our food, have led many in the rich world to embrace organic farming practices. People who choose organic produce because they mistakenly believe that it is healthier, should be aware of the risks of rapid spoilage and the risk of pathogens such as E. coli. Organic farms, however, often support greater biodiversity, and are more likely to use natural pest-control methods, such as encouraging insects that specifically feed on the pests, or spraying an extract of seeds from the Neem tree, as I saw on an Indian organic farm. This natural pesticide is far safer than synthetic alternatives, and Neem trees also provide shade and fertiliser. Other techniques, such as draining rice fields in the middle of the growing season, and keeping fish or ducks (or encouraging animals such as ants and termites) among the crops, can increase yields, reduce greenhouse gas emissions and slash water or pesticide use.

It would be impossible to feed a growing global population using purely organic methods, however. Yields are on average 25% lower (and substantially lower still with some crops) than with standard methods, so vast new tracts of land would need to be used, further depleting our remaining forests and other ecological spaces.[15] In many areas, organic fertilisers simply don't contain the nutrients plants need: with much of Africa's soil severely depleted, animal fodder – and thus manure – is deficient in the nutrients needed by crops. There is no doubt that Africans need access to synthetic fertilisers if they are to catch up with the rest of the world in crop yields. Currently, with very few African countries producing fertilisers, and the high cost of transportation within the continent (because of high fuel prices and terrible roads), African farmers end up paying as much as six times the global average price for fertiliser – when they can find it at all. In the 1980s, the IMF and World Bank imposed conditions on development aid, preventing African governments from subsidising fertilisers. However, after decades in which the population was continuously on the edge of famine, and following a disastrous harvest in 2005 when more than a third of the population needed emergency food aid, Malawi's President Bingu wa Mutharika rebelled, and reintroduced fertiliser subsidies – the first sub-Saharan country to do so. Agricultural production trebled, and by 2007, Malawi was exporting its surplus corn to Zimbabwe and Kenya. A dozen countries are now following suit, and introducing subsidies.

In the Anthropocene, farmlands will have to work harder for us and the goal of food production is evolving beyond famine avoidance and the quest simply to grow enough calories to survive. Now, scientists are figuring out ways to make staple crops more nutritious, especially in places where people have a limited diet. Save the Children estimates that half a billion children will grow up physically and mentally stunted over the next fifteen years because they don't have enough to eat – 300 children die every hour because of malnutrition, more than die of malaria, Aids and tuberculosis combined. People need adequate protein, vitamins and minerals to develop healthy brains and bodies not just to survive, but to live a fulfilling life, contributing to society intellectually and in physical health.

I take a six-hour bus ride to Soroti in northern Uganda to visit one of East Africa's most promising weapons in the battle against hunger. I find him in a low-rise sprawl of rudimentary biology laboratories, surrounded by experimental crop fields.

David Kalule Okello heads a national research programme and yet he earns less than $6,000 per year; he's continually frustrated by the poor lab facilities and paucity of science funding in Africa, yet he won't consider working in the US or Europe; he's lived through three civil wars, yet he's only 33. The guy smiling above his crisp white shirt is remarkable.

'Initially, I wanted to be a doctor or a dental surgeon. But I ended up in the agricultural field and I think that was the best thing that could have happened to me,' says Okello. 'Our professor used to tell us: this is the best faculty. Look at the doctors, they cannot treat patients if they themselves are hungry. Look at the patients, they cannot recover unless they have food. Look at the lawyers . . . Without food nobody can work and you are the guys that can provide it.'

It's a responsibility that Okello shoulders with enthusiasm and grace. As head of Uganda's national peanut research programme at NaSARRI, he is charged with ensuring that the country is adequately supplied with the right quantity and quality of this important, high-protein food crop. 'I have to keep bringing new varieties into the market according to the farmers' needs, which means improving drought tolerance, resistance to pests such as rosette virus, and things like making the shell easier to peel in one go' – the last being important for processing.

With staples like rice and wheat being hit hard by climate change, scientists are working to develop varieties of nutritious alternatives. Peanut is particularly useful because it is not a nut at all, but a legume, so it fixes nitrogen. Peanuts are also a valuable cash crop because they can be processed into a variety of products from peanut butter and pastes to oil. So a large part of Okello's job is advising farmers how best to grow the improved varieties he develops at the institute, for their particular farm conditions. To reach farmers across the country, Okello has become a radio DJ, handing out advice and getting feedback from farmers on air. It's made him a bit of a celebrity, particularly among female farmers, who flock to his stall at nationwide agricultural fairs. It's not just radio – across the developing world, farmers are tapping into the Internet and using mobile-phone applications to get advice on use of fertilisers, pesticides and seed varieties, and up-to-date prices of crops and seeds, as well as the latest weather reports.

Okello is an unlikely success story. Born in Kitgum in the heart of the troubled Acholi province, the twelfth of sixteen children, money was tight growing up, but his parents instilled in him a love of education that saw him through the country's most turbulent years. Okello entered high school just as the civil war peaked. 'From 1986 I was in hiding in the bush. I couldn't study for two years – my school was in the middle of a war zone,' he remembers. 'We were caught in crossfire between government forces and rebels.'

During this time, Okello's father became very sick with bone cancer, and so in 1988, they took refuge in the hospital in Kitgum. 'I went back to school immediately there was a cessation of hostilities. I would sleep beside Dad's bed, wake up, go to school. I got to see a lot of dead bodies. That was the kind of life we were going through: you try to go to school, you get caught up in crossfire, you cannot pass over dead bodies, so you return to Dad's side in the ward. Until 2004, there were continual insurgencies and I never experienced any peace. I lost many of my cousins in that war and most of my friends didn't make it. Part of me died and was dark,' Okello says, his cheerful countenance clouding over.

The cancer claimed Okello's beloved father within the year, but not before he had left his son with a love of science and the ambition to pursue a career in it. 'My dad was a senior medical assistant, almost a doctor, and he was a very popular person in the community. He was answerable to the health needs of over 30,000 people,' Okello explains. 'Seeing him heal people using medicine made me want to go into science. So I went back to school and I studied. I knew one day I was going to make it, and that conviction made me go on, despite the hostilities and the trauma.'

Okello's mother, an infants teacher, was unable to support him through high school and on to university, but he won much-coveted government sponsorship for his studies in agricultural science at Makerere University in Kampala. Okello excelled at university and decided he wanted to be a plant breeder. 'I got interested in the issue of food security in Africa and in Uganda in particular, so I thought if I majored in genetics I could be one of the people who one day helps alleviate the problem of hunger.'

Africa is the continent with the fastest-growing population and the continent with the greatest number of malnourished people. Global warming and drought will hit yields hard, with declines of 43% predicted by the end of the century in the United States under the slowest warming scenario, and 79% under the most rapid warming scenario.[16] Wheat yields, for example, are already about 6% lower than they would have been without climate change, while around the world, entire foods are at risk of disappearing or becoming too expensive to buy because of climate impacts, including coffee and chocolate.[17] One-third of the world's population lives in drylands where land degradation is reducing food supplies, biodiversity, water quality and soil fertility. So the rush is on to develop heat- and drought-resistant crop varieties, as well as improving nutrient levels in developing-world staples, where people lack a varied diet.

In 2003, Okello was awarded a grant from the Rockefeller Foundation for a master's research project working on an improved corn variety called 'quality protein maize' at Makerere University. The maize, bred to contain elevated levels of tryptophan and lysine (the two essential amino acids which are lacking in ordinary cereals), has 82% of the protein value of milk, and is aimed at nursing and pregnant mothers and malnourished children. Okello's task was to produce high-yielding local varieties of the maize. 'The work involved both molecular approaches in the lab and the conventional garden cross-breeding work.' It was a success, and the group released a maize variety called 'Longe5', which is doing well in Uganda. But the end of his master's degree meant the end of funding, and the project was abandoned. In 2006, Okello applied for a government research position working on maize, and ended up getting a job at NaSARRI in the oil crops department, which include sesame, sunflower and peanuts. Two years later, he was awarded a three-year grant from the Alliance for Green Revolution in Africa (the merger between the Rockefeller and the Bill and Melinda Gates foundations) to develop new varieties of high-yielding, drought-tolerant, rosette-resistant peanuts.

He puts on rubber boots and we wander outside to see the new peanut crops undergoing final tests in the field. 'I now have about 500 varieties – until 2016 [when the funding ends] I will be able to release at least two varieties per year.'

His next project will be to investigate the genetic make-up and diversity of the rosette virus, which attacks peanut plants, stunting growth, destroying leaves and causing yield loss of up to 100%. 'I want to try out transgenic varieties,' he says, in which useful genes from one species are put into another. 'In thirty to forty years transgenics will be used everywhere, but I want to see how they can be used in the management of this rosette disease.'

Back in India, I'd met researchers already improving peanuts in this way. At the International Crops Research Institute for the Semi-Arid Tropics (ICRISAT), outside Hyderabad, scientists focus on five different food crops that are highly nutritious and require little water: pigeon pea, chickpea, pearl millet, sorghum and peanut – over 120,000 seed varieties of these crops are stored in cold vaults at the institute, hoping to protect future generations from hunger. Like David Okello, they use genetic markers to identify the section of DNA in a crop that is associated with drought or mildew resistance, say, and selectively breed hybrid varieties that enhance this trait. It's a sophisticated version of the process that rose breeders use to grow bigger blooms, or that dog breeders use to achieve the tiny stature of a Pekinese or the height of a Great Dane.

Scientists at ICRISAT are also improving crops in ways that breeding alone cannot achieve. For example, they have inserted into peanuts a gene from Arabidopsis, a type of cress, that further improves its tolerance to heat and drought. In another, exciting experiment, they fortified peanuts with the gene from maize that codes for beta-carotene, which our bodies convert to vitamin A. Vitamin A deficiency, affecting 250 million children worldwide and more than one-third of children under the age of 5 in Africa, is a leading cause of preventable blindness, disease and premature death. The GM peanuts are yellower but taste exactly the same.

They are not the first crop to be fortified with vitamin A – there has been Golden Rice (expected to be launched as soon as 2015, as the first fortified GM crop on the market), and golden sweet potato. But fortifying peanuts makes more sense because they are so oily – vitamin A dissolves very well in oil but not so well in water, so an oily food delivers the vitamin far better to our bodies. Genetically modifying crops to fortify them with vitamins and minerals, such as zinc and iron, or protein, has the potential to improve the health of millions in the poor world. Even in the rich world, where many people are malnourished because they eat a poor diet, fortifying staple grains could improve health. Scientists are also looking into producing GM fruit and vegetables that have other health benefits, such as reducing the body's accumulation of unhealthy fats in the diet, which could ease the global obesity epidemic. Other possibilities include engineering plants to produce chemicals, fuels and pharmaceuticals.

Even though farmers have been growing GM crops since the 1990s, we're really only just starting to explore their potential. New gene-sequencing tools have made the technology faster and cheaper than could have been imagined even a decade ago. It means that labs in India and Uganda, and in China and Brazil (where such research is well funded), can produce crop varieties tailored to their climates, soils and needs. More than half of the world's GM cropland is now in developing countries.

Governments around the world have been slow and poor communicators about foods that have been genetically modified. And the market for GM plants has been dominated by commercial enterprises such as Monsanto and Syngenta, that jealously guard their patents and prevent their customers from collecting and sowing seeds produced during the harvest, compelling them to buy fresh seeds from the company each time. While farmers in the West do this anyway, poor people can't afford to. Practices like this, coupled with opaque research by the companies into environmental and consumer safety – they do not allow any research on their seeds by independent scientists – have left many people suspicious of genetic modification. One way to make GM foods more palatable would be to openly publish the genetic data so it is available to all – the genetic code of humans and of rice is already freely available to researchers, and a few small biotech firms are now 'open-licensing' their technologies. Fortunately, most of the research into this kind of crop improvement is being carried out at publicly funded universities, where there is little incentive to massage results. And these studies have shown many benefits. For example, Indian and Chinese cotton farmers using plants genetically modified to resist common pests are now using little or no pesticide, benefiting their health.[18] Other GM crops need less toxic herbicides or less fertiliser, thus reducing the hefty emissions involved in their manufacture as well as run-off into waterways. Crops engineered for drought resistance haven't yet shown such impressive results, but here, too, progress seems assured.

Genetically modified crops currently only account for 10% of global agriculture, and much of that is for fuels rather than food, but their potential to help feed the hungry could be great. Despite public distrust, GM products pose no additional health risk, as numerous studies have shown.[19] It's worth noting that transgenic alterations are widespread in nature. For example, the powerhouses that sit inside each of our cells, supplying them with energy – the mitochondria – were originally bacteria that somehow got swallowed inside early organisms. Genetic manipulation of foods is also not new – many crop varieties, including organics, were made by exposing seeds to mutagenic chemicals or radiation. But GM foods are not a panacea – insects will evolve resistance to pesticide-emitting GM crops, superweeds with resistance now plague where herbicide-tolerant GM crops are sown, and even the most drought-tolerant crop requires some rain. Farmers still need access to the seeds, irrigation, fertilisers and good soil, whether they are growing GM or conventional crops.

Ultimately, despite the passionate and often extremist views about whether farming should be small, GM-free and organic or industrial-scale with the efficiencies of synthetic biology tools, the truth is it will increasingly be all of these things. Farmers reap multiple benefits from planting nitrogen-fixing trees (with bacteria in their roots) and hedgerows around their fields, for example, while also planting the most effective crops for the area, climate and soil type – whether that be GM rice or conventionally bred cassava. With an enormous variety of crops, fields and farmers, there will be a variety of different practices used, and we will need this variety into the coming decades.

In terms of environmental impact, the way crops are grown is far more relevant than whether a seed is genetically modified or not. Does the farmer overuse fertilisers and pesticides? Is farmland replacing rainforest? Is a very thirsty crop like cotton being grown in an unsuitably arid area? Is the soil being allowed to recover? The humble plough, for example, can damage soils and speeds the release of large quantities of carbon dioxide as microbes get the oxygen they need to break down organic matter – soils now release ten times the carbon emissions of humans.[20] Instead of ploughing, some scientists argue that seeds should be sown into holes drilled among the plant matter left after harvesting. Leaving the old crop stalks and roots holds the soil and its moisture in place, reducing both erosion and oxidation of minerals (which harms soil fertility), while the slow-rot of vegetation recycles nutrients and attracts microorganisms, worms and insects, which help maintain soil fertility and provide ecosystem services. So-called no-till agriculture is gaining popularity in Brazil and the US, but getting the world's poorest farmers to adopt it will take time, partly because of the equipment cost of drilling holes for planting.

In the Anthropocene, the pressures of increasing population and richer appetites are conspiring to increase the area of wild lands converted for agriculture, so a degree of agricultural intensification is needed in low-yielding countries to feed the hungry and protect biodiversity – yields in western Europe are already close to their physiological limits. If yields in poor countries improved, reaching 95% of their potential, global food production from current farmland would increase by 58% – 2.3 billion tonnes, US ecologist Jonathan Foley calculates. Even if yields were only brought up to 75% of their potential, there would still be a 28% increase in global food production, or 1.1 billion tonnes. As Hardevsinh Jadeja, Winifred Omoding and other farmers show, this can be as simple as improving access to irrigation, quality seeds and fertilisers. Vast mono-cultures can be more efficient, but industrial farms can be polluting, provide less employment, and because they use fewer varieties, a larger proportion of food may be at risk from disease or weather. Dependable native crops such as drought-resistant sorghum, cassava, millets, sweet potatoes, savoury bananas and peanuts, that have hitherto been neglected for more fashion-able grains, are more appropriate for a hotter, drier climate, and should be scaled up.

Longer term, entirely new agricultural technologies will need to be explored and developed – for example, foods based on algae or fungi that can grow in oceans or on racks in factories. Scientists are developing the first perennial grains, which, if successful, will hugely benefit food security and the environment, since they would require far less fertiliser, irrigation, soil tilling and pesticides than the annual crops we use now.

Another potential game-changer would be improving the efficiency of photosynthesis in staple crops. Most plants, including rice and barley, are C_3 plants: when they take carbon dioxide from the air, they 'fix' it into a molecule with three carbon atoms. C_3 plants thrive in temperate climates where there is plenty of water, but become less efficient as temperatures climb. Already, global warming has caused rice yields to plummet by as much as 20%. However, around 3% of species, including maize, sugar cane and sorghum, have evolved a more efficient type of photosynthesis, fixing four carbon atoms per molecule. These C_4 plants need more sunlight than C_3 plants, but cope better with higher temperatures, drought and limited nitrogen. Scientists are working to convert rice to a C_4 plant, potentially increasing yields by a half. The good news is that the C_4 mechanism has evolved naturally on many separate occasions, and many plants, including rice, already possess the necessary enzymes to switch.

In many cases, feeding the hungry in the Anthropocene is as simple – and difficult – as improving roads and storage so that crops can reach market before they rot. When I drove east from Rwanda to Uganda's capital, Kampala, I passed bountiful crops of bananas and vegetables being destroyed in their fields by farmers who couldn't sell them locally, owing to the glut, and couldn't transport them to people suffering famine a few hundred kilometres further north-east because the roads were so bad (which was attributable to corruption and mismanagement of road-maintenance funds). Some 40% of food produced in the developing world is wasted before it gets to market because of refrigeration and transportation issues.[21] The same proportion of food is thrown away in the rich world. Cutting waste would be the fastest and cheapest way of meeting future global food requirements. Food wastage is the third largest emitter of greenhouse gases after the US and China, and uses a third of agricultural land even though 870 million people go hungry every day.[22]

We also need to be more judicious about what we use global arable land for. With hundreds of millions of malnourished people in the world, why would anyone deliberately set fire to food? Yet this is what the biofuels industry is doing, sanctioned and supported by governments and NGOs. Biofuels might seem ideal: crops like corn or sugar cane can be grown for fuel, and burning it releases only the carbon dioxide that the growing plant sucked from the air. However, giant companies have already grabbed more than 5 million hectares of land in developing countries for biofuel production. I have met malnourished farmers in India and Central America who were either growing biofuels (including sunflowers, jatropha and sugar cane), or who had been evicted from their land for these crops. Either way, they and their families have been losing out, while huge tracts of rainforests and savannahs are being cleared for biofuels.

With a significant portion of global crops diverted to use for fuel, the price of food inevitably goes up. And the situation gets far worse if harvests are destroyed by droughts, floods or other disasters. In August 2012, after an emergency G20 meeting over food prices, the director general of the United Nations Food and Agriculture programme called on the US government to suspend its biofuels quota for the maize harvest, after a months-long drought and heatwave destroyed much of the crop. Cattle farmers around the world backed the call, fearing costly animal feed over the winter. But what the US feared more was a rise in the gasoline price: so it maintained its biofuels quota. In that year, at least, the battle between hungry people and thirsty cars was won by the latter.

Worse still, substituting biofuels for oil doesn't significantly reduce greenhouse gas emissions; in fact, emissions are sometimes increased. That's because biofuels require fertilisers (which emit greenhouse gases in their use and manufacture), and because tropical forests, which soak up carbon dioxide, are often cleared to grow the biofuel crops (or the food crops displaced by the biofuels).

So-called second-generation biofuels, derived from waste plant matter or manure, are being used as an alternative. But even this is problematic. Not only does it take more energy to produce fuels from these sources (otherwise our bodies would digest them), but in most parts of the world, this material is not 'waste' at all – it is valuable mulch or fertiliser, or fed to livestock, dried for roofing or burned as fuel for cooking and heating.

Third-generation biofuels are on the horizon, exploiting genetically modified algae or bacteria. So far, experiments have been encouraging. However, bacteria need their own fuel – usually sugar – which will have to be grown. Algae, which can be grown in oceans, lakes and wastelands, also need a supply of nutrients (fertiliser run-off will do) as well as sunlight, but they do not require crops. But even with a very efficient conversion process, algae would need a vast area to perform and could realistically only provide a fraction of our fuel needs.

Even without biofuel production, the agricultural footprint of the Anthropocene is vast and currently unmanageable. Agriculture is transforming the nitrogen cycle, emptying rivers and lakes, and sucking out so much groundwater that cities are sinking into the ocean. The greenhouse gas emissions from fertilisers, cattle burps, paddy fields, and deforestation account for 30% of our total – more than from electricity generation or transport.

The main driver of these planetary changes is our mouths. Some 10,000 years ago, at the beginning of the Holocene, there were only around 5 million of us; now we've passed the 7 billion mark. There have been dire warnings that the population will exceed resource limits, causing horrendous famines for at least two centuries.

But the rate of population growth has already peaked, reaching its zenith around 1968, the year when Paul R. Ehrlich's *The Population Bomb* was published. Since then, the rate of growth has declined by about 50%, with the average woman in developing countries (outside of China) now producing three kids, rather than six. Some countries have actively promoted contraceptive use and delayed motherhood, as well as promoting female education, which reduces family size. China, the most populous nation, introduced a controversial one-child policy in 1978, which has prevented hundreds of millions of potential births. However, social engineering has turned out to be not quite as effective as economic growth in reducing family size: over the same period, Taiwan, in moving from 'developing' to 'developed' status, has seen a slightly bigger reduction in fertility than China.

From now to 2050, population growth will take place almost entirely in the developing world. Sub-Saharan Africa will show the highest growth rate, but most of the extra people will be Asians, with India expected to outrank China as the most populous country by 2020. Rich countries, meanwhile, are seeing a decline in fertility that, in some cases, such as Japan, will actually lead to a smaller population by 2050. The demographic shift, already under way, will force rich societies to rely increasingly on migrants from the developing world, hastening the Anthropocene's mix-up of people around the globe. The decline in population growth is not fast enough for many environmentalists, who see reducing human numbers as vital for a safe future for humanity. However, there are plenty of agronomists and economists who believe there is enough of everything, including food, to go around. Despite the growing impact of climate change, degraded soils and water shortages, they insist that we can feed the extra billions, so long as we improve agricultural efficiency and reduce pre- and post-harvest waste.

As people become wealthy, they switch from rice or cassava to milk and meats – if everyone on Earth was to eat like an average American or to live his lifestyle, there would not be enough to go around. Kilo for kilo, getting our protein from meat rather than plants uses far more land, water and energy, and emits far more greenhouse gases. Currently, around 2 billion people live on a primarily meat-based diet, while 3 billion are malnourished. But as populations increase and some grow richer, the numbers and proportion of meat-eaters will become completely unsustainable. The livestock population of the US already outweighs the human population by five times, and directly consumes seven times more grain (grown on seven times the land area used to feed people) – which could feed 840 million people. Worldwide, just 62% of crops go to feed people directly; one-third go to livestock and another 3% to biofuel or other ends.

Humans kill around 1,600 animals and birds every second for food, and that doesn't even count the marine life.[23] The total biomass of livestock is double that of humans.

In the Anthropocene, if we are to preserve the few remaining forests and other diverse ecosystems around the world, eating meat from a recently slaughtered animal is likely to become increasingly rare. Insects, which are far more efficient than cattle at transforming their feed into protein, are likely to provide a greater proportion of our dietary protein – over 2 billion people already supplement their diets in this way. Livestock will perhaps continue to be farmed for meat, but intensively in sheds, perhaps being fed novel foods like algae (and, as some suggest, removing their cerebral cortex so they are unaware of their uncomfortable lives). Superfarms already exist in Saudi Arabia and the US, housing more than 100,000 live animals. Growing meat synthetically is another option. It would reduce water use by 90%, greenhouse gas emissions by 95% and land use by 99%, scientists have estimated.[24] Theoretically, the entire world's meat demand could be met by culturing just one animal's cells on racks in laboratories – and could be modified to make the meat less fatty, healthier and of different flavours.

To make the biggest impact, humans need to dramatically reduce meat consumption in the Anthropocene. Many communities around the world eat little or no meat for cultural, religious or ethical reasons, and there are plenty of nutritious and tasty meat substitutes; the difficulty is in overturning the almost universally held belief that meat is a sign of wealth. There are already encouraging signs – meat consumption in the US has dropped 10% over the past decade, for example, as people seek healthier choices and the cost of animal feed pushes up the price of meat.[25]

Humanity's relationship to agriculture in the Anthropocene will be different. Worldwide, smallholder farmers like Winifred will be fewer. People will be increasingly urban and more distant from gardens and farms. Food will be bought rather than grown, and increasingly in processed

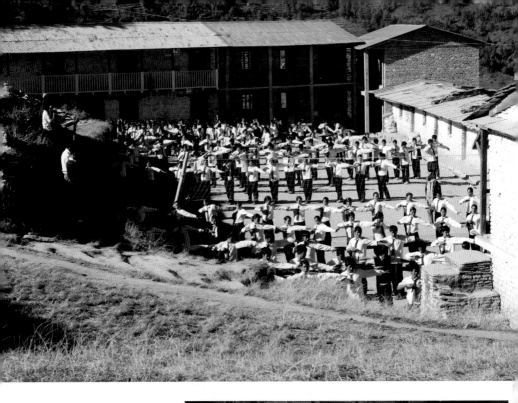

m top: Nangi village
dren exercising in the
oolyard; Mahabir (in
sses) at the WiFi nerve
tre; Nangi kids online in
computer room.

Clockwise from bottom left: The construction of banks and ridges for the artificial glacier above Stakmo; the glacier that forms there melts in the spring; Norphel talks to Tashi outside his Stakmo home; Tashi's bumper harvest; Parco and Torres painting their mountain white in the Peruvian Andes.

Clockwise from bottom left: M
boat along the Mekong in Laos
floating market in Chao Doc; i
Vietnam people still live on the
river; the vast wilderness of
Patagonia is home to gauchos;
glaciers; Fransisco Vio's toddle
enters his warm self-built eco
house near Coyhaique, Patagon

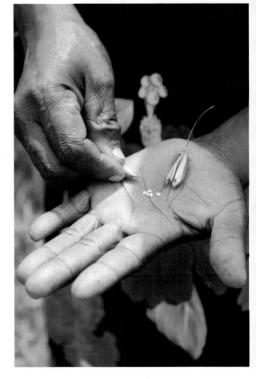

Clockwise from bottom left: Robinah strips her poor harvest of peanuts near Masindi, Uganda; Winifred and her family stand in the doorway of their prospective large house; the sesame that transformed Winifred's fortunes.

Clockwise from bottom: a schoolroom in Raj-Samadhiyala, India; Gulab Givi, a former beggar, in his thriving shop; David Okello wades through his new peanut varieties in Soroti, Uganda.

Clockwise from top left: Malé viewed from a seaplane; Ani Nasheed; Gerald McDougall's house o
Garbage Island, Belize (it was ravaged by a cyclone just days before this shot and he was carrying
repairs).

forms. Farms will be elsewhere. People will become divorced from the daily struggle of growing food from seeds. Whatever we eat, we will surely have to become less numerous over time, and use food more sustainably. We will grow more food on less land using less water and with less pollution. The food we grow will be more nutritious, and we will waste less of it. We will only manage this by using all the tools we have, including ancient practices such as shifting agriculture, nomadic pastoralism and organic conservation, as well as promising new techniques, such as genetic modification. Agricultural transformations can occur in just a few seasons, as Winifred, Jadeja and others have shown, but nurturing these shoots until they are self-sustaining requires an external support framework, including improved infrastructure.

In this way, it should be possible to feed our growing population despite climate change and other pressures. However, there seems little doubt that, into the Anthropocene, farmland will almost certainly spread further, covering perhaps half of Earth's land surface as humanity's quest for food impacts on every part of the planet.

5

OCEANS

*T*he great body of swirling water that flows around Earth distinguishes our blue home from the other planets spinning around the sun. Ours is the only one to have liquid water. But it didn't start that way – there was no water of any kind on our molten planet for millions of years.

As the planet's skin formed, its churning belly gave birth to the first water in volcanic belches of steam that rose into the cooler atmosphere forming clouds. As Earth cooled, it rained – for thousands of years. The low-lying parts of the planet's surface became bathed in water, creating the first oceans. Meanwhile, young Earth was being constantly bombarded with large icy comets. Perhaps as much as half of our planet's water arrived in this way.

The oceans defined and shaped the planet's coastlines, eroding cliffs and channels, battering islands, lapping at beaches and heaving its great mass in a regular tidal rhythm that affected the Earth's spin and increased day length.

Oceans were responsible for the most remarkable experiment in the universe: life. It began around deep sea vents, cracks in the oceanic crust that issue hot, mineral-laden gases, rich in the ingredients for life. Around 3.5 billion years ago, complex molecules capable of replicating, coalesced at these energetic sites, eventually forming living cells. Life would not venture from this oceanic womb for billions of years.

The oceans now cover 70% of our planet, they drive the weather and climate patterns, and are a critical player in the global carbon, nitrogen and water cycles. The majority of water that rains down on our crops and forests, lakes and cities comes from clouds that were created through oceanic evaporation. More water evaporates from the oceans than precipitates on to them, and if it were not for replenishment by rivers and other run-off, sea levels would drop. In fact they are rising, because humans are heating the oceans, causing expansion, and because glaciers are melting and pouring into the seas.

In ancient times, the oceans represented the edge of our world, the limit of human knowledge and experience. Fish were a valuable food source for some of the earliest people, but in many ways the oceans were to be feared. Courageous explorers set forth to explore new worlds separated by the seas, but few ventured into open ocean, most preferring to keep land in sight. The oceans separated lands so effectively that entire civilisations were unaware of each other's existence for thousands of years. Even the smallest channels can keep apart animals and plants long enough for them to evolve distinct differences, as Charles Darwin discovered among the Galapagos Islands. Eventually, adventurous mariners searching for bountiful new fishing grounds crossed the terrifying expanses of dark sea. They returned with tales of impossible creatures, like 'unicorn' whales and colossal squid more than ten metres long, which wouldn't be verified by scientists for hundreds of years. And they found new continents. Their explorations would join together the world's people and begin the process of globalisation.

In the Anthropocene, humanity is transforming oceans. Where the oceans separated landmasses, we have joined them with bridges and tunnels. In the *process, endemic biodiversity has been diluted and dispersed by us as we introduced species from one island or continent to another. Where continents separated oceans, we have joined them with canals in Panama and Suez. Perhaps our biggest changes can be seen in the Arctic, though, a place composed almost entirely of ocean, where vast expanses of ice made entire regions impassable year-round. Now, we are gradually melting the frozen seas so that ship travel, habitation and drilling for oil and minerals is possible.*

The oceans have become our sewer for chemical, biological and material waste. Our pollution has made them more acidic and altered the nutrient concentrations, especially near shorelines. Humanity has heated, swelled and overhunted the oceans. We are selectively culling entire species, upsetting marine ecosystems even as we discover new species and habitats, exploring at previously inaccessible depths.

In the Anthropocene, oceans are our food larder and our transport passage, but increasingly also our power supply and our source of oil and gas. They provide livelihoods and food for billions and are home to the biggest 'forests' on Earth: the phytoplankton that scrub the air of our carbon dioxide emissions. We have only explored 3% of them and yet, as we learn ways to probe their mysterious depths, calamitous happenings are under way at the surface: entire islands are disappearing beneath the waves.

From the rounded cabin windows of the seaplane, the Maldives archipelago resembles a chain of turquoise splashes in the Indian Ocean. Each of these shallow aquamarine lagoons is fringed with a halo of coral that looks green from this height. And many of the bright pools have a white swirl of sand at their heart – the biggest host a jumbled crowd of buildings, while a few of the others contain tiny jungles of coconut, papaya, banyan and drumstick trees.

Nine-tenths of the Maldives is covered by sea and the little that is above water is dispersed into 1,200 tiny islands, spread over hundreds of square kilometres. Around 470,000 people live on these scraps of land, the largest of which is less than eight square kilometres. The Maldives capital Malé is the most densely populated city in the world, with 130,000 people crammed on to an island of just two square kilometres.

It's a precarious existence. The country's tallest 'mountain' is just 2.3 metres above sea level, putting the Maldives' fragile swirls of sand among the most vulnerable lands in the world facing the ravages of climate change. What little land there is, is built from the skeletons of sea creatures – the Maldives is as much a part of the ocean as it is terrestrial. Sea-level rise, increasingly intensive storm surges, monsoon alterations, ocean acidification and associated erosion all threaten to drown this tropical paradise – some scientists predict the islands will become uninhabitable as soon as 2030.

'The islands are so much smaller than they used to be. And the sandbars are shrinking away year by year,' says John Gregory, peering through the seaplane window. John and his wife Sue, from Bath in the UK, are 'returners'. Since 2000, the Gregorys have spent two weeks every March and September in one of the country's luxury resorts. Over the past decade, the couple have noticed several changes. 'Some of the resort bungalows have had to be abandoned because they kept getting flooded. One simply fell into the sea. And the beach where we used to sunbathe has completely gone,' Sue says. 'It's a shame. I suppose soon we'll have to find somewhere else to go.'

The Gregorys' inconvenience is something the government can ill afford – tourists provide at least a third of the country's income. I'm visiting at a time of great physical, social and political upheaval, when nationally there seems to be everything to lose and everything to fight for, and, globally, everything to win with the country now firmly on the international stage. It is a time of great fear for the people of this vulnerable land and, paradoxically, a time of great optimism. After thirty years of brutal dictatorship, there is a new young president, Mohamed Nasheed, full of fresh ideas. But he needs all the tourist dollars he can get to overcome the budget deficit he inherited, to make good on his pledged social and infrastructure improvements, and for his grand plan: a sovereign fund to rescue the nation when the waters get too high to stay. Before the corals and beaches disappear, Nasheed needs to build up sufficient funds from visitors like the Gregorys. He's in a race against time, a race against the human-changed oceans of the Anthropocene.

In many ways the plight of the Maldives offers a snapshot of the challenges we all will face as humanity encroaches on Earth's oceans, the last wild and wet frontier. From overfishing to global warming, they are felt first and most keenly here. I've come to find out what it's like to live in the world's lowest nation, the country most at risk of disappearing. And to meet the remarkable man who is innovating ways to fight the global force of humanity, refusing to let his nation drown silently.

Nasheed became the Maldives' first democratically elected president in late 2008, promising to turn his state into a renewable-energy powerhouse and the world's first carbon-neutral country, even while he develops the economy and infrastructure. In a strong message to other nations, he rejected his predecessor's plans to extract oil from below the country's waters. Never before had a nation so unambiguously stood for environmental policies in spite of the economic hardships that doing so would entail. Then again, never before was a nation so directly and obviously threatened by the invisible pollution produced by humanity's quest for progress. It is an exhilarating and fascinating time to visit.

I meet 'Anni', as Nasheed is affectionately known, after sunset in the modest courtyard garden of his family home in Malé. He has changed out of his shirt and tie, and appears shod in thongs and wearing a comfortable pink collarless shirt in the traditional style. He exchanges a few words in local Dhivehi, through the open terrace doors to his wife Laila who is preparing dinner for their two young children inside, and offers me a drink. Then, switching on a pedestal fan to stir the sluggish evening air, he smiles a proper greeting and sits down opposite me. For the first time, I see how young he is – in his 40s, but slight and boyish, and with a quietly passionate voice that's unusually high-pitched in an occupation in which most have modulated theirs down. We begin chatting and I'm quickly charmed by his openness and unguarded sense of humour. During our conversation, he describes inaction on carbon mitigation by developed economies as 'stupid', reliance on oil as 'very silly' and questions why 'the rest of the world doesn't stop talking and start doing something about the problem by focusing real funds and attention to switching to renewable energy right now'. This is a president who isn't afraid to speak his mind on issues he is passionate about – human rights and climate change – or to pull a publicity stunt. In 2009, he held a televised underwater Cabinet meeting around a submerged desk, persuading his ministers to wear scuba gear over their business suits, to draw attention to the impacts of sea-level rise on the nation. Later that year, Anni emerged 'the real hero' of international climate negotiations in Copenhagen, according to the Danish premier, maintaining intense pressure on other world leaders 'to leave pride aside and adopt this accord for the sake of our grandchildren'. It is widely recognised that his intervention and persistence is the only reason there was any agreement at all at the disastrous meeting.

Anni is refreshingly different from other heads of state. He has a passion and urgency about him – an almost physical desire to get things done quickly. It's as if he knows he hasn't got much time in the presidency (sadly, this is all too prescient). Or that too much time has already been wasted. He has spent most of his life as a human-rights activist and journalist, campaigning for democracy during the previous president Abdul Gayoom's autocratic rule. Anni was imprisoned more than twenty times, often held in solitary confinement, tortured several times (including being forced to eat crushed glass), and missed the births of his two daughters – one of whom was conceived during a rare conjugal visit. He spent his incarcerations studying when able, writing three books on

Maldivian history. In 2003, Anni demanded an investigation after a young inmate was tortured to death, sparking national protests and riots. He fled the country and set up the Maldivian Democratic Party (MDP) while in exile in Britain. When he swept to power in 2008, Anni promised his jubilant supporters a raft of reforms. One of his first moves was to release all political prisoners and destroy the buildings used to torture prisoners.

However, Anni soon realised that he'd inherited a system riddled with corruption. Despite having the highest per capita earnings in South Asia, the Maldives is poverty-stricken with half the population earning less than $2 a day. Most people lack basic services from clean piped water to sewerage, there is next to no health care, inadequate education, ancient infrastructure, and no public transport system between the many islands. Worse, the Maldives has very high heroin dependency among young people – in some areas as many as one in five uses the drug, making it the worst in the world relative to the country's population. The contrast between the indigenous population and the sizeable volume of wealthy holidaymakers could not be starker, and there is little interaction between the two – almost all tourists fly directly to and from their resort island, never visiting an island inhabited by Maldivians. 'We have Third World islands where Maldivians live, next to luxurious European resort islands,' Anni says, losing his easy smile.

Desperately crowded Malé is home to a third of the population in a mass of concrete buildings stacked against and on top of each other. Inside, the youth, who are largely unemployed, battle addiction, poverty and boredom – a dangerous concoction that leaves them vulnerable to fanaticism stoked by Anni's political enemies. 'The Maldives can be an example of how a Muslim nation can go peacefully into a functioning democracy without bloodshed and acrimony,' Anni says. 'We need to educate people and bring some of the more isolated villages into the twenty-first century so that they are better able to deal with the challenges ahead, whether that be moving to another island or country,' he says.

For oceans as much as anywhere, the extent of change in the Anthropocene depends on us – on how poor people develop their economies, on how rich people reduce their environmental impact. Anni is uniquely attempting to decarbonise the country's energy supply and heed environmental limits while industrialising. It's a brave decision, I say; perhaps unnecessarily painful . . . ? Anni interjects sharply: 'No, it is the only way to progress. We have no choice. It is brave – foolhardy even – to build power stations using fossil fuels when we know it is making things worse for the climate, for sea levels. Britain, the United States and the industrial world has been poisoning the atmosphere for the last 300 years. We must show the world there is another way.'

Defending his people from climate change, he says, is a humanitarian challenge on the scale of World War Two. 'Climate change is not just an environmental issue,' Anni says. 'It's a human-rights issue that risks the homes, livelihoods and lives of millions of people. People around the world have a responsibility to help those who are most vulnerable.' International climate negotiations are aiming to limit the global temperature rise to 2°C – something that would condemn the Maldives. 'It is too late for us,' he says. 'It would be unliveable here. But it's not too late for everywhere. And if we aim for 2°C, maybe we would find it easier then to keep within 1.5°C.' (In fact, as I've mentioned, many scientists now think we're headed for a 4°C temperature rise by the end of the century.)

Anni's many ideas are radical, and include geoengineering artificial coastlines and islands within the archipelago, constructing floating islands, and even buying land in a foreign country to relocate the entire Maldivian population, using the sovereign wealth fund. That is something that has never been tried before outside of colonial conquest, and it's difficult to imagine the practicalities of a state within a state existing in the current political landscape. Yet ideas like these will be considered increasingly seriously in the Anthropocene, as each disappearing state seeks to find its own solution to what to do when climate change steals their country. 'We want to stay together. We don't want to lose our culture,' Anni says. 'But we also don't want to be climate refugees living in tents for decades.'

Meanwhile, he is trying to cope with current problems and stave off the day when his fellow citizens will have to leave their home. 'We have water contamination from sea incursion, we have sixteen islands that need relocation, we have fishing problems, poaching of marine life, erosion problems, the coral bleaching issue . . . There's a big list of things that are happening now,' he says. 'We have to repair the natural defences to our islands, by restoring reefs and planting vegetation.'

Anni's plans all cost money, much of which will have to come from tourism. Honeymooners and the monied have for decades flocked to the Maldives, drawn by its serenity, guaranteed sunshine and calm waters swimming with pretty tropical fish. I head to a resort island to see what they are up against. My boat pulls up at a picture-postcard scene: transparent waters kiss crystal white sands in front of tastefully designed timber guest bungalows. But when the beach empties of bikini-clad bathers, the machinery moves in. Twice a day, a large pipe descends into water offshore and retrieves sand, which is pumped on to the beach. Without the pump, there would be no beach. In fact, there might not be much of an island either. The Maldives is as transient and ephemeral as the shifting sands that move with the tide. Many of these islands are mere sandbars that come and go with the changing currents. Local people also call their country 'Wodun adhi Girun', which means 'the nation of appearances and disappearances'.

The country has been treading water since its birth as a string of volcanic islands millennia ago. It was Charles Darwin who first figured out its origins, during his stay in the Maldives in the 1840s. The twenty-three necklaces of islands and lagoons started out as coral reefs fringing volcanic islands, he surmised. In time, as the Earth thawed from the last ice age, glacial melt caused a rise in sea level. Meanwhile, the dormant volcanoes gradually subsided, and these two effects slowly submerged the islands. Reef-building corals prefer shallow waters – most cannot grow easily at depths of more than forty-five metres – so they began constructing their calcium carbonate skeletons on top of one another at a rate fast enough to keep up with the sea-level rise. At the same time the corals at the surface grew laterally to stay abreast of the ever-diminishing coastline. The corals continued to push upward and outward well after the volcanic island was completely submerged, forming a ring encircling a lagoon where the original volcano once was.

It is this gradual process that led to the formation of the jewel-like coral necklaces I saw from the seaplane, which Darwin called 'atolls', using the Maldivians' word for them. As the corals are broken down by other sea creatures, such as parrotfish and starfish, their skeletons are reduced to sand that builds up with other debris to form low-lying islands within the lagoons.

But in the past fifteen years, this delicate balance has been upset – the sea level is rising too fast for the corals to keep pace, and more and more islands are disappearing permanently. Globally the oceans are rising at 3.5 millimetres per year – roughly twice the rate seen during the twentieth century. Most of the sea-level rise in the past decade has come from the thermal expansion of the oceans – as the greenhouse effect warms the water, the agitated molecules shift further apart, taking up a greater volume – but glacial melt is now the leading cause. Until now, sea-level rise has been directly proportional to the rate of global warming, but scientists expect sea-level rise to start accelerating, rising 2.3 metres for every 1°C of warming.[1] Their calculations will help determine whether large parts of the Maldives will be uninhabitable by 2070, 2050 or as soon as 2030. Even a one-metre rise would be devastating for the

Maldives, making a large portion uninhabitable and totally submerging most infrastructure.

Rising waters, combined with coral growing more slowly and dying, have left the reefs less able to shelter the islands from waves and storms, which are rapidly eroding their coastlines. In the Maldives, coastal erosion isn't just an issue of losing a few metres of coastline every year. The country itself is at stake.

Erosion is already having a big impact on the population. Roads and houses are crumbling into the sea, coconut palms are being washed away, and the groundwater has become so polluted with seawater that on many islands it is undrinkable. So far, twenty islands have been abandoned. Whereas the resort islands might be able to stave off their disappearance in the short term using sand pumps twice a day, locally inhabited islands couldn't afford to even if it solved the problem. And resort managers know only too well how temporary the measure is.

The country has already experienced a foretaste of the inundations that lie in store. The Maldives suffered the worst per capita effects of the 2004 Asian tsunami, even though lack of a continental shelf meant that the maximum wave height experienced was just over four metres. On one of the worst-hit islands, Kandholhudhoo, the highest tsunami wave was just 2.5 metres, but when it receded minutes later, three people were dead, not a single house was habitable and residents had to leave the island for good. Of the 150 inhabited islands, fifty-seven had serious damage to critical infrastructure, fourteen had to be totally evacuated, and six were decimated. A further twenty-one resort islands were forced to shut down due to serious damage. The total damage was estimated at $470 million, about 62% of GDP. With rising seas and more frequent storms forecast, it may not take a tsunami to inflict similar damage in future.

I travel with Anni to the official opening of the nation's first storm-safe 'designer island', Dhuvaafaru, in Raa Atoll. Formally an uninhabited jungle, the entire vegetation was razed, and a new village funded and built from scratch by Red Cross workers for the 4,000 tsunami survivors of Kandholhudhoo. The houses are positioned further inland compared with other islands, and each has a traditional well for washing (rising sea levels have turned the groundwater brackish in all but the Maldives' largest island). The school and community centre are both constructed to withstand storm surges and other inundations, are large enough to house the entire island community if needed, and are raised high on stilts to allow floodwaters to gush through and over the island unimpeded.

While Anni shakes hands and chats to community leaders, I have a poke around the new houses. Fatima Mohammed, who has moved here with her husband and three children, loves her new home. 'It is so much bigger and cleaner than our old house, where we were sleeping with our whole family in one room. And it has a fan,' she shows me. The family has spent the past four years living in refugee accommodation, with three other families – fourteen people to one room. But grateful and happy as Fatima is for her new home, in her heart she will always live in Kandholhudhoo, she tells me.

Relocating people to designer islands with emergency buildings high on stilts might work in the short term for a few people – protecting houses against inundations is being tried successfully in other low-lying places like Bangladesh – but it is not a long-term solution to the rising sea level across the nation's 200 inhabited islands: there simply aren't enough suitable new islands to relocate too. There has been some attempt to bolster the defences of existing at-risk islands with concrete, but this has had mixed success. In the 1990s, the Japanese government paid for a $63 million sea wall to shelter Malé from the rising waters. While this protected the city against the tsunami, it has had damaging environmental side effects, and Anni is hesitant about building more on other islands. Sea walls reduce the currents flowing over the coral reefs – living coral needs a flow rate of around ten metres per second to breathe – causing this protective natural barrier to die off, increasing erosion. Without the reef, the islands simply disintegrate. Perhaps most damaging are the slew of island harbours hastily built by Gayoom in the run-up to the 2008 election. More than a dozen were located in positions where they interfere with currents and speed up erosion of coral, sand and shorelines – often causing sand to be washed and dumped inside the harbours, rendering them useless.

Local environmental organisation Bluepeace believes that the nation needs to think bigger than walls and island extensions. It argues that what is really needed is a series of raised artificial islands, dotted around the archipelago. Seven islands, perhaps paid for by the international community by way of compensation for climate change, could allow the entire Maldivian population to stay ahead of the rising waters, the charity suggests. The idea is not entirely far-fetched. One artificial island, Hulhumalé, was built in 2004, north-west of the capital. While it was designed mainly as a commercial port and to reduce pressure on the overcrowded capital, it was built three metres above sea level, enough to ensure it lasts to the mid-century, at least by conservative estimates. However, Hulhumalé has increased erosion in adjacent islands by disrupting natural currents, and dredging the seabed for materials also reduces the resilience of the shores. Several companies are now working on floating-island concepts. One of them, Dutch Docklands, is in negotiations with the government to create four individual ring-shaped floating islands (each with seventy-two water-villas), forty-three floating private islands in an archipelago configuration, the world's first floating eighteen-hole golf course, and, an 800-room floating hotel. According to the plans, the floating islands will be interconnected by underwater tunnels, and the golf course will feature an underwater clubhouse adjoining two luxury hotels. 'The reality of floating islands could start to shape future urban landscapes, with further scope for agriculture, offices, housing and leisure,' the company says. 'This will lead to new economic opportunities where governments can cost-effectively lease islands with flexible solutions instead of investing in static developments.'

At the other end of the scale is Thilafushi, another artificial island, created inside a natural lagoon twenty years ago as a solution to the problem of waste in the Maldives. Each tourist generates 3.5 kg of waste a day on average, which is five times the locals' average, and the island, known locally as Rubbish Island, now receives more than 330 tonnes of garbage a day, by boat, to be piled up in a gargantuan fetid heap swarming with flies. Among the stinking garbage and acrid smoke from rubbish burning, it is just possible to pick out the Bangladeshi migrant workers sorting the waste – among it are oil drums, asbestos, lead, cadmium and batteries, which pollute the ocean during garbage 'landslides'. Every day the island grows another square metre, as the mountains of waste fill in the lagoon.

While Anni is 'open to all options' and isn't ruling out artificial islands, he first wants to improve the country's natural resilience. 'In some places, we are making our situation worse,' he admits. The fate suffered by Kandholhudhoo is a stark warning of how not to manage other islands. Terribly overcrowded and with the coasts eroding fast, Kandholhudhoo's residents had completely dredged the coral reef for island expansion and building materials, meaning there was nothing to slow the devastating waves and reduce their energy. Meanwhile the vegetation had all been chopped down for timber housing, which left no mangroves to slow the waters and maintain the integrity of the coasts, and no trees to hold the topsoil. Properly managed mangrove plantations could restore this barrier, and since some species can mature in five to ten years, it could have relatively quick benefits.

In 2011, the government published the world's first Strategic National Action Plan that integrates disaster-risk reduction and climate-change adaptation. The plan details several ways for the country to build its resilience through improved infrastructure and social investment, and also through bolstering the health of the naturally protective ecosystems, including mangroves and reefs, that can delay erosion. The Maldives coral reef ecosystem is the world's seventh largest, and it contains two of the biggest natural atolls – Thiladhunmathi and Huvadhoo. It is also among the richest in terms of species diversity, hosting over 1,900 species of fish, 187 coral species, 350 species of crustaceans, nine species of whales, fifteen to twenty species of sharks, seven species of dolphins, and five species of turtles. Coral is not just important for the islands' biodiversity, fisheries and tourism, it is integral to the islands – they are, after all, made of coral.

There has been only partial mapping of the corals in the country and much is still unknown about the reefs and how they will behave as temperatures and sea levels rise. But the prognosis is grim. Coral reefs are on track to be the first of the world's ecosystems to be entirely wiped out by humans. Some scientists estimate that reefs will have gone as soon as 2050 – atmospheric carbon dioxide concentration may already be above the levels that will see corals condemned to extinction. 'By 2050, we may still have corals, and things we'll call "reefs", but they will be massive limestone structures that were built in the past, with tiny patches of living coral struggling to survive on them,' says coral ecologist Peter Sale.[2] By 2100 he thinks there will be no reefs visible. 'We're talking here about killing off a whole integrated community of organisms that have been with us throughout our existence and long before there were people of any type on Earth – having it disappear from the universe because of our activities, because we have changed the planet and made it a place that reef communities cannot live on.'

If so, the oceans of the Anthropocene will not only be less colourful, they will also be far poorer in terms of fish diversity – coral reefs support a quarter of all marine life. Reefs also support millions of livelihoods, contribute to at least half of the national economy in many island and coastal countries, create sandy beaches and protect coastlines. And as the sea levels rise, they provide vital protection against storm surges and inundation, particularly in low-lying places like the Maldives.

Coral actually describes a range of primitive organisms called coelentrates, which are related to jellyfish. They consist of a calcium carbonate (limestone) skeleton and they host an algal lodger that can photosynthesise, providing the coral with most of its nutrients, while the algae benefit by being protected from predators. Humanity's carbon dioxide emissions are a double whammy for corals. Higher ocean temperatures force them to expel the colourful algae – during a coral bleaching event, reefs lose so much algae that they become white and experience massive die-offs. Also, as the oceans dissolve more and more of the gas out of the atmosphere, they become more acidic, and this dissolves calcium carbonate. It means that corals must expend far more energy to build their skeletons. The oceans currently swallow one-quarter of our carbon dioxide emissions, dropping 0.1 in pH because of the increase in carbonic acid in the past century.[3] It may seem little, but pH is measured on a logarithmic scale, so that equates to a 30% rise in acidity (15% since the 1990s alone) – a rate of change faster than at any time over the past 65 million years, more than ten times faster than during any of the evolutionary crises in the Earth's history.

At the same time, reefs are under assault by pollution, overfishing (which removes key species and allows profusion of others, such as destructive algae which overwhelm and kill the coral), and boat-anchor damage. A healthy reef can yield up to 1,500 kg of fish per hectare, but once a reef is fished below 300 kg per hectare, it is likely to collapse.[4] It is one of the reasons that the protected resort islands show less reef damage than the local islands; reefs are also far more likely to recover from a global-warming event if they are located in a protected area than if they are exposed to further human-related damage. Reef survival in coming years will depend largely on the next-generation corals – on how well coral larvae settle on rock, metamorphose into feeding polyps, and establish new reefs after extensive damage.

With humans on the verge of exterminating corals forever, the battle is on to try and save them. Here, Anni has been at the forefront, promoting and actively participating in some of the first experiments to grow artificial reefs. The world's first coral cultivation projects aim to restore the ocean bed through underwater gardening, and the dynamic president has been helping scientists, who are transplanting nursery-raised 'nubbins' of coral polyps among their dead limestone cousins, and on specially submerged lattices. The hope is that a living reef structure will arise and eventually form a new island. So far, it looks successful, with tens of thousands of new coral nubbins taking root on proto-reefs, off Nakatcha Fushi, Landaa Giraavaru and other islands, and a corresponding increase in marine life.

Another resort island is going a step further, and I head to Vabbinfaru to visit what may be the future of atoll management in the Anthropocene. Here too they are cultivating coral fragments in a nursery, but as I pull on my snorkelling gear, marine biologist Roberto Tomasetti explains that they are giving corals a helping hand in the acidifying waters. We jump off the jetty, swimming above the barren sandy bottom and occasional polystyrene-like remnants of once-living coral. Then, a few metres further out, an enormous colourful spectacle heaves into view. The researchers have submerged a huge steel cage called the Lotus on the sea floor. The twelve-metre structure, which weighs two tonnes, is connected to a long cable which supplies a low-level electric current. The electricity triggers a chemical reaction, which deposits calcium carbonate on to the structure. Corals seem to find this irresistible, and the Lotus has been so thoroughly colonised by coral that it is difficult now to make out the steel shape beneath all the elaborate shapes and colour.

The El Niño Pacific-warming phenomenon of 1998 killed 98% of the reef around Vabbinfaru, so the scientists here have been able to compare the growth rates for corals grafted on to concrete structures on desert patches of seabed, and those stuck on to the Lotus. Roberto says coral growth on the structure is up to five times faster than elsewhere. The electric reef may also make the corals fitter and better able to withstand warming events and other stress, perhaps because the creatures waste less energy on making their skeletons. A smaller prototype device was in place during the 1998 warming event and more than 80% of its corals survived, compared to just 2% elsewhere on the reef. 'I would like to be able to find out whether we can transplant heat-tolerant ones to parts of the reef where it is more exposed and so build coverage there,' Roberto says. Researchers are trying to selectively breed coral varieties that are tolerant to higher temperatures, ultraviolet light and acidity. One way of doing this might be by exchanging the coral's algae species for ones that are more robust, which some corals do naturally, and early experiments are proving fruitful.[5] Done on a larger scale, this could keep enough of the reef alive to delay erosion and buy the islanders some more time. Several electrical 'biorock' structures have been installed around the world, including a 220-metre-long garden at Pemuteran on Bali, Indonesia, designed by marine scientists Wolf Hilbertz and Thomas Goreau, in an aim to protect reefs from global warming. But the cost and effort involved make it impossible to do except on a small scale. 'There is no current technology that allows thousands of kilometres of reef to be regrown, so it is of very limited value,' Peter Sale says. However, he concedes it is a useful technique for tourist resorts and to help small areas of reefs recover that have been hit by temporary damage, such as an oil spill or dynamite fishing. And artificial reefs can be a physical barrier against illegal trawler fishing – concrete structures are used in this way in the Mediterranean and off the Florida coast – even if they are not always successfully colonised by coral.

The problem of disappearing reefs has attracted plenty of creative solutions – one idea is to erect giant canopies to shade and cool the shallow waters where reefs grow. British conceptual architect Rachel Armstrong envisages using synthetic biology to make a coral-like organism that would create an artificial reef of calcium carbonate using energy from the sun and, through this, support sinking cities like Venice. Unfortunately, whether the process is done by nature or human invention, while the oceans continue to acidify, a calcium carbonate reef is doomed to dissolve.[6] And it is not just corals that are affected. Any animal that builds a calcium carbonate shell, including crustaceans, oysters, plankton and any animal that feeds on such creatures, from whales to humans, will feel the impact. Tiny planktonic pteropods (sea butterflies), an important food source for fish and marine mammals, are already dissolving, as are oysters.[7] The bones of fish and whales are also made of calcium carbonate, and researchers have found that acidification is altering bone formation, particularly in the delicate ear bones, which fish use to navigate and sense speed and direction.[8] Fish are not just disorientated by acidification: biologists have discovered that their neurotransmitters are also disrupted, changing their behaviour.[9] Acidification also changes the way sound waves travel through the water. A 0.3 decrease in pH allows sound to travel 70% further in the water – the ocean is becoming louder in the Anthropocene, perhaps confusing animals that communicate through sound, like cetaceans.[10]

Places naturally high in carbon dioxide, such as deep-sea vents, show a corresponding low level of marine diversity, where corals are replaced by algal slime, offering a window into what the oceans of the Anthropocene might become in coming decades.[11]

One tack would be to reduce ocean acidity by adding bicarbonates or lime to the water.[12] Trials show that it is a successful technique and coral growth improves locally where it is done, but we'd need to add around ten cubic kilometres of lime (around 9 billion tonnes) to the oceans every year to offset the global effects of our carbon emissions.[13] At the moment, we produce around 300 million tonnes of lime annually, mainly for the concrete industry, so the problem quickly becomes apparent. Perhaps sea cucumbers are the answer? Researchers have found that the large invertebrates excrete calcium carbonate and ammonia as part of their digestion process, which raises the pH of the water. But we'd need a lot of cucumbers. The oceans currently absorb about half of our carbon emissions, so if we want to reduce ocean acidity, ultimately we need to reduce those emissions.

The problem of coral reef destruction in the Anthropocene is acute and global. The 1998 mass bleaching event destroyed 16% of reefs world-wide.[14] Since then, there have been at least six major bleaching events, and scientists estimate that at least 20% of Australia's Great Barrier Reef – the world's largest – has been destroyed, and up to 90% of coral has been lost in the Indian Ocean and Caribbean. Nowhere is doing more to combat the problem than the Maldives, but disappearing coral is just one of the sea changes that Anni is facing. With so little cultivable land, people here depend on fish as much as any other marine predator – most meals involve salted, dried or curried tuna caught by a family member – and fisheries are in trouble. For generations, Maldivian fishermen have relied on the Nakaiy, a unique and complex weather calendar, thought to date back to the fourteenth century. It gives precise and detailed information for the year, in fourteen-day intervals, on where to find shoals of tuna and other fish across the archipelago, and how the stars, winds, storms, monsoon and tides change. 'It always worked perfectly,' says Abdul Azeez, a mariner in his sixties. 'But over the past fifteen years it has become increasingly inaccurate. Now we cannot use it any more.' The climate has become unpredictable in the Anthropocene and fishermen have noticed a reduction in the size of fish and their catches. 'The tuna used to be this big,' Abdul says, his arms outstretched. 'There were clams as big as your face and so many types of shark and barracuda.' The government has responded with legislation to ban fishing in certain areas and of certain threatened species. Most fishing is for skipjack tuna and its rarer, larger cousin, yellowfin tuna, and contributes one-tenth of the nation's income. They are caught by pole and line – nets are banned in the Maldives to protect stocks – from motor-powered boats called dhonis, which replace the prettier sail dhonis used here for centuries. It's the most sustainable fishing technique, and yet stocks have declined sharply nonetheless. Catches are down 40% since 2005, fishermen have to travel far further to make a living, and some are tempted into illegal fishing on protected reefs or of endangered species. Meanwhile, the better-conserved stocks around the Maldives attract unscrupulous fishermen from other parts of Asia – and they're not just visiting for tuna.

I join a group of biologists, who are heading off on a three-day expedition to find the world's biggest fish: the whale shark. They swim throughout the tropical oceans, but relatively little is known about their behaviour, their migratory routes, how long they live or their breeding habits, other than that they give birth to live young that hatch out of eggs inside the mother shark. Whale sharks are being hunted for their large dorsal fins, which are used as billboards outside restaurants in Asia to advertise sharkfin soup. Shark numbers have plummeted in the Maldives in recent years because of finning. It can take more than ten years for a shark to reach sexual maturity and their fertility rate is very low, so it is impossible for many populations to recover from overfishing. The scientists on board are hoping to discover more about the threatened species, before it's too late.

We set off to an outer atoll some forty kilometres away, where the sharks are known to congregate. The sun flashes so brightly off the smooth flat ocean that we have to wear polarising lenses to see beneath water almost as transparent as air. A manta ray glides past with a graceful flap of its giant cloak, turtles cruise like grumpy bath toys, and coloured fish flit by in huge neat schools. For me, it's magical, but the guys are getting anxious. Reaching lengths of up to twenty metres, and sporting a dramatic checkerboard pattern of bright polka dots, a whale shark sounds like an easy spot, but we've spent three hours peering into the water, and seen none. We're not the only ones searching. Speedboats whiz past us carrying resort tourists on whale shark spotting trips. The activity has become big business here, and more than 75% of whale sharks recorded in the area have scars or wounds from boat damage, says Richard Rees from the Maldives Whale Shark Research Programme; they are especially vulnerable because they swim slowly and close to the surface.

Suddenly, Richard spots one. I leap into the water after the practised researchers and swim my fastest towards the large shark that's drifting along. Richard passes me one end of the measuring rope and I power up to the shark's nose, looking directly into its eyes as Richard stays at the tail. The shark is a youngster, just 6.5 metres long, but to me, a majestic painted creature, negotiating the water with grace and ease. Unlike other sharks, the massive mouth of a whale shark poses no threat to humans – they are filter feeders, grazing on microscopic animals and algae. Richard swims beneath the fish to sex it; small claspers confirm that it's a male. He snaps the shark on the left and right side with an underwater camera. Another scientist swoops in with a harpoon and tags the shark just beneath the dorsal fin, which will record data to transmit via satellite whenever the shark surfaces.

We are surfacing when Richard spots another and we freedive down again. This poor creature has a badly distorted dorsal fin that is almost completely detached after an unsuccessful finning attempt. A shark fin of this size can go for $10,000 in Taiwan or Hong Kong, which is one reason why their population is expected to decline by as much as 50% over the century.[15]

After four days at sea, I return to land to hear that Anni has banned the hunting of reef sharks throughout Maldivian waters, making it the first nation in the region – and the second in the world – to outlaw the practice. The move is welcomed by marine conservationists who have long campaigned for better protection for the thirty-seven species of shark that frequent Maldivian waters. 'Eight years ago, you could see sixty or seventy black-tip reef sharks swimming off our jetty whenever you went snorkelling,' says Anke Hofmeister, a marine biologist based at the Soneva Fushi resort in Baa Atoll. 'Now you would be lucky to see two,' she says.

I've come to this luxury resort, in the north of the Maldives, on Anni's suggestion – he describes it as an inspiration, and regularly sends ministers to look at the innovations being trialled here. In his own small way, owner Sonu Shivdasani is aiming for the same as Anni: a profitable enterprise that is socially and environmentally progressive. By 2015, Soneva Fushi plans to absorb more carbon dioxide than it produces. To that end, the resort has planted trees and mangroves (a big departure from almost every other resort island, which are bare but for a few palm trees). Each bespoke villa is designed to use natural ventilation, and to supplement this, Sonu experimented with a novel heat-exchange air-conditioning system using pipes that descend into the cool deep ocean water. However, after battling with the design and spending over $1 million trying to get it to work, he had to abandon the system and start again. The new plan is to provide the cooling in the same way as their other electrical needs, using a new solar power plant. Hot water for showers is already being provided by a heat-recovery system, with backup from a biofuel system generated from the restaurant's waste food. These efforts should be sufficient to make the resort's diesel generator obsolete, he hopes. Now, the resort is working towards offsetting the carbon footprint of its visitors' international flights, through tree-planting projects around the world.

The human – social – environment is as important to Sonu as the natural marine one. He employs twice as many local staff as other resorts, and invests in them and their villages through various projects, including a scheme in which tourists spend several days volunteering with Maldivians to earn ten days' free holiday at the resort.

Soneva Fushi, which frequently wins gongs for best as well as greenest hotel, has not sacrificed luxury for sustainability. Its small changes, including recycled elephant dung paper and jute bags, are as important as the big technological innovations, Sonu stresses. For example, plastic bottles are banned on the island. Instead, tourists are given a glass bottle which can be refilled for free with still or sparkling water from the desalination facility, a move that the resort says saves more than 45,000 bottles from landfill each year. Discreetly hidden away in a maintenance area is a recycling system and, beyond that, a couple of gardens that supply fruit, herbs and salads for the resort – quite an achievement in a place where more than 90% of food is imported.

What difference do these efforts make, I wonder, if the islands are going to disappear anyway because of a much greater volume of emissions released thousands of miles away? Back in Malé, I put the question to Anni. 'We need to lead by example,' he replies. 'If we are genuinely asking other countries to cut their emissions, then we must be willing to do the same ourselves. And that applies to all countries rich or poor.'

This is a massive departure from the usual position taken by developing nations, who argue that they need to develop their economies and while they do so they should be allowed free reign to pollute. Since they haven't caused the problem of greenhouse gases in the air, the argument goes, they shouldn't be subject to emissions controls. Anni, however, sees no point in arguing over fairness when there is an environmental emergency to be faced. 'Yes, it is unfair that some countries got to develop their economies in a dirty polluting way, and it's unfair that the poorest people are also going to be the hardest hit by climate change. But we must move beyond this argument and start acting to solve the problem.'

The pioneering attempts by Anni to couple human rights and climate-change action impress me deeply. These islands will disappear, he knows it, but these people, their children and grandchildren will still require a habitable planet, and Anni is thinking about the long term while taking action in the short term. Such a trait is unusual in anyone; in a president, it is particularly rare. Which is why the events that later unfolded on that paradisical island state seem to me especially tragic.

Just as he had feared, Anni's presidency was all too brief. Midway through his term, in early 2012, he was ousted from power in a *coup d'état*. His many projects were put on hold or cancelled as investors pulled out, and in many ways the country went backwards. At the end of 2013, after increasing international pressure, violent street clashes and protests, elections were finally held. Anni lost by a whisker to Abdulla Yameen Gayoom, the brother of the former dictator, who had led an Islamist campaign accusing his rival of being too secular and close to the West. In opposition, Anni continues to fight for action on climate change, development and environmental issues, but I look back on my Maldives visits with bittersweet poignancy, remembering the sense of optimism and hope that I felt so recently. And not a little despair, because, as carbon dioxide levels continue to rise, and the oceans become warmer, more acidic and threatening, there is no other world leader as determined to confront the problem. And that is one of humanity's biggest failings.

Can paradise lost become paradise regained elsewhere? It's an issue that other island peoples living with encroaching oceans, coastal erosion, shrinking groundwater, and worsening reefs and fishing conditions will soon have to face up to. Tuvalu, Bangladesh's Bhola Island, Papua New Guinea's Carteret Islands, Kiribati, Fiji and the Pacific atolls are all on the list.

I caught up with Anote Tong, president of Kiribati (pronounced Kiribass), while we were both in New Delhi. Like Anni, Tong spoke with eloquent passion about the plight of Kiribati's 113,000 citizens. Tong was one of the first world leaders to speak out on global warming and urge action. Perhaps this is because he has a science degree or, more likely, it is because of the increasingly visible effects of climate change on his Pacific island home, which, like the Maldives, is a series of thirty-three low-lying coral atolls and islands. 'We can see the islands disappear, tree by tree, house by house, into the waves,' he told me. 'It is no longer a case of "if the sea levels rise we drown". It is already happening and we have reached the point of no return.'

Kiribati is expected to be the first country in which all land territory disappears due to climate change, and nations such as the Maldives and Bangladesh are watching with fascinated horror. Tong is pragmatic and planning for the inevitable. He has asked Australia and New Zealand to accept I-Kiribati (as nationals are called) migrants. Kiribati is one of the poorest nations on Earth. Most people fish or work in the coconut industry, making things from coir (the hairy outer fibres) or copra (the dried white flesh from which oil is extracted while the remainder is used in animal feed).

Tong is anxious that the populace should be as prepared as possible for their new lives in perhaps radically different countries. 'I don't want our people to be climate refugees, living from day to day with little hope,' he said. 'Many young people have their whole lives in front of them and we must make sure they have the best chance in their new lives, so they can contribute usefully to their new country and not be second-class citizens. It is not their fault that they have to move.' To that end, he has negotiated with the New Zealand government for some I-Kiribati to work in fruit-picking and gain horticultural training. And Australia has agreed to train almost a hundred more as nurses. These are small numbers, and neither country has so far agreed to accept any as new citizens. Indeed, a New Zealand judge in 2013 rejected a Kiribati man's claim for refugee status because of climate change.

The president of Zambia, Levy Mwanawasa, was the only head of state to welcome the islanders, telling Tong that there is 'plenty of room' for them in his country. But in 2008, he died suddenly in office, before the move was agreed. Tong is currently in negotiations with Fiji's military government to buy 5,000 acres of land on the country's second biggest island, Vanua Levu. Several I-Kiribati are already at the University of the South Pacific in Fiji, which is jointly owned by the region's twelve island nations. Tong plans to send the youngest, skilled workers first, so that the new country doesn't become overwhelmed with a migrant burden. Still, it will be a difficult move, not least because while Kiribati is a democracy, Fiji's democratically elected government was ousted in 2006 and the country is being governed under oppressive 'emergency laws'.

If the Fiji plan doesn't work out, Tong has been looking for other potential homes, including the possibility of building giant floating islands or oil-rig-type structures on which to move everyone. 'We have to look at every option,' he says. 'We are pioneers in this process and we don't know what the answer is yet.'

What is it like to know that the country in which you were born will no longer exist? He pauses, and for a few moments the professionalism that enabled him to answer easily the same familiar questions slips.

'It is painful,' he says at last. 'We are talking about the loss of a culture that goes back perhaps 5,000 years, the loss of a unique language, of fairy stories and songs. I would hope that our population can all move together somewhere else, but it is likely that we will be split into small groups here and there wherever we can be accommodated.

'Older people tell me they would rather die under the waves than leave their homes, the place of their ancestors and the graves of their loved ones. It is likely many will die of grief rather than move on,' he says. And the reality of what he is facing – that an entire culture will be extinguished within decades and before our eyes – sinks in. The geographic map of the Anthropocene will be different, but so will the cultural map.

'This is the biggest global emergency of our time – bigger than terrorism. People that continue to emit greenhouse gases, they are climate terrorists,' Tong says, riled up now. 'We are the first to go. But others will come after us.'

The fate of these low-lying islands depends on simple physics happening some 10,000 kilometres away – a place so utterly alien that it's hard to believe it shares the same planet. In the Arctic, where most of the day is hidden in darkness, where howling near-hurricane winds blast the frozen seas and icebergs the size of cities crash into each other, change is under way of a magnitude so extreme that Earth hasn't experienced its like for over 10 million years. The wildest, most protected ocean on our planet is finally succumbing to human influence – the North Pole is gradually weeping its ice.

Climate models predict that melting of ice in polar extremities such as Greenland and Antarctica will reach a tipping point after which the rate will suddenly accelerate. This is because meltwater lubricates the faster break-up of ice sheets, and because there will be less ice and snow (which reflect up to 90% of the sun's rays) and more dark water that better absorbs the sun's heat. The models differ on timescale, though. James Hansen's group at NASA Goddard predicts a global sea-level rise of as much as five metres by 2100; more conservative models, such as that by Stefan Rahmsdorf's group at the Potsdam Institute in Germany, peg it at 1.4 metres by the end of the century. That would be enough to inundate low-lying islands like the Maldives, along with many of the world's major cities. Even Rahmsdorf says that sea-level rise of metres rather than centimetres cannot be ruled out if there is 'non-linear or threshold behaviour of ice sheets' – a tipping point, in other words.

What is certain is that glacial melt is speeding up. The disappearance of ice that I witnessed in the Himalayan mountains is nothing compared to what's happening in the Arctic. The Arctic is warming at least twice as fast – perhaps as much as eight times as fast – as anywhere else on the planet.[16] Average global warming of 2°C, for example, would mean Arctic warming of 3–6°C. This is partly because of hot southerly winds converging on the region, bringing soot pollutants. The darker particles absorb the sun's rays, heating up the atmosphere and, when they are snowed down on to the ice, they darken it, speeding up melting. Almost all the Arctic glaciers are shrinking – the Greenland ice sheet is losing 200 gigatonnes of ice a year, four times more than a decade ago. Researchers studying just one area of Greenland in 2011 found that crevasses in the ice sheet had increased by 13% since 1985, speeding its flow into the sea.[17] The Greenland ice sheet covers 80% of the island's surface and is the second largest in the world after Antarctica's. If it melts completely, it will cause global sea-level rise of seven metres.

The oceans are not rising uniformly – the Pacific, for example, will grow 20% higher than average.[18] And, while tropical islands disappear, new ones are showing up in the Arctic. Uunartoq Qeqertaq (meaning Warming Island) was born in 2006, the result of the melting of an ice bridge that formerly cloaked it and mainland Greenland as one. Atlases will have to be redrawn in the Anthropocene.

The warming seabed and thawing land of the Arctic are releasing ancient stores of hundreds of millions of tonnes of methane, a greenhouse gas that over a century is twenty times more potent than carbon dioxide. Scientists have observed vast volumes of the gas bubbling into the atmosphere, with the potential to utterly transform our world beyond anything habitable. 'This is the first time we've found continuous, powerful and impressive seeping structures more than 1,000 metres in diameter,' Russian scientist Igor Semiletov told the UK *Independent* newspaper in 2011. On land, we are seeing greening of the Arctic tundra, a vast region of permafrost that is thawing, making agriculture newly possible. 'Tundra' means 'treeless', and yet in some regions, researchers have observed a 20% increase in vegetation, including forests of alder and willow.[19] Animal species may also begin migrating there, while others, such as the polar bear, risk extinction – it is already breeding with its north-venturing southern cousin, the grizzly, producing a hybrid called a pizzly or grola.

The greatest and most dramatic regional change is to the sea ice, though. The Arctic is the smallest, shallowest ocean, almost completely surrounded by land and frozen over year round. For a few months in the summer, the ice partially retreats before refreezing. However, the extent of this retreat has been growing, and it's been occurring earlier. In 1980, the Arctic ice in summer made up 2% of Earth's surface, but it has since halved in area and its volume has dropped by 75%. Suddenly, in 2007, it melted at an unprecedented rate. Clear channels appeared, allowing ships passage between the Atlantic and Pacific oceans over the 'top' of the Earth. Subsequently, the melt has continued, with the summer of 2012 so clear of ice that scientists were forced to re-evaluate their models and predictions. The data could be underestimating the melt rate, because scientists can't distinguish pack ice from slushy melted ice on satellite images. What is clear, though, is that the ice is entering a death spiral of accelerated melting and warming. The Arctic Ocean could be ice-free by 2030 – perhaps as soon as 2015. The speed is alarming scientists, who calculate that the financial burden of a warming Arctic could be comparable to the size of the global economy in 2012.[20] Others are recommending urgent geoengineering measures to slash temperatures, such as spraying the atmosphere with seawater to increase the reflectivity by clouds of sunlight.

Unlike meltwater from land-based glaciers, melting sea ice doesn't directly affect ocean levels, it simply displaces its own volume. However, the implications of having a great dark body of water at the planet's north pole in the Anthropocene, rather than a thick sheet of reflective ice, extend far beyond the Arctic. Ocean circulation is largely governed by the temperature difference between the warm tropical oceans and the freezing Arctic waters. Already, scientists are seeing changes, with impacts on climate phenomena around the world, from El Niño to hurricanes. The influx of fresh water into the Arctic, from glacier melt, and its relative warmth, is generating stable layers of water that don't overturn and mix as they should. Fish are migrating further north as the temperatures become less hostile and food availability increases – as the ice melts, more light can penetrate through the water, stimulating great blooms of phytoplankton.[21]

Herein lies one of the oceans' greatest gifts: phytoplankton are photosynthetic – they are the rainforests of the sea, locking up carbon dioxide – and if we could harness their activity, we could simultaneously cool temperatures and neutralise the oceans. That's the thinking behind geoengineers who are hoping to stimulate these phytoplankton blooms using iron – a fertiliser that is naturally in low concentration in the ocean. Various mavericks have carried out their own small-scale trials, but the best hope lies in the freezing, barren Antarctic waters. There, waters rich with krill once supported a large population of whales. Humanity's whaling industry virtually wiped out these large marine mammals, and with them vanished an entire ecosystem, most likely because the whales fertilised the water with their iron-rich faeces. Without the krill, there are few whales. Fertilising the water there with iron dust would perhaps be more acceptable to a geoengineering-sceptical public, because it would be helping restore an ecosystem we destroyed. And at the same time, it may also go some way to rectifying global temperatures.

Important global weather patterns are heavily dependent on polar conditions, and the scale of ice melt in the Arctic is already changing the once-temperate zones of the northern hemisphere. As the temperature difference reduces between the hot southerly winds and the icy northern winds, the two fronts don't churn up with as much violence, but disperse more gently. The energetic winds formed in the churning, also known as the jet stream – a current several hundred kilometres wide that is responsible for the milder climate experienced in northern Europe – are pushed further south. The relative stability – a stagnation of stultified air – means that the usually changeable weather experienced in temperate zones becomes more extreme and lasts for longer. Thus, in the summer, instead of brief rain showers, the region gets monsoon-like flooding, and instead of brief dry periods, it gets month-long droughts. In the winter, the new more southern position of the jet stream exposes northern temperate latitudes in Europe and the US to freezing temperatures, and the increased humidity in the air from evaporation leads to long and heavy snow showers. The north of Europe, in other words, gets colder, snowier winters, while south of the jet stream, drought persists. In the Anthropocene, the planet will no longer have large swathes of continent with a temperate climate.

The Arctic of the Anthropocene will soon be unrecognisable to a visitor from the Holocene. The opening in the ice of the North-East Passage – a long-coveted sea route between Europe and Asia – is bringing new trade possibilities to this forbidding area, boosting marine traffic. The journey from Europe to the Far East takes nearly half the time over the Arctic that it takes through the Suez Canal – Rotterdam to Yokohama is more than 7,000 kilometres shorter via the North-East Passage, and around one-third cheaper too, both in fuel and insurance premiums since it avoids the risk of an encounter with Somali pirates in the Gulf of Aden.

The disappearing ice heralds new opportunities for others too – the very companies, in fact, which are fuelling the warming that's causing the melt. An estimated 50 billion barrels of oil and 7 trillion cubic metres of gas – 15% and 30% respectively of the word's undiscovered reserves – lie beneath the Arctic Ocean, as well as a wealth of other minerals. Companies are jostling to drill into these previously inaccessible fields and the enormous economic potential is generating diplomatic tensions between the five Arctic countries, each of which lays claim to the undersea riches, as well as the massive wind, tidal and geothermal energy possibilities. Each country can claim the portion of continental shelf that protrudes from their coast, and so sovereignty is being decided on the undersea geology. The US claim alone, at 8.6 million square kilometres, is estimated to be the world's largest underwater land extension and bigger than the lower forty-eight states combined. In 2007, President Vladimir Putin pre-empted international agreement, planting a Russian flag on the ocean floor at the North Pole. But despite the bravado, fears of a new, Colder War, seem unfounded for the moment.

Of more concern is that energy companies will destroy the fragile ocean environment in their haste to make a buck. The ecosystem that exists has evolved extreme mechanisms to survive such cold and darkness – for example, fish species have antifreeze running in their veins, and huge mammals like bears can survive without eating for months. Nevertheless, they live on the edge of survival – any oil disaster could push entire species into extinction. Drilling in the Arctic, among hurricane-strength winds, fierce ice storms, crashing icebergs and compression freezing, in the hope of finding enormous wells of oil, is not easy, and the potential for leaks, for vessels to spill or crash like RMS *Titanic* did a century ago, is higher than elsewhere. A spill in the Arctic – even a small one – is also far harder to clean and contain. Oil doesn't naturally disperse in freezing water as it does in the tropics, and the microorganisms that help break up oil in the Gulf of Mexico don't inhabit the Arctic. Using booms and other physical barriers to contain a spill would be impossible.

However, despite environmental concerns, oil and minerals prospectors are heading into the region in large numbers. Even if it is not economically viable to set up drilling operations now – and several have pulled out of the area after realistic evaluation of the costs involved – there will come a time within the decade when such ventures are profitable.

Towns and even cities are already being built to support the new industry, in places previously inhabited only by eskimos and whale hunters. Greenland, the world's largest island, with a population of just 56,000 (85% of whom are Inuit), is in danger of being overwhelmed. Indigenous people are finding life in the region increasingly hard anyway. Many cannot cross the previously frozen ground without getting bogged down in mud or risking falling through cracks in the thin ice. Melting ice is exposing the fragile coastline to the violent battery of Arctic waves, and piece by piece it is crumbling into the sea. In some parts of the Arctic, the rate of erosion has doubled to tens of metres per year due to thawing and the loss of protective ice, which increases wind fetch.[22] Traditional hunting grounds are now inaccessible and new migrants from the modern world are introducing a way of life that is eroding their culture. Greenpeace has vessels deployed in the region, and its activists are trying to obstruct drilling and generate public opposition to the plans. How successful the well-funded NGO will be against the better-funded oil companies remains to be seen.

Meanwhile, the so-called 'doughnut hole' of millions of square kilometres of ocean that lie outside of the Arctic nations' exclusive fishing zones is melting for the first time, exposing Arctic cod populations to a potential international fishing free-for-all. The wild Arctic could fall victim to the scourge of every other ocean: overfishing by humans.

Fish are the last wild animal that we hunt in large numbers and yet we may be the last generation to do so. Globally, 85% of stocks are over-exploited, depleted, fully exploited or in recovery.[23] Stocks of large fish are down 90%. Global annual fish consumption is now approaching 20 kg per person per year (four times as much as in the 1950s), even though global fish stocks have continued to decline. Humans are now the biggest ocean predator, making oceans less biodiverse and disrupting food chains. It is likely the only fish that people will eat, further into the Anthropocene, will be farmed. Entire species will never be seen, let alone tasted. Vast areas of the seabed of the Mediterranean and North Sea already resemble a desert, expunged of fish by Europeans using increasingly efficient methods such as bottom trawling. And now, these heavily subsidised industrial fleets are cleaning up in the tropical oceans too. One-quarter of the EU catch is now made outside European waters, much of it in the previously rich West African seas, where each trawler can scoop up hundreds of thousands of kilos of fish in a day and return to Europe with the catch. All West African fisheries are now over-exploited, so that many local people can no longer support their families through artisanal fishing.

The policy of maintaining vast fishing fleets to go after ever diminishing stocks by means of massive state subsidies is unsustainable in every way – it costs an estimated $50 billion annually. In Spain, for example, one in three fish landed is paid for by subsidy. Artisanal fishing, by contrast, catches half the world's fish, yet provides 90% of the sector's jobs.[24] Clearly, industrialised countries are not about to return to traditional methods; however, it is also obvious that the massive fleets will have to be slashed. Humans have so changed marine ecosystems that in order to retain something of the Holocene fish diversity, we will have to do what no other species has managed: rein in our hunting even while there is opportunity and capability to hunt more.

In the EU alone, restoring stocks could result in catches estimated 3.5 million tonnes greater, worth $4.4 billion a year.[25] Rather than the current system, in which member nations hustle for the biggest quotas, individual governments could set quotas based on local stock levels. Fishermen would have individual tradable catch shares of the quotas – they have a vested interest in seeing stocks improve, after all. This method has been used successfully in countries including Iceland and New Zealand, and helps end the situation whereby everyone grabs as much as they can from the oceans before their rival nets the last fish. Research shows that managing fisheries in this way means they are twice as likely as open-access fisheries to avoid collapse. In severely depleted zones, the only way to restore stocks would be to ban fishing. In other areas, quota compliance would have to be properly monitored – fishing vessels could be licensed and fitted with tracking devices to ensure they don't stray into illegal areas, spot checks on fish could be carried out to ensure size and species, and fish could even be tagged to be sure of their source.

What we're now seeing in the Anthropocene is marine species living not wild in the oceans but in small enclosures: humanity has employed its usual method of dealing with food shortages and moved from hunting to farming. More than half of the fish we eat now comes from farms – in

China, it's 80%. Raising fish alongside rice in a paddy field, as many communities in Asia have done for centuries, is a sustainable way of diversifying agricultural production, ensuring protein and other essential dietary nutrients, and improving yields: the plants shelter the fish from heat and predators; the fish eat pests and their faeces fertilises the rice. Industrial-scale fish farming, however, is far more problematic. Farmed fish, like salmon and tuna, eat as much as twenty times times their weight in smaller fish, like anchovies and herring, which has led to overfishing of these smaller fish. Fish farms are also highly polluting. They produce a slurry of toxic run-off – manure – which fertilises algae in the oceans, reducing the dissolved oxygen and creating dead zones. Scotland's salmon-farming industry, for example, produces the same amount of nitrogen waste as the untreated sewage of 3.2 million people – over half the country's population.[26] And fish farms are also breeding grounds for infection and parasites that then frequently infect wild populations. Nevertheless, fish farms will certainly become more important in coming decades, and their problems can be overcome through better management and innovations – for example, feeding the fish a vegetarian diet supplemented with artificially made omega-3 oils.

Overfishing, combined with pollution, climate change and acidification, is altering the entire marine ecosystem. A whole range of exotic sea creatures are being hunted to extinction, including turtles, manta rays, marine mammals and sharks. Around 100 million sharks are killed each year – their numbers have declined by 80% worldwide, with one-third of shark species now at risk of extinction. Sharks are top predators and play an important role in maintaining the balance of the marine ecosystem. A decline in shark numbers can lead to an increase in fish numbers further down the food chain, which in turn can cause a crash in the population of very small marine life, such as plankton. Without the smallest creatures, the entire ocean food web is threatened.

One repercussion is jellyfish. In the past decade, several regions have seen jellyfish explosions – or 'blooms' – in which the waters turn into gelatinous soups that take over coastlines, harbours and bays, leading to beach closures, nuclear power plant shutdowns, and disrupting industries from tourism to fishing. Refrigerator-sized jelly monsters called Nomura's, weighing 220 kg and measuring two metres in diameter, have swarmed the Japan Sea annually since 2002, costing the Japanese fisheries industry billions of yen in losses. Marine ecologists are warning of worse to come: human changes to the marine environment may lead to a tipping point in which jellyfish rather than fish dominate the oceans of the Anthropocene, much as they did hundreds of millions of years ago in pre-Cambrian times.[27] Jellyfish are simply better prepared than other marine life for Anthropocene conditions – warmer temperatures, salinity changes, ocean acidification and pollution work to their advantage. Jellyfish can cope far better with anoxia (low oxygen) than muscular fish. Even our infrastructure – pontoons, piers and drilling platforms – is a boon for jellyfish, providing an anchor for polyps.

We need to decide what kind of ocean biodiversity we want for the Anthropocene – and then act fast if we are to save species. Solutions can be simple. For example, more than 300,000 seabirds are killed annually when they get caught in fishing lines, pushing populations of albatrosses, petrels and shearwaters to the edge of extinction. But using weighted lines with bird-scaring streamers reduces seabird deaths by more than 90%.[28] Strengthening and expanding protected marine reserves helps conserve marine life, and nation after nation is stepping up to the plate. The Maldives now has more than thirty protected marine areas, and Australia has created the world's biggest marine reserve at over 2.3 million square kilometres. Currently, less than 1% of the ocean is protected, although the international community has agreed to raise this to 10% by 2020. Marine reserves are only effective, however, if governments have the resources to patrol and protect them. Also, many marine creatures, from whale sharks to whales, are migratory – they don't stay in the protected areas, making them easy prey for fishermen. What's needed, many argue, are mobile reserves that follow migratory animals and those that shift habitat due to currents or climate phenomena like El Niño. The zones need to be well targeted and needn't impact on fishermen's livelihoods. One study found that designating just twenty sites – 4% of the world's oceans – as conservation zones could protect 108 species, or 84% of the world's marine mammal species.[29]

None of this will help much if we don't curb our pollution of the oceans. Humans have always used the oceans as an enormous swirling toilet for our waste, but in the past, when our global population was much smaller and everything we produced was made of natural materials that bio-degraded, this habit had little impact on the marine environment. It was a sensible option, stopping human sewage from building up on land and causing disease, for example. But the scale of our rubbish is now trans-forming the oceans. Heavy metals and other toxic waste are consumed by bottom feeders like mussels, and accumulate up the food chain to the extent that top predators like polar bears are becoming sick, and pregnant women are advised not to eat carnivorous fish like tuna to avoid heavy-metal poisoning. We produce millions of tonnes of plastic waste – plastic production has increased 500% over the past thirty years – that doesn't biodegrade and is buoyant, as well as chemical effluent including laundry detergents and fertilisers which feed huge blooms of toxic algae that can be seen from space, and starve other marine life of oxygen, creating vast dead zones. Even laundering synthetic clothes is becoming a problem because of the scale with which we do it. Polyester garments release around 1,900 plastic fibres each time they are washed, which are being eaten by marine animals and getting into the food chain in measurable amounts. Every square kilo-metre of ocean now contains an average of 18,500 pieces of floating plastic, and vast floating islands of garbage coalesce in the currents. In the major ocean gyres, the ratio of plastic to marine life is six to one by weight.[30] While this poses problems for fish and other marine life including turtles and albatrosses which feed plastic to their young, the floating rafts of debris

– covering an area of 5,000 square kilometres in the north Pacific – offer sun shelter and a hard surface for molluscs, crabs and invertebrates, creating a new ecosystem that scientists are calling the plastisphere. In the Anthropocene, the plastic has effectively added hundreds of millions of hard surfaces to the Pacific Ocean (other floating structures, like seaweed, do not naturally occur there).

They're not the only creatures to live on our rubbish. In the Caribbean, I found a man who had created an entire island out of garbage – perhaps pointing to a potential solution for crowded islands and eroding coasts. Westpoint Island, a few kilometres off mainland Belize, contained a couple of timber houses, some fruit trees and pigs all surrounded by mangroves. It was unremarkable except for the fact that five years ago, it didn't exist. Gerald McDougall, a Creole-speaking fisherman, explained how he came up with the idea after noticing 'coconut trees sprouting in the sea'. 'Workers on nearby Saltwater Caye were dumping trash inside a clump of mangroves and, over time, sand built up on the rubbish and coconut trees grew,' he told me. Gerald had been looking for a piece of land to build a house, but the islands and cayes in this prime pool of Caribbean were too expensive.

'I realised that I could make a home for my family out of all the garbage here,' he said, crinkling his black leathery face into a smile of missing teeth. In the Tobacco Caye range – a small group of islands and mangrove clusters – Gerald saw a promising spit of sand surrounded by mangroves. It measured one hundred by seventy metres and was part of a government-owned chain. He approached the Belize authorities and inquired about a lease, explaining what he wanted to do. 'They didn't believe it would work, but they let me have it,' he said.

Over the next few years, Gerald spent every spare day he had on his embryonic island. He convinced all the local islands and resorts to deliver their garbage to his sand spit, and gathered a team of helpers from his many friends. Together, they separated out the trash: the glass beer bottles he sold back to the bottling companies to help fund his island; the cans he burned to make a metallic filler material; he used plastic bags for bulking; and the rest of the plastic and metallic waste became his new land. Growing an island is a layering process. First he dug out the mud and seagrass, then he packed in his sorted garbage, then a course of mud and sand, then another layer of garbage, then sawdust from the Dangriga millers, then more mud and sand. He barricaded the island in from the water using wooden fencing. 'When it rains, the sawdust holds the water. I planted papaya and coconut trees and their roots hold the whole thing in place,' Gerald explained. 'The government inspector came out to see it and he was amazed. It really works!' The results were impressive: Gerald had built two large houses on the site, which already measured two acres when I visited, and cultivated a flourishing fruit garden.

Will we see a range of artificial islands or even floating cities further into the Anthropocene? Perhaps – artificial islands are already being created everywhere from the Maldives to Dubai to accommodate bigger populations on purpose-built land, and it is changing the geography of our planet. To date, most have been low-lying and so, like their natural counterparts, they are vulnerable to sea-level rise. Floating islands, though, would rise with the sea. Billionaire Peter Thiel, a co-founder of PayPal, is funding the Seasteading Institute, which hopes to create entirely floating autonomous city states.

In the Anthropocene, we need to radically rethink our relationship with the ocean. Until recently, the vast, wild seas, with their seemingly inexhaustible buffet of fish, were considered too powerful for human adulteration. Perhaps the first clue that they were not was back in the 1800s, when the whaling industry nearly exterminated from the planet the biggest creatures Earth has ever known. Now, humanity's influence can be seen throughout the oceans and we are in danger of losing key species we depend on. We land mammals have never been more part of the marine environment: most of us are coastal and, like Maldivians, many of us live at sea level – we will need to decide when to retreat and when to protect. It is neither desirable nor practical to separate humans from the oceans in order to 'save' them. But we need urgently to curb our destructive impact.

Oceans have fundamentally changed and we must engineer new ways of living with these changes. With sea levels rising, we need to carry out the geology of creating our own islands and coasts; with fish stocks plummeting, we will need to farm our own domesticates; under warmer more acid conditions, create our own artificial reefs, and perhaps even garden the oceans with photosynthesising phytoplankton.

Those at the forefront of this change, such as Anni Nasheed and Anote Tong, have shown inspirational leadership in approaching these challenges. Now the rest of the world must step up.

6

DESERTS

Dry, barren and windswept, deserts are the most hostile landscapes on Earth, and the places that most closely resemble the surfaces of other planets. One-third of our planet's land surface is covered by desert, places that receive so little precipitation that water evaporates from the surface faster than it falls from the sky. Half of the world's deserts are in polar regions, frozen expanses of land, much of it devoid of life.

The rest are close to the equator, often abutting the most lush and dripping tropical rainforests. These arid lands are vast expanses of sand and dust, with such little vegetation that winds race unhindered across them, scooping up the eroded earth and dumping it thousands of kilometres away. Minerals from the Sahara, the largest desert on Earth, are responsible for fertilising the rich Amazon.

During the day, the sun beats down on the harsh desert surface, raising temperatures as high as 50°C; at night, the heat escapes into the cloudless sky, *plunging temperatures into the minus figures. And yet, plants, animals and humans have found ways of inhabiting deserts, occupying their fringes and venturing across their deadly expanses. Grazing herds and the nomadic human pastoralists that learned to follow them have managed to survive for generations in seemingly impossible drylands.*

In the distant past, many of the areas that are now deserts were wet. The Sahara was once a savannah teeming with life. Rock paintings made by humans thousands of years ago depict large antelopes and domestic cattle. A series of massive lakes once flooded many deserts, including the Sahara, and their remnants are now being discovered, buried beneath the dry surface. This fossil water has the potential to regreen the desert.

In the Anthropocene, the world's deserts are growing. Global warming is increasing the Earth's surface temperature so that water evaporates faster, and climate change is reducing precipitation in semi-arid regions.[1] In addition, the way humanity is using the land is converting formerly vegetated areas to desert. Intensive farming and population expansion on fragile drylands are hastening desert encroachment. The spread of deserts is ruining former arable lands, chasing away villagers to city slums and increasing poverty. Across the planet, desertification degrades more than 12 million hectares of arable land every year.[2] Some 40% of Africa is already affected and analysts warn that two-thirds of the continent's arable land could be lost by 2025.

People are trying various strategies to beat back the dust. One of the most remarkable is an attempt to grow a Great Green Wall, a tree belt fifteen kilometres wide and 7,775 kilometres long, along the southern edge of the Sahara from Senegal to Ethiopia and Djibouti, across eleven nations. Another green wall in China hopes to stop the Kubuqi Desert spreading east from Inner Mongolia.

Humans are exploiting deserts in the Anthropocene, finding new uses for the relentless sun and wind. Some of the poorest nations on the planet are starting to benefit from schemes to trap this solar and wind energy for electricity. Vast energy belts are planned across deserts in Australia, the US, China and Africa, with some of the electricity being used to help make deserts more hospitable. Solar-powered desalination, for example, has the potential to enable agriculture in places where it was previously impossible.

In the Anthropocene, humankind has invaded and transformed deserts, turning these alien landscapes into cities and parks, green fields and massive lakes. From space, such places are unrecognisable. Circles of wheat and yellow flowers bloom out of the grey sand, under the giant arm of a turning irrigator. Cities like Phoenix, Las Vegas and Dubai emerge pristinely from the arid sand, complete with spurting fountains and floral displays.

Deserts have always been places of little water and vegetation, extreme temperatures and harsh winds. But in the Anthropocene more and more people are trying to live there. They are learning new ways to use this alien landscape – and transforming it in the process.

'I would like to visit the tribes living around Lake Turkana,' I tell my ever-helpful guest-house owner, Sally. 'Do you know which buses go there from Nairobi and whether it's popular and we need to book?'

She breaks into peals of laughter. 'Buses don't go there,' she laughs. 'Hee, hee, which bus to Lake Turkana?' she chuckles again.

The desert area in the far north of Kenya, bordering Ethiopia, is a lawless zone, troubled by frequent tribal conflict, aggressive bandits and reached by some of the worst roads on the continent. Years of crippling drought have only stoked up tensions. I want to go there to discover how people are coping as the environment becomes yet more hostile, and to learn about some major changes heading their way.

Eventually, Sally stops laughing and takes pity on me. After establishing that I really am keen to go there, she tells me that she has a cousin who's a Catholic priest at the lake mission, and she promises to arrange it. By happy coincidence, another of the lake's priests is in Isiolo awaiting a spare part for his vehicle. He can drive me to the lake from there. The opportunity is too good to pass up, and by 6.30 the next morning I'm at the *matatu* (minivan) station in the city, waiting to leave for Isiolo. It is late afternoon, after an uncomfortable journey involving multiple police stop-checks, when the *matatu* finally rolls into the town. I scan every billboard and building for Isiolo's Catholic mission, but it's a difficult task – 400 million Africans are born-again Christians and the various sects of Christianity are well represented in this small town, from the Church of Wonderful Miracles to the Church of the Best Future, and there is also a large Muslim community. People give around 10% of their meagre incomes to these groups – that's far more than the government takes in taxation. And it's made some of Africa's church leaders multimillionaires with private jets, megachurches, video productions and publishing houses all preying on the desperate.

I spot the mission eventually, and there, waiting patiently in an off-road vehicle, with its engine running, is Father Fabio-Miguel Chaparro, a 35-year-old from Colombia. I am more than three hours late, but he couldn't be more understanding. He makes final preparations for the journey, as girls from the mission's high school file past. There are plenty of longing glances directed at the handsome young priest, but he is oblivious, checking the engine and politely replying to unnecessary questions from schoolgirls, who hitch their skirts as short as they dare under the scrupulous gaze of passing nuns. Within minutes we're on the road, driving a little too fast through the town, past faces that glare at us in open hostility.

And then we see it: a great blackened field in the heart of the town. It is still smoking, but people are already attempting reconstruction using timber A-frames. The marketplace was burned to the ground yesterday, Fabio tells me. Everything was destroyed – all the shops, stalls, produce (much of it desperately needed food), people's houses and livelihoods gone.

It started with the shootings, he explains. Yesterday there were gun battles in the streets here between the Turkana and Borana pastoral tribes. More than ten people died, including a 14-year-old schoolgirl from the mission school. She was travelling in a truck when the Borana ordered everybody off, separated out the Turkana, including her, and shot them all.

The conflict, like all conflicts here, began with one tribe stealing cattle from the other. In this increasingly barren desert, nomadic tribesmen are abandoning a way of life that has remained unchanged for centuries. Their animals dead of starvation, their children sick with hunger and cholera, and their government deaf to their cries for help, tribes such as the Turkana, Samburu and Borana are fighting each other over what animals are still alive and over the few water sources that have not dried up. Whereas past intertribal conflict was fought out with sticks and spears, the neighbouring violence in Somalia and Sudan has provided cheap and easily available guns. Much of the region is now a no-go area, plagued by banditry and violence and left to its climate-weakened fate by the government and police.

'A little bit of drought forces the different tribes to live closely and cooperate over the few water sources. But if a drought is too severe or lasts a long time, then animals start dying and tribes steal cattle from each other, leading to violent conflict,' says Fabio. 'When the drought is really bad, people get their animals back more easily, because the raiding tribes and the animals they take are too sick and weak to go very far,' he adds.

The drought is really bad – the worst in over sixty years. The deserts of northern Kenya and the Horn of Africa were experiencing some of the most desperate suffering, when I visited, with more than 13 million people hungry and on emergency assistance. Average temperatures have risen by 1.3°C and rainfall has declined by some 20% over the past fifty years, causing vegetation zones to shift and one in five trees to die.[3] The Sahara Desert is advancing more than a kilometre per year in places, causing hunger and mass human migrations, and exacerbating tribal and regional conflict. Three-quarters of African countries – including Kenya – are in zones where small reductions in rainfall can cause large reductions in river flow. The drought is affecting livelihoods, animals and vegetation and hydroelectric production.

It's not that the people here are unused to dry conditions – they live in a desert – but droughts are becoming more frequent, lasting for longer, and the rains between them are becoming shorter. Droughts used to occur every nine or ten years, followed by rains. The past decade has seen severe droughts occurring every other year, and it hasn't rained at all in some parts for the past four years. Traditional drought-management strategies no longer work. There is not time for people and animals to recover from one drought and build up their strength and resilience, and breed more animals, before the next one hits. Global warming has brought higher temperatures that evaporate what water there is faster from wells and reservoirs, and the little vegetation is heavily overgrazed by herds from miles around, making it even scarcer.

Kenya is an uneasy conglomerate of tribes, stirred up and manipulated by Arab slave traders, then British colonialists. Following a brief period of post-independence calm and prosperity, intertribal conflict has resumed over land, animals and water. The country has slipped twenty places on the Human Development Index since 2002, the government is corrupt and making little progress towards democratic reform, frustrating conflict-weary Kenyans and alarming the international community. Meanwhile, the population, more than half of whom are children, has doubled in the past twenty years.

More than 80% of this equatorial nation is arid, populated by nomadic people who cover vast areas with cattle, goats and camels. It is difficult to overemphasise the importance that animals hold. Without animals, a man cannot marry, but with a large herd, he may have six or seven wives. Animals are a measure of wealth and security, and they provide the staple diet of blood, milk and meat. Often, people would rather spend money on medicine for a sick goat than their own sick child. As they lose animals, the compulsion becomes greater to raid a neighbouring tribe for theirs. And so the deadly cycle continues.

As we speed out of the town, I am tensely watchful. Squashed next to

Fabio, I am on ambush alert. Every man, woman and child is a potential killer. Many hold guns, and corrupt police officers provide the tribes with plenty of bullets, and then vanish when trouble kicks off. Fabio points out a patch of desert where planes carrying weaponry land at night. 'Insurgents live in that mountain, there,' he points, as we crash past on the dirt track that passes for a road. We are thrown left and right, up and down, painfully against the car's hard interior, but Fabio doesn't slow the pace. Large boulders and high mud embankments appear at the side of the road – perfect concealment for men with guns. We breathe slightly easier on flatter, more open landscapes.

Turkana people flag him down for a lift, but he motions the Kenyan sign for 'full' and speeds past. 'A few months ago, I was carrying Turkana in the back and Samburu tribespeople opened fire at us. Three bullets hit the car: one went past me here [motioning from his door, past his stomach to the other door], one hit the engine and one punctured the front tyre, but I drove on a couple of hundred metres until I lost control,' he says. 'That was the fastest tyre change I've ever done,' he smiles. 'Since then, I don't give lifts to a tribe in a mixed area.'

Fabio's mobile phone keeps ringing with people asking him where he is. Fabio is deliberately vague, not answering and faking signal outage in a way I'm sure priests aren't supposed to. He catches my quizzical look. 'If I tell people where I am, the message will spread and we will be ambushed,' he explains.

Ostriches stand at the side of the road like cartoon birds, and we pass zebras and tiny, delicate fairy deer called dik-diks. It's exciting for me, having only seen these animals on television or in a zoo, but it's too dangerous even to slow, so they flit past, while I hope that we won't run them over at least. We pass no other vehicles except army patrol jeeps and a couple of trucks. 'They travel in convoys to survive banditry,' Fabio says.

The hours roll past. This part of the road will be tarmacked by the

Chinese, Fabio says, but people here are not enthusiastic about the project. The Chinese have a reputation for eating everything near their work sites, he explains. They eat the precious goats, causing conflict with the tribal owners, they eat dogs, so that people either sell their dogs to the Chinese, or steal another's dog to sell to the Chinese. And they eat the wildlife too, everything from deer and antelope to big cats, which are killed by the tribes and sold to the Chinese. Most upsetting is the healthy and lucrative trade in ivory, for which local people shoot elephants. Then there is the concern about prostitution, which clings to construction sites in a region where at least one in ten people has HIV.

Five hours later and the road abruptly disappears into a small river. Unlike the seasonal rivers, which only run after rain, this flows year-round. But it's struggling against the drought. I can see in the sand scars how broad and deep it should be at this time of year. Nevertheless, it is an important source of water. In the villages it passes through, several sand dams have been built. These simple constructions consist of a concrete wall built against the flow of a stream or seasonal river, which traps silt and sand behind it. The sand acts like a raised riverbed, so that the water continues to flow downstream over the concrete wall. But in time, the sand soaks up so much of the water – millions of litres – that as much as 40% of the trapped volume is actually stored water. Even when the river ceases to flow, shallow wells dug into the sand dam continue to supply water, providing lifelines in the desert. Some even have a tap that pipes clean, sand-filtered water from the base of the concrete wall to the surface for people to use.

The opposite bank is high and steep. We plunge forward into the river and with spinning wheels make it to the other side, but the heavy vehicle cannot mount the bank. Again and again Fabio revs the engine and tries to lift us out of the sandy mud, but with the clutch burning and the wheels spinning freely, we go nowhere. He tries to reverse us back out, but succeeds only in sinking us deeper. Around us the gloom descends. We decide to cut branches from thorn bushes and lay them under the tyres. Removing my shoes and socks, I wade into the river, the slimy mud oozing between my toes. We battle on with our branches and spade until eventually, Fabio frees the car. It is now utterly dark, the kind of complete darkness that your eyes can't quite believe, opening themselves wider and wider, hoping to drink light from somewhere. We are exhausted, clogged from head to foot in mud. I have thorns in my feet and scratches on my arms and legs, but we are free of the river and ready to continue. Our destination, the village of Loiyangalani on the shore of Lake Turkana, is at least seven hours away. We pick up speed and career along, glimpsing eyeshine from big cats – perhaps leopard – in the blackness.

Fabio decides we should break the journey at a nearby mission. It's a beautiful house built by Italian missionaries, with a huge veranda that runs right the way around it. The two fathers greet us warmly and welcome us inside. The three priests are great friends, I see, all young and drawn closer together in solidarity over the harsh lives they lead. All have been shot at, faced angry tribal groups and the frustration of seeing their labours destroyed, from the schools they built being burned down, to children they have helped nurture and educate being killed by conflict, disease or malnutrition. We are shown to our rooms – 'careful of scorpions in the bathroom' – and warned to use the mosquito nets. A couple of years ago, an Italian priest staying here alone contracted cerebral malaria and, with his car broken down and no way of calling for help, died horribly. Fabio sleeps with his ghost – just one more of the many casualties of life out here.

Next morning, the priests are up by five o'clock, preparing to head to their distant communities to deliver cholera medication – there is an epidemic around here at the moment, killing babies and old people mainly. We share bread with a bowl of the delicious creamy acacia honey. There's a small bird here known as the honey bird, which likes to drink the honey. But it can't get past the fierce bees to get its tipple, so it has developed a relationship with the honey badger, a small, very aggressive mammal that also likes honey but which doesn't know where it is. The bird shows the badger where to find the honey, flying a little way ahead of it to guide it. The badger then raids the hive with its big claws, driving off the bees, so that both bird and badger can drink their fill. Some local people have learned to watch the badger and the honey bird to find the honey. And the bird has realised that people are faster and smarter than the badger, driving the bees off with smoke, and so the bird now calls to these people and takes them directly to the honey, bypassing the poor badger. The mission priests here are working with the locals to encourage honey collecting so the villagers can trade it in Nairobi for what is becoming a growing liveli-hood, helping families whose animals have died in the drought.

By 6 a.m., Fabio and I are back in the car. It's already warm and we're driving slower to conserve our little fuel. At this speed, we pick out animals more easily. Dik-diks stand in our path, fleeing at the last minute with exaggeratedly startled eyes. 'Mmm, good to eat,' says Fabio. We pass two beautiful, ornate sort of grouse, indigo-coloured with a small comb. 'They're lovely,' I say. 'Yes, very nice,' Fabio agrees. 'Better even than chicken.'

Most deserts are surprisingly full of life, home to the largest, smallest and strangest land animals, from desert elephants to chameleons, geckos, glow-in-the-dark scorpions and upside-down insects. All have evolved adaptations to allow them to survive the harsh heat and aridity of the desert environment. But the persistent drought is affecting the animals as well as the people living here. Migration patterns have shifted southwards, and animals are sickening and dying in large numbers. Every few miles, we are overcome by the terrible stench of rotting flesh. At the edge of the road, a camel, cow or goat will be there, identifiable only by its broad outline, jawbone shape or a hoof, in a slick of buzzing, purple and green liquid. The smell is indescribably bad, we actually retch just passing in the car, although it's clearly perfume to the vultures. Some carcasses are already white skeletons. The animals have been dying since April or May, Fabio says, starving to death or dying of thirst, their owners waiting for rains that never came. The loss of animals has been worse this year than people can remember, whole herds are dying as they grow weaker in search of grazing.

After some time, a village appears in the desert. Like most here, it is just a stopping place – a settlement of perhaps twenty roundhouses, constructed from sticks with skins thrown over the top. The tribes, all nomadic, move frequently, so nothing is permanent. Such wanderings have been practised throughout the Holocene, and are the most sustainable way of living in a desert. All that the Turkana pastoralists take with them is what they can load easily on to a donkey or two, so usually the houses will be left and new ones constructed at the next place, using more sticks from several days' walk away. There are very few trees here. The forest that carpeted this area in the early 1970s has long ago been cut for housing, charcoal burning or firewood. It's a tragic cycle of degradation that worsens as the population increases.

We pull into the village, called Sarima, stopping by the hut of the schoolteacher, Lowoi. He is one of just two men here, I notice, the others are all women and children, who gather excitedly around, tugging at our fingers and repeating the few words of English they know. 'How do you do? How do you do?' The men, it transpires, are all off looking for the goats. Last night, there was a raid by another tribe, and most of their goats were stolen. Suspicion falls on the Samburu. This tiny Turkana village would have a hard time taking on the Samburu, but there is a combatant faction of Turkana, a rebel group that live in the mountains and are trained to survive anything. Last year, the Samburu stole some goats from the Turkana, who called on this military faction. The rebels responded by going into the Samburu lands and stealing thousands of goats and cattle. 'You could see these vast herds of animals were taken, and then distributed among Turkana villages,' Fabio says. 'The army came with helicopters and guns, but they were no match for the rebels and had to return to base with dead soldiers.'

We're in the village to do business. These people produce frankincense – many of them are chewing it, like gum – which they collect by tapping the aromatic resin from a specific tree, more than twenty-five kilometres away. The schoolteacher shows me the three small bags of grain he has to feed all the children, donated by the mission. There are few alternative sources of income for pastoralists, so incense and charcoal-making are useful survival mechanisms to help families buy food during the desperate drought. Unfortunately, charcoal production involves chopping down the few remaining trees, which further diminishes shade and water reserves. And *Boswellia* trees, the source of frankincense, are heading for extinction as the saplings get eaten by the increasingly desperate cattle. The drought is also making fires more frequent, leaving adult trees more vulnerable to attack by long-horned beetles. Fabio is buying frankincense for use by the missionaries during Mass. The entire village crowds around two women who are measuring out cupfuls into a sack.

At this point, I should admit a strong prejudice, a dislike even, of missionaries 'infecting' the poor world with their beliefs and rules. Historically, missionaries have been the cause of countless deaths, and the zealous North American couples I frequently meet in Africa, clutching bibles and doctrine, do not fill me with joy. This journey has shaken me out of lazy judgement, however. Fabio is nothing short of a hero. Without the practical interventions of this dedicated network of missionaries, people here could not survive this changing environment. Through the violence that sends other charities packing, the grim mundanity of life in the harshest landscape, and the political prejudice that sees the army rescuing only select tribes, Fabio still visits, still helps, still feeds and schools Muslims and Christians alike.

We load two large sacks of the frankincense into the car, pay the village schoolteacher, and we're ready to continue. But before we can leave, there's some more negotiating to be done. Several of the women want a ride in our car to the oasis thirty-five-kilometres away, to fill their jerrycans and save them the walk they usually make every two or three days. There's not enough room for all of them and Fabio wants to be fair. Eventually, a compromise is reached, whereby we take six women and a couple of kids. The burden of finding water in dryland communities almost always falls to women, many of whom must make laborious journeys through remote and dangerous regions and then return with their heavy loads – even when sick or pregnant. Water fetching is a leading cause of miscarriage and stillbirth in desert villages, and women often give birth during the arduous journeys. Such babies are called Mwanzia in Kenya, meaning 'born on the way'. Villagers here must walk for several hours to reach known watering holes during the severe drought and, with the daytime temperature now approaching 50°C, the relief on the faces of our passengers is clear. These women may soon get more permanent relief. A United Nations programme called GRIDMAP, launched in 2013, aims to use advanced satellite imagery to map the underground aquifers in Turkana – already, more than 250 billion cubic metres of groundwater have been discovered, with some just a few metres below the surface.

We set off again, and everyone in the back bursts into song. We are driving through uninterrupted desert now. There are no trees, nothing. The high salt levels here render this strange volcanic landscape utterly barren. We're in the Rift Valley. It is hot and there are balls of pillow larva everywhere, and fossils – evidence of a time when this would have been under the sea. We climb a slope and there, revealed below us in this stark rocky landscape, is the lake. It is amazingly green – they call it the Jade Sea – and it stretches all the way to the Omo Valley in Ethiopia. This place has been inhabited by humans since before modern man evolved. A 1.6-million-year-old 'Turkana Boy', a near-complete *Homo erectus* child's skeleton discovered here with other finds, reveals there was a thriving lake community in the early Pleistocene. The shoreline is longer than Kenya's Indian Ocean coastline, but in the Anthropocene, humanity is shrinking the lake year by year. At the beginning of the Holocene, 10,000 years ago, the water was a hundred metres higher and used to feed the Nile. Now, siltation, dams and irrigation in the Omo Valley have reduced its inflow considerably. Ethiopia is also building a major dam on the Omo River for hydropower, which will cut flow further, lowering the level by eight metres by 2024.

More time passes, we mount yet another hill and suddenly behold the most wonderful sight: seemingly out of nowhere, the town of Loiyangalani appears like a mirage before us. It is nothing but a few hundred stick-and-palm huts, but it is a beautiful relief. We arrive at the Catholic mission, fifteen hours late and cloaked in the desert.

When the priests disappear for fatherly duties, I wander around. The mission is set in a cool, shady courtyard with trees and well-tended flowers. An outbuilding at the back contains hammocks and a television, powered by solar panels and a wind turbine. It's cool and breezy here because, unlike the main concrete building, it's made using local materials – like a larger version of the stick-and-palm huts. Rounding a palm grove, I come across the mission's swimming pool! It's heated naturally from volcanic spring water and I remove my shoes and jump in fully clothed, letting the past days' journey wash off.

After lunch, I explore the strange desert town. Almost all of the residents are Turkana, with some Somalis, who are the businessmen here, some Samburu and a few other tribal groups. People are generally friendly and colourfully decorated, the women wearing elaborate beaded necklace collars, and there is a lot of red about – the Turkana's favourite shade. A recent cholera epidemic is contained now, but the priests still distribute antibiotics and the much-needed food parcels. The mission has built a number of schools and is trying to educate girls in particular, knowing that, at the very least, an educated woman ensures that her children attend school. One way they try to encourage attendance is by providing at least one free meal to every child. As elsewhere, though, many girls can't attend because their parents won't let them, fearing that they will become 'too educated for marriage', jeopardising a dowry. The mission also runs evening classes for married girls, some as young as 10, with a free meal, but here too it's difficult for many to attend.

I am generally fairly sceptical about the benefits of education projects in places like this. It seems to me that a Eurocentric concept of education – where kids are removed from traditional learning environments (where they attain skills in, say, animal husbandry or farming) to teach them outdated, inane passages from a tired textbook on subjects that bear no relevance to their lives – is socially and individually damaging. It equips young people for a life in urban slums and shanty towns, even if a tiny few may make it into a rewarding profession. However, here, by the shores of Lake Turkana, I feel the opposite. It feels as if I am watching the beginnings of the end of a way of life that goes back centuries, and these desert people need an alternative.

Pastoralism is actually the most economic and productive use of drylands and a vital source of food during crises like the current drought. The type of nomadic herding practised by these tribespeople uses ancient knowledge and techniques to find water and grazing in such seemingly barren lands. Covering vast areas in a sustainable way, so as not to over-exploit any area in the way that settled farming and ranching does, has allowed pastoralists to continue an uninterrupted lifestyle for millennia. There are more than 250 million pastoralists in Africa, roaming over 43% of the continent's landmass and contributing 10–44% of GDP in the countries they live in, despite being among the poorest people on Earth.[4] Judicious grazing by cattle can recycle the carbon in the soil, making it soak up more water and reversing desertification. Around 90% of East Africa's meat and 80% of the milk comes from pastoralists, and yet they are facing such a barrage of pressures that nomadic herding may go extinct – a 2006 report by Christian Aid found that more than one-third of northern Kenya's pastoralists had been forced to abandon their way of life; by 2008, that number had doubled to 1 million people. The problems are multiple. Pastoralists are losing access to their traditional lands, as farmers, parkland conservationists and mining or industry concessions fence off huge tracts. They are thus forced to rely on diminished reserves of what is usually the harshest and driest land, with few watering holes. The increasingly frequent and severe droughts have been the final straw. With their cattle dying and continual conflict with neighbouring tribes over water and grazing resources, many pastoralists have been forced to migrate to villages and rely on food aid, scavenging and hunting wildlife. In some places, pastoralists are trying settled agriculture, with mixed results. Cattle rustling, which is a fairly consistent activity, becomes a more serious business for hungry desperate people, who, as I've seen, are now armed with modern weaponry. Loiyangalani is full of widows of those killed during such raids.

Anything that helps prepare people for a different lifestyle, empowers women and gives them an opportunity for another livelihood has got to be a good thing. Turkana children who learn to read and write here are much less likely to be married before they are 20, are more likely to earn something for the family and are more likely to have children who wash their hands. Once educated, Turkana have found jobs as nurses, teachers, mechanics, in the army and business. I meet Isabella, who speaks excellent English and went to secondary school for two years until she could no longer afford it (only primary school is free in Kenya). She is 20 years old and has two small children. She practises her English with me and tells me about her plans – she wanted to become a teacher, but that will probably be impossible, she says. Now she hopes to visit Nairobi one day.

Back at the mission, Fabio is fixing the spare vehicle part he collected in Isiolo. He crawls out from under the car and tells me that the widows are going fishing, if I want to join them. These are women who have lost their husbands to tribal conflict and have to feed their kids somehow. 'At first,' Fabio says, 'they would come to the mission every day and beg for food, So we decided to give them fishing nets and teach them how to fish.'

It was a shaky start – fishing is looked down on in Turkana culture – but, with a few false starts, the women were not just successful, they were producing more than they could eat. They stopped begging from the mission, and instead came for advice on how to sell their fish. They are now making a nice living trading their fish as far away as Lake Victoria. The fisherwomen's catch is dried and salted and taken to a holding shed, where it waits for the weekly truck that travels down from Ethiopia to Lake Victoria, via the village.

Fishing has proved itself a very secure livelihood compared to cattle. The fisherwomen don't go hungry, nobody can steal their fish because they don't need to wander to remote areas to graze, and fish (unlike sick or dead goats) can be sold in the city for cash to buy other things. So, a subtle change is occurring: widows, who have long tolerated their lowly position in Turkana society, are gradually seeing an improvement in status. Being more settled also means they can take advantage of education and health-care facilities. But, like the herders who now sell charcoal at the side of the road, it is spelling the end of a nomadic way of life.

Our car, full of singing Turkana once more, makes a bumpy journey down to the shore over boulders that seem insurmountable. In the end, fishing with the widows is postponed, because when we get to the lake it turns out that the nets are all broken from crocodile attacks – a legacy of when the Nile fed this lake. But the widows, who range in age from 16 to 35, lose no time in stripping off and going into the lake for a swim. With crocodiles and hippos sharing the lake, swimming is exciting. The fearless widows splash Fabio mercilessly and then we head back to the mission, where Father Andrew is disappointed in our lack of a catch. No matter, we still have dried tilapia for dinner, and it's delicious. While the Turkana locals survive on meat, milk and blood – they pierce an artery in an animal's neck and drain it into a cup before pinching the vessel closed until it clots into a seal – the migrant priests enjoy a more varied diet of fruits and vegetables brought from the outskirts of Nairobi. The local people won't even eat cheese 'because it is like soap', Andrew says, while professing that tilapia is the only fish one can eat in Kenya. Mombasa and the rest of the coast also provide a lot of fish, I say. 'Oh, but that is fish from the sea. It is not good to eat,' he replies. I ask him what is wrong with it, and he says that it comes from the sea so it is salty. I point out that the fish from Turkana is salted. Then he says in ominous tones: 'Some fish in the sea have both eyes on one side of their head.' And that ends the matter.

On Sunday, we go to the next village so Fabio can conduct Mass in the church there. The village belongs to the smallest tribe in Kenya, the El Moro, who number about fifty, although so many have intermarried with Samburu, it's hard to know whether there are any 'true' El Moro left. This is one of the only traditional fishing tribes in Kenya – the men take log rafts out and pull in Nile perch and tilapia weighing as much as 100 kg. Like the widows of Loiyangalani, they dry and salt the fish for sale in Lake Victoria. All the transactions can be made using the M-Pesa mobile money service.

Mass is conducted with much beautiful, harmonious singing, wafting of incense and plenty of curious glances my way. The children and adults have deformed bones and teeth. Fabio thinks it is because they are drinking water straight from the lake. It could be that the minerals dissolved in the water are preventing them from absorbing calcium. Child mortality is high here. Women have six or seven children and losing one or two is normal and an accepted part of life – children 'don't count' as permanent people in the fragile years up to 5 years old, after which they more securely join the living.

However, life here is about to be shaken up beyond anything these villagers – let alone Turkana Boy – could have experienced. Fabio drives me up from the lake's depression, back on to the desert plains. It's incredibly windy here, the sort of wind that knocks down a motorcyclist, he tells me ruefully, having been flattened himself. The area is traversed by the Turkana Corridor Low Level Jet Stream, caused by daily temperature fluctuations that generate strong and constant wind streams between Lake Turkana and the desert interior. It makes it one of the best sites in the world for wind-energy production. And it is about to be home to the continent's biggest wind farm, a giant 300-megawatt construction by a Dutch–Kenyan consortium, meeting around 20% of the national electricity need. The reliable strong winds mean the same turbines can produce double the power of those in Europe. More than 360 turbines will be installed here over an area of 7.5 square kilometres, with the electricity being sent hundreds of kilometres south on overhead cables to Nairobi and other heavily populated areas. Currently, some 80% of the national power is supposed to come from hydroelectric plants, but frequent droughts make them unreliable, dropping their contribution to 30% and forcing businesses that can afford it to rely on expensive diesel generators.

Wind turbines are ideal for deserts because the landscape allows uninterrupted sources of wind with little population to be disturbed by the turbines, although this also means that the population they serve is often located some distance away, requiring long cables that lose energy en route. Nevertheless, deserts around the world, including the Mojave in California, the Thar in Rajasthan and the Gobi in Inner Mongolia, are busily being planted with wind turbines. Some projects are combining with other generation, such as joint solar–wind projects, where the turbines provide energy during the night. A plant in Abu Dhabi uses wind turbines to generate electricity to condense water out of the desert air vapour. The invention, by French company Eolewater, extracts 500–800 litres of water a day out of the air, using just one thirty-kilowatt turbine and a condenser, something that could be scaled up to support communities elsewhere.

Wind turbines could provide a significant proportion of global electricity in the Anthropocene – in theory, 4 million turbines around the globe could easily produce 7.5 terawatts annually (nearly half the eighteen terawatts humans use) – but they are controversial because they kill birds and require a vast amount of land or ocean compared to conventional power generation.[5] One way around this might be to harvest wind from higher in the atmosphere, where it can be more powerful.[6] Several designs of kite and aircraft turbines are being developed to use winds as high as 500 metres in the sky. Offshore wind farms face less resistance from land users, but they are more difficult and costly to install and can interfere with shipping lanes. One potential solution is to have tethered, floating islands of wind farms that wouldn't need to be driven deeply into the ocean bed and could be moved to the best (windiest) locations in deep waters, while being kept out of shipping lanes.

The Turkana wind farm will generate around 2,500 jobs during the first three years of construction, and require new infrastructure and a road system to link Nairobi to Turkana – currently, executives visit the area by helicopter. Fabio worries that the isolated tribes living here will lose their culture and way of life with the influx of so many outsiders to the area. Local pastoralists are already feeling the effects of the recent discovery by Anglo-Irish firm Tullow Oil of 143 metres of oil at a well in Turkana. Since oil was found, parts of their tribal land have become no-go areas given over by the government to oil prospectors without the communities' consent. But the wind farm's project's leader, Carlo van Wageningen, insists his development will bring opportunities, including jobs, infrastructure and electricity, to the north-east for the first time. He says that the consortium will establish fisheries at Lake Turkana, and clean-water projects that could end the interethnic conflicts over water. A promised new clinic could give these isolated people real health improvements.

Some people here are already benefiting from electricity for the first time, thanks to the desert's other major power source: the sun. A couple of the El Moro men have solar-powered mobile phones, which are sold cheaply in Nairobi, and one of the straw huts in the village has two solar panels on its roof. And there is a proposal to build a large 250-megawatt solar power station here. Standing on the edge of this desolate desert lake, peopled by nomadic tribes, it's hard to believe that this place could become an energy hub. But, as humanity realises the dream of reaping boundless free energy from the sun, deserts like this one will be transformed in the coming decades as we bring electricity to the one-quarter of the world's population currently lacking it.

Even in deserts where the wind doesn't blow, the sun continues to shine unhindered by clouds, providing a reliable source of electricity, which is particularly useful because desert communities are often remote and far from power stations. Indeed, solar power proves far more reliable than grid power in many developing countries. During India's frequent blackouts, it is often the poorest off-grid villages with recently installed solar panels that have uninterrupted power, enabling them to continue to pump water for crop irrigation, for example. The sun supplies our planet with a constant flow of 120,000 terawatts of energy, which powers our entire world from bacteria to plants and animals, to weather systems and chemical cycles. Humanity uses around the same amount of energy in a year as the sun provides in an hour. The problem is that the sun's energy is dispersed, whereas we need to use it in a concentrated form. Desert solar power is one way of doing that.

Companies have been quick to see the market in off-grid homes that need phone juice, and a variety of pay-as-you-go solar options have sprung up across Africa and the Indian subcontinent. They are aimed at the very poor, who cannot afford the capital investment of solar panels, and who pay the most for energy because, without access to cheap grid electricity, they are forced to buy kerosene or diesel to power their homes. In some pay-as-you-go solar deals, users are issued a large battery for a small fee, from which to charge their appliances, and when it runs down, they bring it to the community solar charging centre and pay a small amount to swap it for a recharged one – the tariff can often be paid for using M-Pesa mobile phone credit. In Shanghai, I met an Israeli entrepreneur who has set up a bulk-buying scheme in Africa for Chinese solar panels. Currently, Africans and other developing-world consumers have to pay ten times more for solar PV panels because they purchase them individually or in small quantities in remote locations. Yotam Ariel's company, Bennu Solar, is working to reach 1 billion rural villagers, advise them on the most appropriate solar systems, and then coordinate a group purchase order for cheap solar panels on their behalf. Other schemes work more like hire purchase, in which the solar panel, batteries and other paraphernalia required to charge devices are rented at a low cost over several months, until the user ends up owning the equipment. UK-based start-up Eight19 has such a scheme in which users pay a $10 deposit and get a solar panel, a solar lighting kit for two rooms with LED lamps, and a rechargeable lithium battery with connections for mobile-phone and laptop charging. They then buy a weekly scratchcard for $1, text the card's number to receive a code that must be punched into the solar system's battery to keep it working. After eighteen months, the users own the gear. Customers can pay to add to their kit, incorporating more panels or batteries to power more rooms or a TV or sewing machine, for example.

In the deserts of the world, people are scaling up solar power production. Some use arrays of solar photovoltaic panels to produce electricity directly from the sunlight hitting the panel. The most impressive of these include a 200-megawatt plant in Arizona, another in Golmud in the Gobi Desert, China, which produces more than 300 gigawatt hours of electricity, and the world's largest at 600 megawatts in Gujarat, India. In the Atacama Desert, Chile has already achieved grid-parity with solar power plants (in which the cost of electricity produced is equivalent to conventionally generated fossil fuel power), without the need for government subsidies. But the desert is a dusty, sandy place and currently tens of thousands of litres of water are needed to keep solar panels and mirrors clean and efficient in even small solar plants. And water is scarce and expensive in such places. Engineers are working to design ways around the problem using dirt- or water-repellent coatings, and an engineer based in Saudi Arabia has invented a battery-powered automated dry-cleaner for solar panels, which is being trialled.

Rather than generating power via panels, another option is to use the sun's rays to heat up a material, such as water or oil, and generate electricity by getting the vapour to drive a turbine. By syphoning off some of the steam during the day into a steam accumulator, the generators can run during the night as well. It is very similar to the way a conventional power plant works, except instead of using fossil fuels to provide the heat, the sun is used. An array of mirrors concentrates the sun's rays into a trough or a heating tower, where the fluid is stored. The International Energy Agency estimates that concentrated solar power could provide as much as one-quarter of the world's electricity by 2050. The most ambitious concentrated solar power plants include the world's biggest in the Mojave Desert in California at 400 megawatts, and an interesting one in southern Spain that uses a spiral of mirrors arranged like a sunflower to direct the sun's rays on to a tower of molten salts. The salts store the sun's heat so effectively that the plant can provide twenty-four-hour power, 540 gigawatt hours annually, by releasing the heat during the night to drive turbines.

These projects are dwarfed by the most ambitious transnational solar venture, Desertec, a 2,000-megawatt plan to generate electricity from the North African desert sun – using a range of solar photovoltaic and concentrated collectors, as well as desert wind power – and send it across Europe. The Sahara receives as much energy in six hours as the world uses in a year. When it is complete, Desertec aims to provide 15% of Europe's electricity, transmitted from the Sahara north via efficient, high-voltage direct current (HVDC) lines, conserving 90% of the energy over long distance. Importantly, the project won't only power Europe – African countries will also benefit, including the host solar generators in Morocco, Tunisia and Egypt. HVDC transmission is only one option – with the increasing advances made in high-temperature superconductors, efficient transmission of electricity can also be made over larger distances using submarine cables. Geothermal energy produced in Iceland, and hydroelectricity produced in

Norway, can be transmitted in this way to Britain, for example. Another option is to store and transport the energy in hydrogen – the solar energy is used to split water into oxygen and hydrogen, and the reaction reversed using the hydrogen in a fuel cell to produce electricity where it is needed. Solar power could also be used to generate the heat needed to react chemical agents, such as carbon dioxide and water, to produce hydrocarbon fuels including ethanol, diesel and methane. That is essentially what the chlorophyll in plants does during photosynthesis, so it is the reaction which underpins all life on Earth. The trick is finding a catalyst that is better than chlorophyll for the reaction – and making the process cost-effective, which rules out using the best contender: platinum.

Currently, some of the most enthusiastic adopters of solar energy have been in Europe (where 80% of panels are Chinese imports), particularly Germany, thanks to government subsidies. While solar can contribute to northern Europe's overall energy supply, it is one of the least efficient ways of generating power there. Far better to concentrate solar generation in deserts and rely on other non-carbon sources, such as nuclear and wind, in the cloudy north.

As vast, unexploited tracts of sun- and wind-exposed lands, deserts will be key to future energy production in the Anthropocene. But are they really as empty as they seem? The pastoralists I meet argue that this is their land and the basis of their livelihoods. The fragile desert ecosystems are also feeling the effects of our need for industrial-scale low-carbon energy, with large areas of desert around the world being effectively scraped clear of wildlife. In some places, the only inhabitants able to tolerate the changed conditions are rodents, snakes and small shrubs – birds, tortoises and larger wildlife disappear as we undertake massive remodelling of our desert landscapes. In the Mojave Desert in California, the company building a giant solar farm was forced to relocate endangered desert tortoises from the site before construction. Whether other desert animals will be treated with similar sympathy is unclear.

In fact, the world's deserts are becoming less and less empty. Partly as a result of desertification of previously arable land, and partly because of expanding populations everywhere, deserts are increasingly places humans call home. Finding water is the most important occupation for all desert-dwellers but particularly for humans, who can survive without water for a shorter time than most animals and plants that live there. However, what we lack in biological adaptations, we more than make up for with our ingenuity – and we're going to need it on our drying planet. As our drinking water, agricultural and hydroenergy demands on freshwater reserves become ever greater, we are intensifying efforts to conjure water from every source. Many desert nations have exhausted their underground aquifers and they are not being replenished. These aquifers were laid down many thousands of years ago during a time when the climate was different, there was more rainfall and rivers flowed in the region. Recently, geologists have discovered vast reservoirs of water – a hundred times the volume of Africa's surface water and the world's third largest freshwater store after the Greenland and Antarctic ice sheets – in aquifers beneath the Sahara Desert and below Namibia. The water, which was last deposited some 5,000 years ago, could be accessed through boreholes, but in many areas it lies more than 150 metres below ground, making it expensive to pump out. Nevertheless, it offers hope to many water-stressed countries trying to develop irrigated agriculture as other desert nations have done. Libya's Great Man-made River project, the world's largest irrigation operation, taps 6.5 million cubic metres of fossil water a day from the Nubian Sandstone Aquifer – which contains some 150,000 cubic kilometres of water – through wells descending 500 metres below ground. The water is piped to 70% of the country's population as well as irrigating crops, but experts predict the aquifer – which was laid down during the last ice age – will dry up within the century. Groundwater levels in neighbouring Egypt are already dropping as a result of the extractions. Despite Libya's extraordinary artificial river, most desert countries remain desperately short of water.

Theoretically, water shortage is something humanity shouldn't suffer on this blue planet. In the Anthropocene, we have the technology to desalinate the oceans; the only thing holding us back is energy. Desalination plants are already being built across the planet, with water pumped to cities hundreds of kilometres inland. The process simply relies on evaporating filtered seawater and then condensing it back to a liquid elsewhere, leaving the salts behind. But the energy at the moment comes from burning fossil fuels. However, even oil-rich countries are increasingly looking to solar energy to power desalination – Saudi Arabia is building three solar-powered plants, for example, and is predicted to become a net importer of oil by 2030. Efficient, large-scale desalination powered by wind or sun has the potential to transform the deserts of the Holocene from barren fearsome places to agricultural hubs, supporting thriving cities – the sustainable Las Vegases of the Anthropocene.

In the Anthropocene, clearly not everywhere can become a Las Vegas, though. Poor desert communities will continue to rely on local solutions, such as sand dams and rainwater harvesting and storage, which can be highly effective at conserving rainfall, as I've seen. But what if it almost never rains?

Lima, in Peru, is a strange location for the country's capital. Not only is it shrouded by leaden skies for more than half of the year – something of which the Spanish conquistadors were likely ignorant during the sunny January of 1535 when they chose the spot for their capital – but it receives less than 1.5 centimetres of rain per year. Lima is the world's largest desert city after Cairo, and is entirely dependent for its water and all of its electricity on the drought-prone Rimac River, which originates hundreds of kilometres away in the glaciers of the Andes. More than 70% of the world's tropical glaciers are in Peru, and one-quarter of them have melted since the 1970s – including two-thirds of the glaciers supplying the Rimac. 'The predictions are that almost all our glaciers will have gone by 2030,' says

Elizabeth Silvestre, scientific director at the National Meteorological and Hydrological Service in Lima. 'But it's happening so much faster.'

Ironically, drought in the Andes is one of the major reasons that people migrate to the city. More than 80% of Peru's population now lives on the desert coastal strip – which receives 1.8% of the country's water – 9 million of them in Lima. Even in the wealthiest parts of downtown Lima, businesses and residences suffer frequent water rationing, during which turning on a tap produces nothing. But the problem is far worse in the 1,800 euphemistically termed *asentamientos humanos* (commonly known as AAHH, loosely translated as 'human settlements'), slum communities of migrants that are home to some 2 million people. It is, therefore, nothing short of astonishing that the residents of one dirt-poor shanty town are attempting to grow a forest on their sand dune. Javier Torres Luna is hoping that reforesting the dunes above his hastily constructed plywood home in Bellavista AAHH will provide a long-term solution to an urgent problem: there is no water in his community for drinking, washing or sanitation. And there is certainly none for forestry irrigation. But the plan is not as crazy as it sounds. Before the sixteenth century, these hills were wooded. Once the Spanish came, the trees were cut for timber.

It's not that there's no water in Lima, it's just that there's no rain.

Climbing up the small hill that rises above Luna's home, it proves difficult to breathe because of the high humidity. A thick, grey fog, known locally as the *garúa*, hangs perpetually over the city between May and November, hiding the sun and promoting bronchitis and pneumonia among the city's poor. The fog arises because of a thermal inversion, whereby the chilly Humboldt current cools the water-laden air coming from the Pacific Ocean, preventing rain along the coast.

Fog is a ground-touching cloud of water vapour that is common on coastlines, where spray from crashing waves provides the salt nuclei that help water droplets form in the air. Over the past couple of years, Luna and his neighbours have erected a series of large nets at the top of Bellavista hill to harvest precious drops from the wet air. The nets work by condensing the fog vapour into liquid water: trapping airborne droplets so that they coalesce into a heavier stream. It's a technique perfected in nature – the Californian redwood tree, for example, secures up to half its water from fog; and the shiny black Namibian tenebrionid desert beetle condenses fog on to its corrugated carapace and then performs a headstand to channel the water into its mouth.

Luna and his neighbours have built reservoirs and tanks, and constructed large plant-shade nets forty metres square, using steel cables at strategic points on the top of a hill to trap the precious fog. Fog-carrying wind from the ocean hits the nets and the results are audible: it sounds like a gushing fountain. 'We had no idea it would work so well,' Luna says. 'The only problem is protecting the nets from vandals in neighbouring communities who would destroy our hard work and steal the cables.'

A gutter under the nets catches the water and pipes it through a sand filter, and into storage tanks. The water is being used for washing, drinking and to irrigate the tree saplings. In as little as four years, the trees will be large enough to trap the fog themselves, producing a self-sustaining run-off that will replenish ancient wells and provide water for the community here for the first time in 500 years. After experimenting with various tree types, tara was planted, a native that produces an economically important fruit used to make gallic acid for the tyre, tanning and herbal medicine industries. 'Without being irrigated, it would take 400 years for a tara tree to grow to full height in this desert, but with four years' worth of watering, it will fruit,' says Luis Marquez Cano, a forestry engineer at the National Agrarian University at Molina, who advised on the project. After repeated experimentation, German biologists Kai Tiedemann and Anne Lummerich, who designed the Bellavista nets, have come up with their 'best yet' net in the neighbouring AAHH of Virgen de Chapi B. It's a three-sided structure dubbed the Eiffel Tower, which produces as much as 2,500 litres of water a day, while only taking up the space of the regular rectangular one. Other organisations are now also proposing fog-harvesting projects, with plans for as many as twenty nets in some Lima communities.

So does fog harvesting offer a way out of Lima's water crisis? The technique is incredibly useful for small communities of between about 500 and 1,000 people, says Bob Shemenauer, who pioneered the technique in Chile's Atacama Desert. But it will only provide for people's drinking, washing and small-scale agricultural needs. Once a community gets large enough to support businesses, such as cafés, swimming pools, and so on, the per capita water use goes up from forty to fifty litres per person, to hundreds. Then, governments need to invest in piped water supply. 'But for small rural villages and isolated communities, it can be a lifeline,' he says.

What if this fog-net principle were combined with desalination? That's the genius invention with the potential to transform deserts in the Anthropocene, even turning them into new agricultural zones. Farmers in Australia are irrigating vegetables in the coastal desert in a greenhouse with walls made from sheets of corrugated cardboard. The honeycomb-shaped wall is sprayed with seawater (using a solar-powered pump and spray) and the wind evaporates out the salt, leaving the humid air to condense on to the greenhouse crops. If there is no wind, solar fans are activated. The system is being used by Sundrop Farms, who have built a giant greenhouse in the arid coastal scrub outside Port Augusta in South Australia, growing tonnes of juicy vegetables. The seawater greenhouse is now being expanded to twenty acres to grow millions of vegetables hydroponically. Parabolic mirrors follow the sun all day long, heating a steam-driven turbine that pumps and desalinates seawater, heats the greenhouse at night and cools it during the day. Around 10,000 litres of pure irrigation water are produced daily and sprayed with nutrients on the plants' roots. Because the system is sealed in and soil-less, there is no need for pesticides, although 'friendly' orius (pirate bugs) are released to munch any spider mites or thrips that would eat the crops. The system is now being built in Qatar too, and there are plans to use it in other deserts. If such greenhouses were built in large numbers, they could even alter the local climate, producing a regional cooling effect similar to the greenhouses of Almería. In this way, the desert landscape could perhaps shift to savannah in places.

There may no longer be a place for the nomadic pastoralists in the growing deserts of the Anthropocene. But we are learning other ways of using these vast barren expanses, as bountiful sources of low-carbon energy and, incredibly, as new land for agriculture.

7

SAVANNAHS

*T*he first vast, treeless plains on the planet were very similar to the savannahs of today, with one major exception: grass had yet to evolve, so these ecosystems were covered in mosses, lichens, ferns and other prehistoric plants. During the Mesozoic era, herds of herbivorous dinosaurs and other reptiles roamed the savannahs, munching these plants. It wasn't until after the Cretaceous and the extinction of the dinosaurs that grasses truly took over – they now cover 20% of Earth's land surface.

Over time, as grasses became the dominant plant, animals evolved various adaptations to make use of this plentiful food source. Ruminants, which include cattle, antelope, sheep, camels and llamas, have four compartments in their stomachs that allow them to digest the fibrous cellulose in grass in several stages, through regurgitating. The mammals still rely on a horde of gut bacteria to crack the cellulose, and their specialised digestive tract includes fermentation chambers where the microflora do their work. Like grass, ruminants are now found on every continent bar Antarctica – wildebeest and bison chew the cud where dinosaurs once grazed.

If humans could be described as having a natural habitat it would be the savannahs. Like all predators, our ancestors were drawn to these big open grasslands, thronging with herbivores. Savannahs proved an irresistible hunting ground, with enough trees to provide shelter and firewood, but not enough cover to hide large numbers of grazing animals. It was on the African savannahs that human bipedalism emerged – our upright-walking species, with two hands free for all-important tool use and carrying capacity, owes its extraordinary success in part to changes evolved on the windswept grassy plains, millions of years ago.

Humans began 'managing' savannahs tens of thousands of years ago, enlarging and creating new ones by regularly burning vegetation, and cutting trees. We learned to track and outrun faster and stronger animals; and, when climate change sent the ruminants on long migrations we followed them out of Africa and across the globe. Journeying after those beasts gave humans a source of food over long distances, allowing us to trade and explore new pastures. When the world became warmer at the end of the last ice age, it was the grasses growing on savannahs that inspired the first farmers to begin cultivating crops. Settlements grew up around their fields and human civilisation emerged.

In the Anthropocene, few savannahs remain wild – the ones that do represent some of the last places on Earth of minimal human interference, where it is possible to imagine what a world without humans might look like. Most of the world's savannahs have been converted to farmland, built upon or tamed into submissive parkland. The big wild mammals that once roamed the world's savannahs have been hunted to extinction. In their place we have created new herbivores. Domesticated cattle, bred in just a few varieties, now graze the tamed savannahs. Death no longer comes at the jaws of a hungry predator, but is orchestrated regularly, at scale and by humans. In the Anthropocene, there is no true wilderness. Humans no longer experience the fear and *unpredictability of hunting and being hunted. We are a superspecies and we decide the fate of all other animals. When we invented the gun, our prowess was assured: we have now killed so many creatures that we are creating a planetary mass-extinction event – only the sixth in the Earth's history, and the first to be caused by living species.*

We have travelled the globe, cherry-picking the plants and animals we like, and bringing them to new locations, creating new ecosystems. We have created a pastiche of the wild savannahs in our suburban gardens, golf courses and parks. In some of these, we are reintroducing native herbivores, with regular culls replacing the missing predators.

In the Anthropocene, now that we have tamed the natural world, filled the wilderness with our few chosen creatures, and hunted wild ones to extinction, we must decide what sort of nature we want – and how humans might continue to coexist with wild species on this shared planet.

The Hadzabe people have nothing; no animals, no land, just the clothes on their backs. And, crucially, the skills and resourcefulness to produce everything they need from their environment.

I travel four hours west from the cosmopolitan city of Arusha in northern Tanzania to Lake Eyasi, a 1,000-square-kilometre seasonal soda lake scooped out of the Great Rift Valley on the savannah of the Serengeti Plateau, to meet this ancient tribe of hunter-gatherers. The Hadzabe offer an insight into our world before humanity's takeover, a time when the relatively few people in existence were just another – albeit cleverer – species on Earth. Until recently, this part of the world was virtually untouched by the Anthropocene, and the Serengeti region is still the place that most closely resembles the Pleistocene, a time in the Earth's past when large mammals roamed the planet in vast numbers and humans hunted by following their migrations.

Lake Eyasi is just forty kilometres from Laetoli, where three early humans went for a stroll some 3.6 million years ago. When palaeoanthropologist

Mary Leakey discovered their footprints, fossilised by the soft ground they trod, she realised that the two adults and a child had been upright walkers like us, walking on two feet with the big toe in line with the other toes, unlike apes.

Approaching the Hadzabe camp, I have the sensation of treading the same ground as my ancestors, of returning to our family home to meet the people who never left.

I find them sitting in gender-segregated groups, with the men playing small lute-like stringed instruments and applying a poisonous tree resin to their metal arrowheads. They light a small fire by rapidly twisting a hard-wood twig into a softwood stick from the local commiphora (myrrh) tree. It soon smoulders and, intrigued, I have a go too. It's surprisingly difficult, but with their help, I get it to smoke eventually.

The Hadzabe bushmen live in groups of about fifteen people and speak a unique language of clicks that is very different even from other click languages elsewhere in Africa. There are now fewer than 400 Hadzabe left, practising a way of life that was widespread in eastern and southern Africa for millennia. The tribe is believed to have been continuously living in this area for at least 50,000 years, but their ancestral range has been shrinking as their land is swallowed up by the modern world: farmers, government-designated conservation areas and private game reserves. Indigenous-rights organisations are legally challenging the government to provide a hunting area for the Hadzabe, but so far nothing has been done.

A relationship has developed between a couple of Hadzabe groups and a local Eyasi village of Datoogas, an agricultural tribe that was pushed out of its traditional grazing lands in the Ngorongoro Crater by invading Maasai 300 years ago. Edward, a local Datoogan, has introduced me to one of these groups, and he acts as translator.

More than half of Hadzabe children die before age 5, from malaria usually, but also from other treatable diseases. A large proportion of women die giving birth, haemorrhaging in the bush or succumbing to any of the other dangers out here, including sleeping sickness transmitted by the tsetse fly. Edward's village is trying to get labouring Hadzabe women to hospital in time, and children to schools, but it's a struggle. The tribal people are culturally scared of buildings – 'Being under a roof is thought to be fatal and so children only stay in school a month or so, and women are scared of giving birth in a hospital and refuse to come,' Edward says. The Hadzabe live in simple twig-and-skin shelters, under a tree canopy or in caves. British colonial rulers also tried, in the early twentieth century, to civilise the Hadzabe and settle them in villages, but they resisted and returned to their traditional way of life.

It's not that the Hadzabe 'don't know any better'. They have had contact with pastoralists and settled agriculturists ever since the first of those tribes migrated down from the Horn of Africa in the fifth century. Although the Hadzabe are highly secretive, usually only marrying within their tribe and suspicious of outsiders, who have historically captured them as slaves, killed them or brought disease, the communities have traded with other tribes over the past centuries, exchanging meat and honey for metal arrowheads. It seems clear to me that the Hadzabe have deliberately and consciously chosen their ancient way of life over the alternatives they have seen. It is an extremely egalitarian and accepting society without hierarchy and with gender equality not found elsewhere.

The group of five men and boys stands up as one and begins to walk off, carrying bows and arrows. One of the men, with a head decoration made from baboon fur, is clearly the leader. He carries the longest, most decorated bow, the string of which is made from giraffe tendons. Edward and I follow the group, trotting and sometimes running to keep up as they search the tree foliage and bushes for animal prey. A bushbaby is spotted and there is excitement as the men take aim and fire arrows into the branches. I hold my breath. I'm not sure I want to eat anything with 'baby' in its name. It gets away, I am secretly relieved, and we move on. The thorny bushes grab at us as we hurry past, snagging our clothes and snaring our hair. The Hadzabe are all bare-footed. We pass trees and shrubs that provide supplementary food, including sour, juicy tamarind berries, carbohydrate-rich tubers and plants whose roots can be crushed for juice or eaten for medicine and vitamins.

Honey is gathered from the baobab trees using a ladder of sticks implanted into the trunk like pegs. And, like the people of far northern Kenya, the Hadzabe have developed a relationship with the honey bird, which leads them and the honey badger to a sweet hive to harvest honey and wax. One of the men demonstrates a special birdcall they use – a mimic of the honey-bird's call, which it responds to.

They eat a largely vegetarian diet consisting mainly of honey and fruit, which is predominantly gathered by women. But meat is plentiful here at the moment, now the rains have arrived. The group mostly catches small creatures like birds and rodents, but a baboon is good. The baboons here are understandably skittish around these people, though. Sometimes, a buffalo 'strays from the reserve', we are told, which makes a delicious and long-lasting feast.

Ahead, the men have spotted something in the trees. They surround it with their bows taught, arrows poised. This time they are successful. One of the boys climbs a tree to retrieve the arrow and a small bird, its wings still fluttering pitifully. One of the men pulls the bird off the arrow, sticks its tiny head in his mouth and bites its neck to sever its spine. Then the kill is stuffed into his belt and we continue onwards.

An hour or so later, there are two birds and a squirrel in his belt and we make our way back to the camp, where the creatures are cooked on a fire and we all share the meal – the women join us but eat separately, several metres away.

This Stone Age existence is under great threat in the Anthropocene. For the last hundred years, the Hadzabe have been pushed into ever-decreasing patches. Large tracts of their land have been given over to Tanzanian onion farmers, who have flooded into the area – 50,000 people have arrived since

2000. The Hadzabe's western lands have become a private hunting reserve, in which the tribe is not allowed to hunt. The Datoogas have now expanded their occupation across the Hadzabe's Yaeda Valley, hunted the animals there, cleared the land for farming (which removes the fruit, tubers and honey), and sunk watering holes for their cattle that cause the Hadzabe's ones to dry up. Another 6,500 square kilometres of Hadzabe land was leased by the Tanzanian government to the United Arab Emirates royal family in 2007, as a private hunting reserve, with all Hadzabe and Datooga evicted. However, following international outcry, the deal was rescinded. And, since popular documentaries on the tribe aired on PBS and the BBC in the early 2000s, tourists have descended on the remote tribes. Tour groups bring money, and money buys alcohol. Now the Hadzabe are plagued by alcohol addiction.

I leave the group as I found them, sitting on the ground, strumming their instruments, and travel north across the savannah to perhaps the greatest wildlife show on Earth, the Serengeti National Park.

When President Theodore Roosevelt came here in 1910, spending a year on a slaughter binge that furnished the Smithsonian and similar institutes with more than 10,000 large carcasses, he described to the world a place that time had forgotten – a hunting ground of a richness not seen in America or Europe for centuries. Such a productive wild landscape should be preserved for future generations to hunt, he argued.

Europeans and Americans headed for the new fantasy land where they could act out their 'man versus beast' Pleistocene scenes – albeit with modern weapons, a team of local porters and a handy motor car – returning home with at least one photograph posed astride a recently shot lion. As more and more tourists arrived from around the world to bag their own Big Five – lion pelt, rhino, leopard, buffalo and elephant – it became clear that the animals here were no more inexhaustible than they had been on the great American or Indian plains. Lion numbers crashed, regular hunters discovered an interest in the animals beyond simply killing them, so hunter turned gamekeeper and the national park was born.

I head first into the world's largest complete volcanic crater, at Ngorongoro. It's an incredible site: a vast, perfect circle of steep-sided earth surrounding a flat grassy plain with lakes. Africa's celebrated savannah mammals are all here, except giraffes, which can't make it down the steep sides. Giant tuskers stand in groups – so much larger than their Asian counterparts. The pink soup at the edge of the alkaline lakes turns out to be gatherings of flamingos, and among them walk jackals and hyenas. The grass is black with buffalo and wildebeest that have migrated south from the Maasai Mara in Kenya. They move in crowds or strangely ordered lines, with their stripy necks and golden beards shimmering in the sun. Zebras graze among them like nicely decorated circus horses.

My jeep driver is hunting for another creature altogether: more jeeps. He finds a gathering of half a dozen vehicles and we speed over to them, clinging to the car's metal skeleton as we race around. The attraction is soon visible, a decidedly lazy pride of lions, lolling around regally in the grass. As we approach, the large male heaves his big head up a few inches, before dropping back down. He rolls on to his back, enormous back paws inelegantly skywards and one front paw resting on his massive full stomach. And there he remains for the next hour. Lions that have recently fed are not prone to energetic displays.

Across the plain, two black rhinos, mother and calf, walk towards us like prehistoric tanks of muscle. It's amazing seeing these huge animals so close – they are such strange and different shapes and sizes compared to the cows and horses I'm most familiar with. It's also slightly unreal: they are completely wild and we are in their habitat, but something about the perfect circle of the crater and the other tourists makes it feel set up. I finally realise what is missing – fear. Many of these animals are close enough to kill the lot of us, yet I feel no fear because I know they won't. It's a bizarre sensation – I wouldn't enjoy the experience half as much if I thought I was going to be eaten, but . . .

Crossing into the Serengeti, this feeling of unreality evaporates. The Serengeti is simply vast, with the world's biggest collection of large mammals. Against a backdrop of thousands of migrating wildebeest, our jeep is small – it's nothing in a landscape that stretches eye-achingly far, with a horizon that goes on forever. Serengeti means 'never-ending plains' in the language of the Maasai.

Unlike in a rainforest, savannahs let you see the animals you're hunting – whether by weapons or camera – and it's also easier to see what is hunting us. Perhaps that's why we humans first conquered the plains so many thousands of years ago, and why it's the environment we feel most at home in and the landscape we most often recreate for ourselves. In a few magical days, I find prides of lions and come across a leopard, slung smugly over a branch with its impala kill tucked into a crevice above. I see a cheetah stalking and hunting its prey in an exhilarating flash of muscle and tendon. Families of warthogs grunt around and splash happily in muddy pools on their stumpy legs. Panic-stricken dik-diks skip past, big-eyed – they are every predator's favourite dinner. I see ugly yet majestic marabou stork, waiting like the vultures for the opportunity to steal from a carcass. Bat-eared foxes dart out of burrows in the rusty termite mounds and hippos wallow in magnificent blubbery ugliness, bunched up together in too-small pools. At night, tucked up in a sleeping bag in my flimsy tent, I hear buffalo crashing past, and the heart-stopping roar of a lion.

The park was modelled on the back-to-nature, prehuman conservation strategy of the world's first national park – established in 1872 at Yellowstone in Wyoming. The idea was to return an area of wilderness that had been sullied by modern humans back to its pristine state. Work started on the Serengeti national park in the 1950s. One of the first things the new conservationists did was evict all the people, houses and domesticated animals. The 10,000 Maasai, who had been living here, grazing their animals and hunting for 200 years before the Europeans discovered the place's abundant wildlife, were relocated outside the park.

Banned initially also from the Ngorongoro Crater, they were eventually allowed to return there and cultivated maize to supplement their diet. But agriculture in the Ngorongoro Crater was outlawed a couple of decades later, and further Maasai evictions from the crater of nearly 60,000 people began in 2009 after the government ruled that the human population of 64,800 people – swollen from the 1959 total of 8,000 – was unsustainable and their grazing cattle were impacting wildlife.

The Maasai have been repeatedly uprooted and shifted as the government has created new conservation areas or given land concessions to private initiatives for conservation or game hunting. In 2009, around 3,500 Maasai were evicted and their homes burned when the government introduced a new Wildlife Conservation Act, and in 2013, at least 30,000 Maasai were threatened with eviction when another large tract of land was leased to the United Arab Emirates as a royal hunting ground.[1] The UAE already operates game reserves on thousands of acres of formerly Maasai land, generating sizeable funds for the Tanzanian government.

The Maasai used to range across most of East Africa, grazing their cattle and practising some cultivation, with seasonal migrations in what many scientists consider to be the most environmentally sustainable use of this land. But reduced to ever-smaller patches, banned from hunting, and in many cases from cultivation, the Maasai are finding it a struggle to survive. Devastating droughts, decades of inbreeding and disease halved the numbers of their cattle in 2000 and 2009. Many rely on handouts and are seen as a primitive, unsuccessful relic of the past. In 2005, Tanzania's newly elected president Jakaya Mrisho Kikwete declared: 'We must abandon altogether nomadic pastoralism. The cattle are bony, and the pastoralists are sacks of skeletons. We cannot move forward with this type of pastoralism in the twenty-first century.' Like the nomadic pastoralists of the northern deserts of the Horn of Africa and Turkana, the Maasai are facing an end to their culture and a way of life that goes back millennia. In the Anthropocene, it is likely that people will talk of there once being a Maasai culture, and perhaps theme parks and museums will bear testimony to their traditions, but many anthropologists argue that there are no longer any Maasai living truly 'authentic' lion-hunting lifestyles, just as there are no longer any Viking warriors.

I travel through the overgrazed plains, where Maasai herd their cattle and goats. I pass trees hung with beehives, small mud-hut villages and people planting crops from bananas to rice. The Maasai are striking to look at. Tall and graceful as the Turkana, they wear decorative beads and earrings of silver, copper and turquoise that distend their lobes, and carry a long, iron-tipped spear and often a bow and arrow or dagger. Most characteristic are the red tartan blankets – shukas – that they wear either draped over their shoulders, or wrapped into a dress. They are either barefoot or shod in simple rubber sandals made from tyres. Their warrior heritage leads many Maasai to be employed as guards and doormen in the cities, flipping mobile phones out from the folds of their shukas, but otherwise unchanged in appearance.

When I see the unparalleled Serengeti, with its lush grasses not reduced by livestock, I have some sympathy for what the Tanzanian government is trying to do by banning the pastoralists, preserving the best fodder for endangered animals. But there is an enormous cultural disjoint between what Westerners perceive as 'nature' or 'natural' and the reality of our world in the Anthropocene. Tourists who pay thousands of dollars to come on holiday to the Serengeti certainly don't want to look through their viewfinder and see Maasai herding cows among the zebra and wildebeest, a tour guide tells me. How would they feel, I wonder, if they watched a Hadzabe guy shooting one of those Bambi-like dik-diks on the Serengeti? Many of us would rather continue with the illusion offered by these artificially human-free parks of a world where 'nature' exists in isolation from the corrupt influence of man. It's a very Victorian idea that has become increasingly fashionable, despite the clear evidence of people coexisting with wildlife for millions of years, exposed by Mary Leakey here in the 1970s.

The problem with recreating the pristine is that it doesn't exist, and never did. The natural world is constantly evolving as the balance of predator–prey species changes, the climate alters and disturbances such as extreme weather events, diseases, volcanoes and other phenomena occur. Humans have always been part of this evolution, especially here in the Rift Valley, the so-called 'cradle of humanity'.

The entire concept of 'us' versus 'nature' is flawed since we too are part of nature. But why then is a bird's nest or a beaver's lodge and dam considered natural, whereas human houses are not? If we are part of the natural world, why is what we produce – the artificial or cultural – deemed any more or less worthy of admiration, awe and respect than what we deem naturally produced by planetary forces or evolution? When people make that distinction between humans and the natural world, surely what they mean is that we are capable of dramatically and deliberately choosing and altering the environment in a way that other species cannot. We can change and enhance nature – even the biology we are born with – and exceed natural physical limitations in a way no other species can. Are we still 'natural', then, in the Anthropocene?

Throughout civilisation, humans have worried about our progress away from nature, while scorning or fetishising societies who live most basically with nature. In the Anthropocene, as the last of these 'natural' cultures now face extinction, it perhaps confirms that humans are not just changing the natural environment, but that we now operate on the planet as a very different animal.

It is occurring too suddenly for many indigenous peoples. Further south in Africa, I come across a family of San people living pitiful lives as alcoholics on the streets of Botswana's second city. I am introduced by Martin Flattery, a white Botswanan, who grew up with the Bushmen – he had to have an interpreter when he started school because he only spoke the San click language. The San bushmen of the Kalahari are, like the Hadzabe, among the world's most ancient people, having lived in the region for tens of thousands of years.

Like the Hadzabe of Tanzania, the San – who exist in five tribal groups speaking seven different click dialects – are true hunter-gatherers. They hunt game in various ways, including by bow and arrow and by setting innovative corkscrew traps for underground-dwelling animals like the spring hare. Their most unusual hunting tactic is to outrun an animal, such as an antelope, by pursuing it ceaselessly at a pace that just exceeds the animal's comfort speed. In two to three hours, a group of two or three bushmen can run down a large and fast mammal like a kudu until it collapses from exhaustion. A television crew from the UK filmed such a hunt here the year before my visit, and on the first attempt to shoot it, the crew in their 4WD vehicle couldn't keep up with the hunters on foot.

The San traditionally live together in small, sustainable, family groups of five to twenty members, building temporary grass huts that can be easily dismantled to suit their nomadic lifestyle. Each group will move according to the seasonal availability of game, hunting only what is required to feed their families and supply materials for their clothes and weapons. If a family member should die in a camp, that place will be abandoned and the group will never stay there again. Although the groups live and hunt separately from each other, they do come together a few times a year during healing ceremonies led by the *sangoma* (medicine men). It is during these meetings that marriages between groups can be arranged.

In the north-west of Botswana is a place called Tsodilo Hills, a kind of Mecca or pilgrimage site for the San. Thousands of cave and rock paintings there testify to the extraordinarily ancient culture of the bushmen, including artwork depicting seals and whales in a landlocked nation hundreds of kilometres from the ocean. They were painted by San from Angola and Namibia who walked the long journey there.

As it was for the nomadic herders I visited around Turkana, the San bushmen's way of life is under serious threat. But whereas drought induced by climate change was to blame for the hardship faced by Turkana and Samburu tribes, in Botswana the government is orchestrating the San's decline. It is to Botswana's eternal shame that over the past decade, the government here has been systematically removing the San from their ancestral lands, destroying their culture and livelihoods in the process, to make way for farmlands, the diamond-mining industry and wildlife conservation. During three major clear-outs in 1997, 2002 and 2005, these indigenous people were forced to live in reservations, where the animals they depend on were soon depleted. Bushmen living in reservations are unable to hunt and so are supplied with government handouts of maize and sugar, which they convert into alcohol to relieve their boredom and depression. Room for manoeuvre is limited in a reservation, so people whose entire way of life is built on a roaming relationship with the land and its wildlife are reduced to a fixed-roof existence that's as alien as asking a suburban Londoner to live in a two-man tent forever. Alcoholism, HIV and TB rates have soared. There is conflict between the reservation people and locals, because of incidents like, for example, local shops selling things on credit to bushmen, who have never dealt with money transactions before and cannot pay. So they poach or steal goats to make payment.

Some bushmen will never be reservationised – they hide in the bush – but the majority are seeing their culture and way of life eroded faster in the past ten years than in the millennia before. Various concerned NGOs and agencies are trying to help (although some of their efforts seem inappropriate – Survival International, for example, has reportedly been giving Toyota Land Cruisers to the bushmen). In 2006, the bushmen won the right in the International High Court to return to their ancestral lands. The problem now, though, is that the San have acquired guns and vehicles, and become accustomed to borehole water supplies and a handy clinic – all provided either by the government or well-meaning NGOs. The bushmen want to return to their hunting grounds in the national parks with all their new comforts, including the goats they've bought. The government has said they are allowed to return, but not with accessories. They are permitted to hunt only with bows and arrows, and must live in their former sustainable way, building grass huts and travelling on foot. Some have returned to this way of life, but for others, the transition back is just too difficult – around fifty bushmen have been arrested for hunting in the national parks with guns. The mess has decimated this ancient people – a loss for Botswana and for the world. 'We're seeing the destruction of a unique way of human life. I've lost so many friends who gave up the will to live once they were moved off their land,' Martin Flattery says.

It's a pattern that has been repeated across the world's savannahs from the North American prairies to the sheep ranches and wheat fields of Australia. Humans who have been a part of these ancient ecosystems for millennia have been eliminated from the developing Anthropocene landscapes in a matter of decades. The land use that follows is almost always more intensive, poorer in biodiversity and less sustainable in terms of water use. In Kenya, for example, the government is encircling the area around Mount Kenya with an electric fence 400 kilometres long, two metres high and set a metre into the ground, ostensibly to prevent wildlife from straying on to cropland, but which will also serve to curtail tribespeople from moving around and grazing their animals.

Shifting indigenous people off their land for farming or mining may be ethically questionable, but in those cases there is no pretence that it is for the improvement of 'nature'. However, displacing indigenous tribes for conservation purposes – to render land pristine – puts conservationists on a dubious footing. The indigenous tribes that were evicted for the creation of Yellowstone National Park – 300 people were killed in a single day – were an intrinsic part of the Holocene ecosystem, and when they were removed, it meant there was no top predator and no one to manage the vegetation through regular burnings which prevent devastating wildfires. As a result, ungulates overbred, eating saplings unsustainably, and the area suffered biodiversity loss. Park managers – for without a functioning ecosystem, these 'natural' areas must be managed by humans – have since tried re-introducing wolves to replace the human top predator and keep the ungulates in check. But here they have run into trouble, because while the park might be a human-free attempt at the Pleistocene, the rest of the country is very much living in the populated Anthropocene, and farmers don't appreciate their cattle being eaten by predators that their ancestors successfully eliminated centuries ago. In some places, conservation practice has moved on from the idea of a natural ecosystem requiring the absence of man – Namibia and South Africa, for example, have successfully managed national parks run by local communities who live and work alongside the wildlife they protect. Even with their best efforts, conservationists wouldn't be able to keep a park entirely free from humanity's influence. Yellowstone's ecology is being transformed by the Anthropocene's climate change, for example – more than half of the conifers have been damaged by pine beetles, which thrive over warmer winters.

Nevertheless, the dream of the pristine persists as a conservation goal. In Montana, former Silicon Valley entrepreneur Sean Gerrity is attempting to expand the rewilding concept across some 3.5 million contiguous acres of native prairie, with a goal of restoring the wildlife abundance the landscape once contained. He wants to introduce 25,000 'genetically pure' bison, multiple packs of wolves, as well as elk and bighorn sheep, and has already bought up much of the land for his dream park. Even in Europe, a continent that lost much of its native fauna centuries ago, conservationists are smitten. There are plans to reintroduce wolves into Scotland, and in the Netherlands an entire Pleistocene landscape has been reconstructed on reclaimed land at Oostvaardersplassen, albeit with proxy species for all the extinct versions that might once have roamed there (had it not been covered by sea). So, Konik ponies from Poland are a stand-in for the extinct tarpans (wild horses), and specially bred German Heck cattle represent the extinct aurochs and European bison, or wisent. In the absence of predators, such as bears and wolves, the park keepers control numbers by shooting as many as 60% of the animals, making a mockery of the park's stated aims of seeing how a 'natural' ecosystem functions. The concept of returning an area 'back to nature' is also saddled with a temporal problem – conservationists have to choose which *time* of nature they want to return it to. For Yellowstone, for example, would it be before the arrival of white Americans in the area – around the mid-1800s? Or is it before any human incursion, some 11,000 years back? In that case, they have their work cut out because the area was so radically different that most of the animals around then are extinct now.

Humans have always had an impact on natural ecosystems because we are a part of them, like any other top predator species. The Pleistocene-like abundance of wildlife on the Serengeti is remarkable because – outside of Africa – it hasn't been seen for some 10,000 years (since the end of the Pleistocene, in fact). Prior to that, many parts of the world had mega mammals. Australia was home to thunderbirds and giant kangaroos three metres tall (some of which were carnivorous) and tapirs, massive koalas, marsupial lions, sheep-sized echidnas and two-tonne wombats the size of an SUV. North America was even more exciting, a Super Serengeti with mammoths and mastodons, vultures with five-metre wingspans, sloths the size of hippos, cheetahs, giant jaguars and sabre-toothed cats, antelope, camels and other huge beasts that ranged the savannahs. And then, quite abruptly, they all went extinct – 75% of the world's large mammals disappeared at the end of Pleistocene, and each continent's wave of mega-fauna extinction occurred with the arrival there of humans.[2] In some places, such as Madagascar and New Zealand, this happened as recently as 2,000 and 900 years ago respectively.

However, after our initial wildlife battering, the human populations – like other migrant predators – mostly settled into a new equilibrium with the ecosystem, hunting sustainably and ensuring enough recovery for future seasons. Those that didn't – such as the inhabitants of remote islands, where there were no alternative resources – paid dearly as their communities collapsed. Such places then reached a new human-free equilibrium, or adapted to another society of people.

There have been a few times in the deep history of mankind when we all nearly died out as a species. Anthropologists call these events 'bottle-necks', times when the population of humans shrank – perhaps to as few as 2,000 people. At those levels, we would be categorised as an endangered species, existing in even fewer numbers than, say, wild tigers do today. But we pulled through, even while our Neanderthal relatives went extinct. In the Holocene, we set about adapting our planet to our needs as a species.

In some places, humans dominated the ecosystems of the Holocene, utterly transforming the biodiversity through farming, burning, building cities and hunting large animals. Many savannahs around the world owe their entire existence to humans, who cleared forests for agriculture and kept the grasslands maintained through regular fires. Around 3,000 years ago, the Bantu farming tribe turned large areas of the Congo rainforest into savannah, for example. And it's thought that thousands of years before that, hunter-gatherer communities burned large areas to encourage grazers and their predators to the area for easier hunting. But not until the Anthropocene has the far-reaching, global impact of humanity come into play. Now, the sheer number of people, their spread and incredible harvesting of natural resources, means that Holocene ecosystems throughout the world are tipping into new Anthropocene states that have never been seen before.

The megafauna extinctions that occurred at the end of the Pleistocene were small, insignificant happenings compared to what is occurring in the Anthropocene – we've lost a third of wild vertebrates just since 1970. In around 300 years, 75% of all current mammal species on Earth will be extinct – that's the startling calculation made by palaeobiologist Anthony Barnosky, at the University of California, Berkeley, based on the current rates of extinction, and assuming the animals already threatened or endangered are wiped out this century.[3] Renowned biologist E. O. Wilson goes further, predicting that half of all species will be extinct by the end of the century. One-third to a half of amphibians are threatened with extinction, a crisis so severe that researchers are creating 'arks' where wild frogs and their relatives can be bred protected from external threats, and theoretically released if and when those threats pass.[4]

Since life first evolved, flourished, diversified and made our planet truly distinct, there have been five mass extinctions. Each was triggered by a cataclysmic event, resulting in at least 75% of all species vanishing. The last of these occurred 65 million years ago, when a meteorite slammed into Earth, throwing up persistent clouds of debris that darkened the sky for years. The climate change that followed led to the extinction of the dinosaurs, as well as three-quarters of all other animals.

Now, Barnosky calculates, humanity is creating a sixth mass extinction on the same scale through a combination of habitat encroachment and fragmentation, hunting, climate change, pollution, and the spread of disease and introduced species. Some 30% of all species may be lost over only the next four decades, conservationists estimate. 'I like to compare what's going on today to the asteroid that wiped out the dinosaurs – except that now we are the asteroid,' Barnosky tells me.

Extinction is actually a natural and common phenomenon – of the roughly 4 billion species estimated to have evolved on Earth, 99% are gone – however, the extinction rate is usually balanced by the evolution of new species. The current, human-caused extinction is happening so fast that evolution cannot keep pace. Barnosky estimates that the current rate is 1,000 to 10,000 times the natural rate, putting it easily on a par with the so-called 'Big 5' mass-extinction events. If we want to retain the wildlife we have at the start of the Anthropocene, we first have to understand how and why we are destroying it.

The world's remaining big-cat species are all threatened with extinction, from the lions of Africa to their relatives in other continents. The different reasons behind their threatened status offer a snapshot into the way humanity is driving three-quarters of all animals into extinction.

The Pantanal is South America's Serengeti – nearly 200,000 square kilometres of grassland and swamp straddling Brazil, Bolivia and Paraguay – and it is extraordinarily abundant in wildlife. I head out into the night to search for America's biggest cat, the jaguar. My powerful torch ignites an array of hundreds of red 'lights' in drainage ditches and rivers, as the eyes of caimans (a type of alligator) reflect the torchlight. Scissor-tailed night hawks take flight and a couple of giant anteaters bumble past, one carrying two babies on her back – a very rare occurrence. Small grey crab-eating foxes play and I see a tapir, the horse that thinks it's a pig. With its funny little trunk twitching in the air, it stumbles around a bit and then crashes out of sight in the bush. A little further along the road, I spot an ocelot, known as the 'painted leopard' for its wonderful markings that are similar to those of Asia's clouded leopard. It stalks the bank like a very large domestic cat, ears flickering in response to tiny sounds I don't hear and its tail held aloft. I'm nearly back to my lodgings, when I hear an unmistakable growl: jaguar. I stop by a thick clump of bush and wait, peering into the blackness, but it remains concealed by the night.

Jaguars are majestic, beautiful creatures. The world's third biggest cat, they have the most powerful bite of any of them in proportion to their weight, and kill prey tiger-style with a massive bite to the back of the head that pierces the brain. They are deadly accurate and superbly adapted to their environment, swimming easily in water and climbing trees despite their considerable weight. Genetically, they are closest to lions and, although very few remain in the US, they were originally North American cats. Jaguars crossed south when the American continents joined at Panama and they were responsible for many of the extinctions that took place in South America after the last ice age. Now, though, jaguars are endangered. Deforestation, expanding farmland and other human encroachment have reduced their habitat to a few pockets of which the biggest are in the Amazon and here, in the Pantanal.

I've come to visit researchers looking at the conflict between the continent's biggest cat and humankind's voracious appetite for farmed food – one of the largest threats to wildlife globally as people around the world consume more meat. In Britain, for example, badgers are being shot because of their threat to dairy cows. Around 95% of the Pantanal is privately owned, used for ranching some 8 million cattle (Brazil is the world's biggest exporter of beef). Many ranchers illegally employ jaguar hunters who chase the animal out of the jungle with dogs and shoot it at close range with shotguns. But in a region bubbling with capybara (the world's largest rodent), caiman and deer, just how much of a threat to domestic cattle are jaguars?

That's what Henrique Concone, a researcher with the Brazilian Instituto Pro-Carnivores, is trying to find out. He's been based at San Francisco Farm in the southern Pantanal for the past seven years analysing jaguar faeces to determine what the big cats are eating. And it turns out their favourite food is capybara, closely followed by caiman. 'We're still analysing the data, but it seems jaguars are responsible for just 0.8% of cattle deaths here – around twenty-four head per year, which is far lower than many ranchers claim,' Henrique says. 'If a farmer has a hundred cows and expects a hundred calves but only gets fifty, then of the multitude of reasons for this, the easiest to blame and the easiest to resolve is the jaguar. But most of the deaths are due to poor management: spontaneous abortion because the cows are not vaccinated, injuries that are not treated, infections and so on.'

Henrique is also looking at data from genuine cattle kills by jaguars to see if there is a pattern that can be avoided, so the jaguar–rancher conflict can be reduced. Again, he says, poor management is an issue. 'A well-managed ranch will expect to lose, say, five head per 1,000 cattle a year and absorb that,' he says. 'Poorly managed ranches allow their cattle to roam over a massive 5,000-hectare range, where they are a regular presence at a watering hole or an easy catch near forest for a jaguar. It is better to keep cattle contained to smaller 200-hectare plots that are rotated, so that jaguars can't be certain of finding them in a specific place.' Other recommendations are ensuring that bulls are kept separate so that vulnerable calves are all born together at certain times of the year where they can be protected more easily, and maintaining the general health of the animals so that the farmers turn a bigger profit and can afford to absorb the loss of a few cows to jaguars. And farmers could use biodefences. Turkana smallholders in Kenya, for example, found that surrounding their plots with beehive fences was much more effective against elephant invasions (and provides a useful harvest of honey) than thornbush fences – the beehives are strung at ten-metre intervals around the fields, leaving a wide margin for elephants to pass, and if an elephant even brushes a fencepost, the bees swarm out. One option being trialled in the Pantanal is grazing water buffalo or a type of long-horned Spanish cow – which aggressively defend their vulnerable calves against jaguars – alongside the regular cattle.

Protecting jaguars from extinction is in the farmers' economic interests, Henrique says, not just because they might earn a few bob from wildlife tourists staying on their land, but because when you remove a keystone top predator from an ecosystem, everything else multiplies with unhealthy consequences. 'Top predators keep the ecosystem healthy by killing off the sick and weak animals, which keeps disease in check,' he says. 'Near São Paulo, where they have killed off all their jaguars and pumas, there is an explosion of capybara and, two years ago, cattle and people there suffered an outbreak of spotted fever spread by a tick that is carried by cabybara. In Europe and North America, you suffer Lyme disease because you have got rid of your top predator, the wolf.'

For all of the ways our civilisation insulates us from the natural world, we remain an integral part of the global ecosystem, and dependent on it. Removing a top predator – be they sharks, wolves or jaguars – has consequences for humans. In the Anthropocene, we will either have to retain our top predators, or artificially manage their roles in the ecosystem. In the case of the jaguar, this means treating lethal tick diseases that spread to humans. We will need to decide as a global community how we value these animals: whether the costs of losing the jaguar are worth incurring relative to the gains to farming. In the jaguar's case, it may be a simple equation, but in the case of, say, bees, it becomes far more complex. Bees are in decline across the world for multiple reasons – some to do with farming practices (pesticide use and removal of plant species), some unknown – with alarming consequences for our food crops. Farmers are now paying for pollination services by companies that truck European honeybee hives to their fields during spring, but it would be cheaper, less stressful for the bees, and better for the local ecosystem, to maintain a farmland environment where bees can naturally thrive.

In Asia, the world's biggest cat is threatened with extinction because, paradoxically, humans have inflated its worth. I visit Bardia National Park in the far-western lowlands of Nepal, home to the country's largest remaining population of Bengal tigers. There are thought to be just eighteen tigers here, a number the government intends to double by 2022, by improving the grassland habitat of their grazing prey, and cracking down on poachers.

Like most wasp-striped animals, tigers are dangerous beasts – they have killed more people than any other big cat – and it occurs to me, while I'm stealthily stalking one on foot, that this is not the cleverest thing to be doing. The wooden stick my local Nepali guide, Siteram, is carrying does little to reassure me as we track the enormous cat. Judging by the prints we're following, its claws alone are big enough to rip my face off. Nevertheless, we press on through the towering elephant grass, following prints to the shore of a stream and across the other side to the edge of dry woodland.

We pause to examine some recent scat on the path. It's full of animal fur and bones – and still warm. Then we hear the unmistakable sound of bones being crunched by a hefty jaw. Siteram motions for me to remain still, while he tiptoes forward. I can smell the tiger now and I start to sweat and my knees tremble a little. Five metres ahead, Siteram silently beckons for me and I will my legs forward towards that terrifying crunching sound. As I round the bush, a dry twig snaps loudly beneath my foot and I catch sight of a blur of powerful movement. With a massive bound, the enormous flame-coloured beast leaps away, disappearing deep inside the forest.

It is still warm where he's been lying, with fresh blood and chewed deer bones. It takes a while for my heart to return to its usual pace, but it takes longer for the grin to fade.

It was a glimpse of one of perhaps 3,000 tigers left in the world, down from 100,000 in 1900. They used to be found from Siberia to Bali to Turkey, but hunting and habitat encroachment by humans have, over the past century, reduced their range by more than 90% and caused the extinction of three subspecies, including the Balinese. There are now six subspecies left, all of which are classed as endangered or critically endangered, including the impressive Siberian tiger, which weighs up to 300 kg. But their habitat is fragmented into increasingly smaller pockets where they are ever more vulnerable – tiger numbers have fallen by at least 40% in the past decade alone, and conservationists fear they may go extinct in the wild within two decades. Most wild tigers are found in India, which is also doing the most to protect them. In the past five years, for example, India has increased its tiger habitats by removing human populations from conservation zones, rehousing them and compensating them for livestock that are eaten by tigers. In 2012, India's Supreme Court took the unprecedented step of temporarily banning all tourism in the central 'core zones' of tiger reserves, in response to conservationists' concerns about unsustainable numbers of people crowding sensitive areas. Despite its considerable efforts, India says that the biggest threat its tigers face is from poachers. The past decade saw 644 seizures of tiger parts, accounting for a minimum of 1,296 tigers, over half of which were from India, according to WWF figures. And the problem is escalating, with more than 200 tiger seizures annually since 2009.

The biggest market for tiger products is China, according to TRAFFIC, which monitors trade in wildlife. Trade in tiger parts was banned in 1992, but a thriving black market openly persists, with tiger skins and bone being sold for hundreds of dollars in China, Taiwan and Korea. The bone is used as a remedy for everything from ulcers to typhoid. And tiger bone ground into a wine – costing upwards of $120 a bottle – is an increasingly popular high-status symbol.

The Chinese lust for endangered species is driving the extinction of everything from tigers to pangolins to manta rays. It has echoes in the past: ancient Egypt's vast mummification industry contributed to the extinction of species such as the sacred ibis and Egyptian baboon, for example. Countless other species have been driven to extinction by humans fetishising their fur, skin or other parts. The difference in the Anthropocene with the Chinese extinction drive is that it is happening on a global scale and affecting so many different species.

The global illegal trade in wildlife is now worth some £12 billion a year and threatens the stability of governments as well as human health – around 70% of infectious diseases have zoonotic origins, including HIV and ebola.[5] Animals and their parts have become the latest conflict resource in Africa. 'It is no longer a case of a few small-scale hunters trying to sell animal parts at local markets. Now it is a big organised activity with grenades and machine guns,' Jean-Baptiste Mamang-Kanga, director of animal protection at Central African Republic's wildlife ministry told me. In recent years, several African countries have experienced heavily armed invasions by organised gangs including the Lord's Resistance Army, the Shabab and Janjaweed and other groups from Sudan to Uganda, poaching everything from apes to rhinos to elephants to fund their other activities. In March 2012, for example, up to 400 elephants were killed in a few hours in a park in Cameroon by a Sudanese gang on horseback bearing machine guns.[6] And a few weeks later, thirty elephants were massacred in Chad by a group of armed militia on horseback, leaving a two-week-old baby elephant fighting for its life. Tens of thousands of elephants are killed each year, with 2012 recording record numbers of ivory seizures, especially in poor countries with weak governance. In the Ituri Forest of the eastern Democratic Republic of Congo, for example, where some of the bloodiest fighting seen in Africa in recent decades has occurred, around 1,000 elephants were slaughtered between 2010 and 2013 – there are now fewer than 7,000 in the country, from around 100,000 twenty years ago. Now, elephant poachers have begun poisoning watering holes with cyanide, a far more indiscriminate way of killing, and conservationists warn that with elephants in Africa being killed now at a rate of one every fifteen minutes, they could be wiped out in the wild in a decade. Rhino poaching, too, has increased 3,000% between 2007 and 2011. In 2014, 100 South African rhinos were relocated to Botswana in a bid to save the species.

Illegal wildlife trade is often conducted by well-organised criminal networks that are undermining governments' efforts to halt other illicit trades, such as arms and drug trafficking, and helps finance regional conflicts. Traders are increasingly sophisticated, using Internet sites like eBay to sell products, which are described as 'captive bred' or with euphemisms such as 'ox bone' for elephant ivory to evade detection. More than half of all illegally traded wildlife ends up in China, where forums offer guidance on how to smuggle endangered species and ivory, and provide a marketplace for such items. 'People might give a bottle of tiger-bone wine to their boss to encourage a promotion, and it's used to sweeten business deals,' says Sabri Zain of TRAFFIC, who's been investigating ways of reducing demand for tiger parts. 'Public education doesn't work,' he says. 'You can put up billboards of slaughtered tigers, but it won't stop people buying tiger wine.' Instead, Zain is working with corporate leaders who, they hope, will come out and openly denounce tiger wine. 'We need to make it socially unacceptable to drink tiger wine. And we need to provide high-status, tiger-free alternatives. China has a 1.3 billion population, which is getting richer. If only a small per cent of those desire tiger parts, the increase in poaching to satisfy them has a huge impact on just 3,000 tigers.'

Aside from poaching, the biggest problem with conserving tigers is that the cats typically inhabit the regions of the world most heavily populated by humans, including India. The answer in the Anthropocene may be assisted migration – to create a new habitat for them in foreign parts. Endangered Tasmanian devils have been relocated to a remote Australian island to save them from extinction, and British entrepreneur Richard Branson has made a Caribbean island refuge for Madagascan lemurs. That's what Li Quan, a former fashion executive, is trying to do for tigers. Li took two South China tigers, which are extinct in the wild and number less than sixty in captivity – from zoos in China to a reserve she's created in South Africa. There, she hopes to successfully breed them, allow them to learn the skills of a wild tiger, and eventually introduce them into reserves back in China.

Does it matter if we 'run out' of tigers – or jaguars, gorillas, koalas or many of the other thousands of endangered species? After all, most people never get to see them in the wild anyway, and for those who want to, we have ample footage and photographs of them in their natural habitat. Moreover, all the alarming figures above relate only to tigers in the wild – they are far from endangered in captivity. More than five times as many tigers live in captivity as in the wild, perhaps 20,000. As we bring about the sixth mass extinction in the Earth's 4.5 billion-year history, we are inevitably losing different species. Saving biodiversity would cost $300 billion a year, according to the chief of the UN Convention on Biological Diversity. So, in the Anthropocene, if we are not willing to spend that, we are going to have to prioritise conservation efforts towards the species we want to save. It's a choice: are wild tigers worth the $350 million global fund set up to rescue them?

We do not depend on tigers for meat or skins. They don't provide transport or help plough our fields. Unlike their far smaller cousins, you can't keep one at home to stroke and pet. The WWF justifies protecting tigers by pointing out the co-benefits for other wildlife. Individual tigers have such a big range that by protecting each tiger, around a hundred square kilometres of savannah or forest is conserved, including other endangered animals such as rhinos, as well as the vital ecosystem services humans rely on from food to water management. Tigers, like jaguars, are a top predator, providing similar ecosystem benefits. Compared to bees, though, or cows or rice, they're useless. And that's largely been humanity's conservation decider to date: if it's useful, we'll keep it; if not, well, I don't fancy its chances. It applies to everything from individual animals and plants to ecosystems and natural landscapes.

Tigers are so big and strong and beautiful, they've become an iconic animal, culturally important in several nations, although that's not been enough to protect other iconic animals, including the Asiatic lion, which once ranged from northern Europe down to South Asia and now persists in tiny numbers in one small reserve in India. Or the Barbary lion, the biggest and heaviest lion that was used by the Romans to fight gladiators, and which became extinct in the mid-twentieth century. But the fact that we like tigers could give them a commercial value in the wild through tourism. A plan being mooted to save the orang-utan, an endangered ape that shares territory with tigers, is to charge tourists a hefty 'conservation fee' to see them. Permits to view mountain gorillas in Rwanda cost at least $500, for example. In the Anthropocene, it is likely that we will have to pay large sums to visit rare creatures that were once more numerous than us.

Nature's 'value' is increasingly being used as a conservation strategy. Scientists are under greater pressure to stress the cancer-treating potential of an Amazon plant, or the financial benefits of ecosystem 'services' provided by soils or forests. One study in 2010 commissioned by the United Nations Environment Programme put the damage humanity had done to the natural world in one year (2008) at $2–4.5 trillion. Another study calculated that conserving biodiversity hotspots could benefit the world's poor to the tune of $500 billion a year.[7] Commercialising nature in this way might stop society from taking for granted the many tasks that the biosphere performs – we could never afford to filter our air, recycle our water, manufacture the proteins, drugs and materials that are derived from the natural world even if it were technically feasible – and value it for conservation. But putting any sort of price on nature is, I think, a nonsense. It is worth noting that while nature performs services for humanity, it also costs us billions of dollars. The natural world is not benign – we spend time and money defending our human habitat from its onslaught, including fighting disease, weather incursion, tree roots, weeds and pests, dangerous animals, crumbling coastlines, and so on. As with any war, it is easy to demonise or revere the opposition. In the Anthropocene, we need to find a way to coexist with ecosystems as we continue to create an artificial order of living systems across the planet. And we need to acknowledge that aside from the services the natural world performs for us, nature has an intrinsic, non-quantifiable value and that we like having it there. Ultimately, the reason for conserving tigers may be less to do with their ecosystem or tourist economy benefits, but simply because they are magnificent creatures. 'We have a moral and ethical imperative to save them in the wild,' Barnosky argues. 'I don't want to be part of the generation that destroyed the last wild tiger,' he says. And I too like knowing that there are still tigers living in the wild. That heart-thudding glimpse of copper fur burning bright will stain my memory for ever.

The Anthropocene extinction will have long-lasting effects. It isn't just the individual creatures that are vanishing, but also their descendants on the evolutionary tree – whole lines of phyla will prematurely cease, so the biodiversity on the planet in a few thousand or millions of years will look very different to how it would have been if humanity hadn't taken over.

One of the most remarkable ways that humans are altering biodiversity in the Anthropocene is through the great homogenisation of ecosystems – what Barnosky describes as the 'McDonaldisation of nature'. Many animals and plants have evolved to occupy specific geographical niches, such as islands or mountain lakes. As a result, across the planet it is possible to find endemic species that exist nowhere else on Earth, such as the giant tortoises of the Galapagos, the lemurs of Madagascar or the koalas of Australia. Occasionally, during the history of the Earth, shifting tectonic plates have forced land masses together, enabling the separate biodiversity to mix for the first time, such as when the North and South American continents joined around 3 million years ago.

Humanity has been orchestrating its own tectonic-scale species migrations, by deliberately or accidentally introducing and spreading species around formerly distinct ecosystems. As a result, some species including rats, goats, rhododendron, wheat and eucalyptus are found worldwide. Meanwhile, many others have become rare or vanished. Many of the introduced species are invasive – or 'weeds' – which out-compete the natives for food, light and habitat, or simply eat them to extinction. And we've also been spreading pests and diseases from one place to another, often causing local extinctions. Isolated human populations have even been wiped out in this way, when we've introduced diseases such as flu or measles to places where the local people haven't developed immunity.

Meanwhile, we've been artificially boosting the populations of certain select species, such as cows, dogs, rice, maize and chickens – most of which are bred varieties that are radically different from their wild ancestors. So, while dairy cows are far from endangered, the last aurochs, their wild ancestor, went extinct in Poland in 1627. Likewise, there are plenty of dogs in Britain, whereas the wolf was eliminated from the island state by the eighteenth century. In fact, more than 95% of the weight of all the land vertebrates is now made up of humans and the animals we've domesticated, according to Vaclav Smil, a Canadian professor and innovative environmental analyst. At the beginning of the Holocene 10,000 years ago, that figure was just 0.1%. Megafaunal biomass (that's anything weighing more than 44 kg) is greater now than at any time since humans evolved 200,000 years ago, despite our voracious hunting of everything from giant sloths to

North American bison to Serengeti elephants. And that's because of the recent huge population expansion of humans and the creatures we've created, whose sole purpose is to be our food. Human biomass alone is greater by a hundred times that of any other large animal species that's existed, Smil calculates.[8]

So humankind is making a pretty distinctive mark on the living planet. Nowhere on Earth is truly wild or pristine any more – everywhere has been touched by humans in some way, from the newly polluted atmosphere that contains different concentrations and isotopes of carbon dioxide, to the wheat monocultures that lie where savannah ecosystems once thrived. If we are to take a more intelligent, intentional, managerial role over life on Earth, rather than blundering through from one trashed ecosystem to another, then we have to accept that the Holocene is gone now. It may be better to shift focus from individual species to creating new types of holistic ecosystems, which may look unlike any that have existed before in their species make-up, but which function better and with more resilience in our human-dominated world.

In the Galapagos Islands, Ecuador, celebrated as a 'living museum', scientists are battling the ecological – and emotional – challenges of the Anthropocene, as they work out how best to conserve one of the world's most unique ecosystems. 'As scientists and conservationists, we need to recognise that we've failed: Galapagos will never be pristine,' says Mark Gardner, head of restoration at the Charles Darwin Research Station. 'It's time to embrace the aliens.'

We're driving uphill on Santa Cruz, one of the four inhabited islands. The glossy mangroves of the coast and golden prickly pear cacti of the arid lowlands are soon replaced by moist cloud forest. Gardner pauses to let a lumbering 400-kg giant tortoise heave itself across the tarmac from one grassy verge to the other. Above, Galapagos finches chatter in a pale-barked, evergreen scalesia tree, its branches dripping with epiphytes and lichens. Despite Gardner and his team's best efforts, the humid highlands are a hotchpotch woodland of non-natives, such as guava and passion fruit, and endemics, such as scalesia and guaybillo, all broken up and intruded on by regular-shaped plots of agriculture. Further up, the fog-laden air becomes wetter and the undergrowth denser. Among a waist-high tangle of brambles and ferns, a rare rusty-leafed miconia shrub fights through, holding up its lilac flowers like a flag of defiance against the introduced weeds that threaten its once-ubiquitous existence.

The sun-aged Australian researcher, with his russet beard, open sandals and shell-pendant necklace, makes an unlikely 'anti-conservationist', as he clambers long-limbed over the blackberries. But that's exactly how he is seen by some at the Charles Darwin Research Station, which coordinates most of the conservation and research on the Galapagos. Gardner has spent the past two decades trying to 'purify' the islands ecosystem, which is celebrated as one of the world's most uniquely pristine reserves. Now he's changing tack. 'Ten years ago, I was much more idealistic and my vision of restoration was to restore nature to its pristine, prehuman state. With the wisdom of hindsight, I now realise that this vision was unrealistic,' he says. As he holds up a thorny creeper with its small, sour fruit, Gardner smiles sadly: 'Blackberries now cover more than 30,000 hectares here, and our studies show that island biodiversity is reduced by at least 50% when it's present. But as far as I am concerned, it's now a Galapagos native and it's time we accepted it as such.'

This remote, relatively young Pacific archipelago is famously home to a vast number of endemic species. Each island nurtures its own distinct ecosystems producing divergent species of the same plant or animal within a few miles of each other in a way that's unique on the planet. But human interaction has taken its toll. Charles Darwin visited four of the eleven main islands in the Galapagos archipelago during his five-week-long *Beagle* stop-over here in 1835, and noted seventeen introduced species just three years after humans first started permanently living on the islands. In the past 150 years – within the lifetime of one of these big old tortoises – the islands have experienced the greatest and fastest changes in their 5-million-year history, with hunting, land clearance, agriculture and the introduction of new species.

In the battle for survival, the alien species have been winning and now dominate island wildernesses. Rallying to the endemics' cause, conservationists, led by scientists at the research station, have spent the past fifty years attempting to remove introduced species and restore the island ecosystem to its prehuman days. There have been some successes: goats have been eliminated from several islands and tortoise populations have been boosted by captive breeding programmes. But the islands' unique flora is proving more of a challenge to restore. The 238 endemic and 376 native plant species have been overwhelmed by 888 introduced species, and the impact is easy to see.[9] *Scalesia pedunculata* forest, for instance, used to cover a hundred hectares; now the species has been reduced by 97% on Santa Cruz alone. The most invasive and problematic of these aliens – blackberry and guava – have developed into forests where nothing else grows, the birds cannot nest and even insects are rare. The battle by conservationists to eradicate thirty-six invasive plant species has cost more than $1 million and succeeded in eliminating just four.

The main reason for this failure, Gardner says, is that invasive plants are far more competitive than native plants. Seeds of invasive species, such as blackberries – which were originally introduced from Asia – are long-lived and accumulate in high numbers in the soil. 'Many restoration activities fail because the disturbance they create actually stimulates these seeds to germinate, so we are stuck in a vicious cycle,' he says.

Now, Gardner is saying: enough. He is proposing a controversial paradigm shift, whereby conservationists embrace introduced species as part of what a new generation of ecologists is terming a 'novel', 'hybrid' or 'emerging' ecosystem. Gardner invited a pioneer of novel-ecosystem management, ecologist Richard Hobbs, of the University of Western

Australia, to the Galapagos in 2010, to help devise this new conservation strategy. But his decision has caused shockwaves among the venerable members of the Charles Darwin Foundation, the fifty-year-old organisation that runs the Charles Darwin Research Station, with many of the old guard 'very upset by the idea'.

'People are upset because it's a real affront to the standard ways of dealing with conservation. The prevailing conservation paradigm has been to preserve "good" and get rid of "bad". But the distinction of what is good and bad has been blurred in ecosystems around the world, due to the impacts of climate change and humans,' says Hobbs. 'A large part of the world is in a hybrid state – around 80% of the world's land surface is moderated by humans.'

While the rhetoric surrounding introduced non-endemics may be negative and imply a morality not usual to scientific discussion – they are invariably described as 'weeds', 'degraded' or 'invaded' ecosystems – a mixed forest of endemics and non-natives needn't be a bad thing. They can provide habitats for animals, ecosystem services, shade, and support even greater numbers of species, for example. Yet until recently, the ecological contribution of these mixed-up wastelands has been ignored. That changed in 2006, with the publication of the first studies to examine such systems by a group of ecologists, who concluded that novel ecosystems are worthy of conservation and in many cases actually contain greater biodiversity than native ecosystems, because introduced species outnumber the native extinctions.[10] Humans have actually caused a net increase in plant-species richness, they contend. 'Biologists will walk for two hours through "trashy" secondary forest, looking neither left nor right, before reaching the pristine forest where they carry out their research,' says Ariel Lugo, director of the International Institute of Tropical Forestry (USDA Forestry Service) in Puerto Rico. 'More than 30% of the globe is covered by novel ecosystems – only a fraction of that is pristine – so if you care about global ecology, you need to look at these systems too.'

However, Bill Laurance, a conservation ecologist at James Cook University in Cairns, Australia, refutes the idea that primary and novel ecosystems have equivalent biodiversity. 'Secondary or novel ecosystems might have a greater number of species, but they are the junk species that are not endangered – you can see them thriving anywhere around the globe. We need to protect the rare plants and animals that are found in primary forests,' Laurance says. 'If people want to resign themselves to managing novel ecosystems then what we're doing is homogenising the world's biota; setting the world on a geological epoch: the Homogocene,' he says. 'I believe in fighting to maintain pristine ecosystems as national parks.'

But Laurance is losing the battle against the 'McDonaldisation' of nature. From the Galapagos to Hawaii, conservationists are switching tack and starting to embrace the introduced species of Anthropocene ecosystems, while focusing their efforts on routing out the more harmful invasives that out-compete unique flora or fauna. Thus, in the Galapagos, the plagues of blackberry bushes originally from the Himalayas are simply being controlled in Gardner's experiment, whereas rats and goats that eat the food of rare tortoises are being eliminated by poison or shot from a helicopter.

In some places, invasives have enhanced the landscapes, reducing erosion, providing handy cash crops or food and habitat for other wild-life. In Cuba, for example, Marabu weed introduced from Africa, brought originally to prettify the island state, was overtaking the landscape. But scientists have now found a productive use for it – the woody shrub makes an excellent activated carbon filter for the rum industry, and when it is ground down and combined with aluminium, it makes light-weight electrodes that can be used in efficient batteries and supercapaci-tors. In Australia's Macquarie Island, conservationists culled all the invasive cats, which were eating native birds. As a result, rabbits (another introduced species) proliferated, devastating native plant life.[11] Removing invasives, which create their own novel ecosystems, needs to be a carefully considered process, if it is to avoid this cascade effect. In the Serengeti, alien plants are also thriving, and there, conservationists are leaving the 'harmless ones' to focus on problematic ones, such as prickly pear, which forms impenetrable barriers to mammals. The cacti were probably introduced in shipments of gravel or carried on trucks passing through the savannahs.

Because humanity has become such a dominant force on the biosphere, we can no longer just expect natural ecosystems to 'get on with it', to maintain the rough equilibrium between species that existed in the Holocene. In the Anthropocene we have to decide how best to manage the rapid extinction rate and homogenisation we're creating. It is certain that the Anthropocene will be a time of much poorer biodiversity in which once-common species will be extinct or exist only in man-made environments like zoos or private breeding colonies far from their natural habitat, such as Richard Branson's lemur sanctuary. In other places our impact is being felt in surprising ways: I've heard wild parrots in Australia who have learned to speak from pet parrots that have escaped captivity, call to people from the trees with human voices; and seen coconut crabs in Indonesia crawl the beach wearing shells made of cans or yoghurt cartons. Many of the places I visited as a child, or even as recently as ten years ago, have utterly transformed and now no longer support the biodiversity they did. On Thai islands, the little fishing villages where once I slept to the hiccuping love calls of tokay geckos now shake to electronic music and the screech of traffic – the tokays are long gone.

When humans decide to actively rescue a species, though, we can have remarkable success. The Arabian oryx, for example, was hunted to extinction in the wild in 1972, but after a successful breeding programme of the antelope by zoos, individuals were released into the wild and now number more than 1,000. Other comeback species include the American bison and bald eagle, which were perilously close to extinction at one time. We can decide on the ecosystem make-up of the Anthropocene, and choose which elements of the Holocene's wildlife we want to preserve, and where. Is it sensible to spend billions of dollars on captive-breeding the Chinese panda, which has insufficient habitat left outside of zoos and has never been released into the wild, while ignoring the plight of the obscure black jumping salamander or the elliptical star coral, for example? The Zoological Society of London has tried to answer this conservation question about which species to prioritise by setting up a scientific framework to identify the world's most Evolutionarily Distinct and Globally Endangered (EDGE) species. This conservation effort focuses specifically on threatened species that represent a significant amount of unique evolutionary history. Pandas fit into this category, but so does the weird and non-cute purple frog, for example. With limited funds and with time running out to protect the 75% of species threatened with extinction, we will have to use strategies like this to choose between survivors and losers.

And we need to decide how we would like to experience nature. For example, will we portion off large tracts of land, such as parts of Antarctica, some tropical islands or rainforest regions, and forbid all human incursion? Is knowing that wildlife exists there enough, or do we need to visit and see it? We have to decide in the Anthropocene what we value about the natural world. Is it something that can be recreated in a 'Holocene theme park', in which the ecosystems of the world can be brought together in one area, with the charismatic species of the African savannah roaming below an artificial mountain home to bears and condor, perhaps?

In other places, such as the vast monocultures we create through agriculture, efforts are already being made to restore native ecosystems, or in some cases plant non-native trees and grasses or introduce animals to artificially restore the functions that the prehuman ecosystem once provided, such as reducing soil erosion, pollinating flowers or helping reduce water loss. Environmental biologist David Bowman has proposed introducing elephants and rhinos to Australia to help eat introduced grasses and curb wildfires that are no longer being controlled by regular burnings by Aboriginal people, and reintroducing predator dingoes (wild dogs) to help control cat and fox populations that are eating their way through the native creatures.

For remaining 'wild' areas, we could try to better control humanity's impact on biodiversity by reducing the areas needed for farmland, curbing poaching, logging and hunting, and reconnecting the patches of natural landscape to create the migratory corridors that larger species rely on to breed and hunt. There have never been so many areas of conservation – national parks and protected zones – so for some endangered species, humans may be acting in time to save them from extinction. Around 13% of the world's land surface is currently under some sort of protection, but many of these areas are protected in name only – the parts of the world with the greatest biodiversity are often in the poorest and most troubled regions.[12] Even famous, well-funded and internationally supported sites like the Serengeti face huge risk: a 480-kilometre road is planned, slicing in two the route of the world's greatest migration of wildebeest to the Maasai Mara in Kenya. The Tanzanian president wants to link economically important areas and bring development. Conservationists argue that animals will be killed by cars and trucks passing through the reserve and that a road will irrevocably harm the biodiversity here. Scientists calculate that if wildebeest were to be cut off from their critical dry season areas, the population would likely decline from 1.3 million animals to about 200,000 – a collapse that would most likely end the great migration.[13] If the wildebeest population declines by even 50% it could increase fire frequency in the park, as less grass would be eaten. In 2011, the World Bank and a consortium of governments and donors offered the Tanzanian government full funds to pay for a rerouted road running south of the fragile savannah, which would serve five times as many people while still linking the major hubs, but President Jakaya Kikwete refused to change the route. Eventually, he agreed to an upgraded, but unpaved road through the national park section, which will be monitored for traffic. Whether this road will remain unpaved in the future, is far from certain. And if we can't protect this most celebrated part of the world from humanity's onslaught, what hope for the rest?

Paradoxically, just as we approach a tipping point for extinctions, we are beginning to understand how to bring extinct animals back from the dead. Scientists are hopeful of cloning mammoths and other animals from the Pleistocene, and even restoring our own extinct cousin, the Neanderthal.

Nineteenth-century flocks of passenger pigeon were described as vast, mile-wide accumulations that took many hours to pass overhead, and were once so common that they would blacken the American skies. However, within fifty years of the invention of the shotgun, the last wild passenger pigeon was reportedly shot by a boy in 1900 (a zoo survivor hung on until 1914). Now, San Francisco-based Restore and Revive is trying to 'de-extinct' it by making a hybrid chick using fragments of DNA from the passenger pigeon and its closest living relative the mourning dove. From this bird, they could 'backbreed' or 'backclone' with more passenger-pigeon DNA, or edit the creature's genome, until they arrive at a 'pure' passenger pigeon.

It's just one of several projects around the world that are collecting DNA from endangered and lost species. In Britain, the University of Nottingham's Frozen Ark project has collected frozen DNA samples of more than 28,000 living species, including 7,000 endangered ones. Meanwhile, the original 'frozen zoo' DNA collections in San Diego Zoo and at the Audubon Center for Research of Endangered Species in New Orleans are being used by researchers to attempt interspecies surrogacy, in which cloned embryos (derived from the DNA in skin cells or other fragments) of extinct animals are placed inside the uteruses of their living relatives. So far, Martha Gomez and her colleagues have been working with the endangered South African black-footed cat, by breeding it using IVF and gestating the embryos inside domestic cats. Next they plan to attempt Tasmanian tigers, extinct since the 1930s. Meanwhile, scientists at Brazil's frozen zoo at the Embrapa research institute are trying to clone endangered species including the maned wolf and jaguar. Other biologists have managed to generate pluripotent stem cells from the endangered snow leopard, theoretically enabling them to create the sex cells that give rise to embryos.

The possibilities are immense. For example, if a species is endangered or extinct because of a disease, a cloned variety could be brought back with its genome tweaked to be resistant to that disease. Or, if the animal is dangerous or competitive with humans, like wolves, it could be genetically modified to not like the taste of sheep and cattle. Domestic cats have been bred that are hypoallergenic, so as not to irritate their owners. However, cloning existing species where scientists have access to their full genetic material is only successful 5–7% of the time – cloning from extinct species with fragmented DNA has only ever been achieved once. In 2003, scientists cloned the Pyrenean ibex, a subspecies that went extinct in 2000, by using frozen skin samples and implanting the embryos into a closely related ibex species.[14] The biggest hurdle in cloning extinct species is that the DNA is too degraded to be able to use. However, modern DNA-reading machines can theoretically recreate the genome of animals from several DNA fragments. Scientists can then use this information to tweak the DNA of a living relative, such as an African elephant. That's what the mammoth cloners are hoping to do, unless they happen upon the dream find of a perfectly frozen mammoth testicle, in which case IVF may be possible.

Sadly, expensive techniques like this, even if successful for individual animals, could not be applied practically to restore the intricate diversity of life that existed before humanity's planetary plunder. Species go extinct for a reason, such as habitat destruction or climate change, and bringing them back won't eliminate that problem. Most of the scientists working in this field are only attempting to clone the animals for a captive future in zoos.

Instead, in our human world, we must decide what type of ecosystems we would collectively like, and set about creating and protecting them. In the Anthropocene, we are no longer just another part of the natural world, we are the planet's gardeners and that requires nurturing skills. But as we try to negotiate a path between the competing demands of the natural world and our human world, we'd do well to remember that, in truth, there is only the one living world.

8

FORESTS

For most of its history, Earth's surface has been barren rock, not unlike the surface of other planets. It took nearly 4 billion years, and a period of global warming, for plants to conquer land. Their arrival was transformative, greening Earth, changing the composition of the atmosphere, and laying a welcome mat for the first terrestrial creatures.

Around 100 million years later, plants had developed the vascular systems to allow them to establish far from wetlands and grow tall – they went from thirty centimetres to as much as thirty metres high by the end of the Devonian period. And the first forests appeared.

Those earliest rainforests would be otherworldly to a visitor from today. For a start, they were silent. The resonant forest hum, the croaking, buzzing, chirping, whistling calls of frogs and insects and birds was absent – nothing flew at all, and the first canopies would rustle and shake only by the wind's breath. Dinosaurs had yet to evolve, and nothing crawled, snaked, walked or clambered through the undergrowth. There was nothing around in possession of a nose, but the forest's scents were very different then. There were no fragrant flowers or sweet-fruiting bodies, no sour animal dung or aromatic leaves.

These quiet, still, simple forests were cushioned with mosses and liverworts, interspersed with the vertical stems of horsetails and ferns. Towering above them were the first 'trees', reaching about nine metres high, called wattieza. They looked much like modern palm trees, with a trunk that branched into a graceful umbrella. Wattieza were not true trees – related to ferns, they were flowerless and reproduced using spores not seeds, and they had fronds rather than leaves – but despite their primitive nature, they transformed our planet. For the first time, plants had deep roots, capable of penetrating and binding soil, and even crumbling and breaking up rock. Their roots helped channel water, creating rivers and banking up lakes and wetlands where aquatic life flourished. The new forests created the first canopies, and produced a wealth of micro-environments that would become habitats for plants, insects and other organisms.

The photosynthetic activity of these green pioneers released enormous volumes of oxygen, creating an environment in which giant insects evolved with metre-long wingspans. The first seed-generating plants and reptiles also emerged into these steamy forests. The vibrant lush jungle grew and died at a tremendous rate, and the swampy ground became a mortuary for fallen trees and plant matter. This wealth of partially decayed trunks of organic carbon in time turned to peat, and sowed the seed of the eventual Anthropocene – for, over time, this peat became the rich coal seams that drove the Industrial Revolution, launching the Age of Man.

At times, the prodigious photosynthesis of those early forests absorbed so much carbon dioxide from the atmosphere, that the global green-house effectively lost its panes, and the planet plunged into ice ages. Massive climate change caused this time by tectonic movements, finally spelled the end of these first forests, which broke up and collapsed under *ice sheets, leaving their energy-rich skeletons for our finding, and a dry desert landscape.*

It would take millions of years – well into the Triassic – for biodiversity levels to return. And the forests that took root were entirely different, seed plants were dominant and conifers were abundant. It wasn't until the end of the Cretaceous, some 70 million years ago, however, that forests truly resembled today's, with the emergence of the first leafy plants and flowering trees, including magnolia and figs. Flowering plants evolved with the insects that pollinated them, including bees. By this time, forests flew with birds and insects, crawled with early mammals and were dominated by the dinosaurs.

In contrast, the world into which humans evolved was dry and cold – the few remaining forests had been banished to refugia at the edges of ice sheets. The last ice age ended some 10,000 years ago, and the warm Holocene ushered in a new green: forests covered half of the world's land surface from the poles to the tropics. Our species took full advantage of this rich resource, gathering fruits, berries, leaves, medicines, oils, bushmeat, firewood, timber and other building materials in forests. They harbour our closest ape cousins and are still used for shelter – and ambush – by humans.

Around 70% of all land creatures and plants live in forests. Tropical rainforests, especially those in the Amazon, Congo, Borneo, Sumatra, Madagascar and New Guinea, support unparalleled biodiversity including many species as yet unknown to science. Many of these forests are home to indigenous people with unique cultures and intimate knowledge of the local plants, which could benefit a wider population.

Forests play an important role in local and global climate. The world's forests absorb 8.8 billion tonnes of carbon dioxide each year through photosynthesis – about one-third of humanity's greenhouse gas emissions. And, like all living organisms, trees respire, taking in oxygen and emitting water vapour from their leaves. At the same time, their canopies provide shelter from the sun and wind, making forests much wetter, cooler environments than surrounding treeless areas. *This nurtures streams and rivers, provides habitat for a range of amphibians and other life, helps cool the regional and global atmosphere, and recycles water. This water is vital to maintaining the trees that need to drink for photosynthesis. Although forests help create the climate, they are also exquisitely sensitive to it – and the smaller a forest gets, the less resilient it is. When trees are chopped down, sunlight enters in the gap and dries the soils. Drought upsets the forests' delicate water cycle – trees start to die and the entire ecosystem can tip from rainforest to grass-dominated savannah.*

Over recent centuries, we have burned and chopped our way through forests, particularly in Europe, the Middle East and China, with periods of intensive deforestation – such as when Britain built a fleet of wooden ships to defeat the Spanish Armada – and of local forestry recovery, such as when local human populations were wiped out by disease. The most dramatic and global deforestation, however, has occurred since the 1850s, with over 40% of the world's forests cleared for timber and, increasingly, to provide agricultural land.[1] Just a quarter of the world's ice-free land is forest now.

Deforestation emits carbon dioxide from soils and decaying plant matter, and is responsible for around 20% of all carbon dioxide emissions.[2] In recent years, deforestation in the tropics has resulted in the same amount of carbon being emitted as was absorbed from the atmosphere. At the same time, warmer temperatures have led to an increase in beetle infestation and fires in boreal forests, so they have become net emitters of carbon dioxide.[3] Only the new regrowth forests in the abandoned farmlands of Europe and the US have tipped the balance in favour of soaking up carbon dioxide.

In the Anthropocene, forests have never been more threatened, and yet, in the wake of global warming, biodiversity loss and water shortages, they have never been more necessary.

The laid-back settlement of Rurrenabaque, in the Amazon lowlands of Bolivia, is surrounded by lush karsts, the foothills of the Andes, and perched on the edge of the rushing Beni River. Once the daytime scooters and market chitter-chatter have ceased, the damp air saturates with the noise of frogs and cicadas, bird whistles and bat clicks – the sounds of the Amazon rainforest.

The sky is thick with El Chaqueo (the Big Smoke), producing spectacular sunsets, but reducing visibility such that the opposite bank of the Beni is impossible to see. The smoke issues from the seasonal burning of rainforest to clear fields for agriculture and grazing – around 300,000 hectares of Bolivia's forest are lost annually in this way. Local farmers believe the fug produces rain clouds, ensuring a good harvest, but in reality the opposite occurs: the deforestation worsens the effects of the drought farmers are already suffering, so they need to burn more forest to increase their yields. And so the cycle continues. Globally, more than 70% of forest loss is for agriculture.

I've come to the Amazon to learn more about humanity's relationship with the world's greatest rainforest as we enter the Anthropocene, and to meet an Amazonian warrior risking her life to save it.

Rosa Maria Ruiz is a petite woman in her sixties with large features, luxuriant long black hair and a noticeable limp. She meets me at the jetty in Rurrenabaque with a hesitant smile, and together we board a motorised dug-out canoe to her home deep in the jungle. The Beni, a major tributary of the Amazon, is wide and fast, banked on either side by forest that becomes more pristine the further we travel, with more expansive tree-trunks and fewer planted species of banana and mango. We pass the wood-and-palm lean-tos of nomadic tribes, who are camping close to the bank to fish during the dry season, when the forest is relatively poor in fruit and other foods. Children with distended bellies trawl the banks for snacks, easy prey for the many caiman, stingrays and piranhas that lurk beneath the rushing surface.

We are two hours into our journey, and finally chatting in a relaxed fashion, when I notice a strange movement under Rosa's loose, taupe shirt.

I try not to stare, but when a stringy black, hairy arm with long slender fingers emerges from between her shirt buttons, it's impossible to stay silent.

Rosa, do you have a monkey under your shirt?

She reaches in and pulls out a four-week-old spider monkey with huge round eyes and a tail that's immediately wrapped snugly around Rosa's arm. 'The hunters injured him because he was clinging to his mother's stomach when they shot her for bushmeat,' she says. 'His chest was split open but I'm nursing him back to health and when he's ready, we'll release him into the reserve.'

Here, in Bolivia's Amazon rainforest, hunting and trafficking of wild animals is a growing problem that Rosa has spent much of her life trying to counter. In 1995, she was instrumental in setting up the Madidi National Park, which, at nearly 19,000 square kilometres, is one of the largest protected areas in the world. Her efforts cost her her home, her business and very nearly her life, yet the future of this important patch of the Amazon Basin is still far from certain.

About half of the forests that once covered the Earth have already gone because of humans and each year, another 16 million hectares disappear.[4] More than half of the world's forests are in the tropics, but, as we head into the Anthropocene, rainforests are being cleared at a rate of 1.5 acres per second, with deforestation penetrating even the interiors of the last remaining virgin rainforests.[5] Less than a quarter of the world's old-growth, original forest cover remains 'intact' – most of this is in the northern coniferous boreal forests, and the tropical forest of the Amazon Basin and the Guyana Shield – and the quality of the remaining forests is declining. If deforestation continues at this speed, all of the world's rainforests could be wiped out entirely this century. The world's biggest and, arguably, most important remaining rainforest, the vast tropical Amazon, which sets the climate and rainfall patterns for entire continents, is now under such threat that scientists fear it will switch to a savannah, and then desert, within decades.

Rosa is one of the few individuals fighting to save this unique biome from the onslaught of humanity. Other similar brave souls have died in the attempt, so vehement is the human desire to strip the forests of trees for timber, mined minerals and the agricultural soils that lie beneath. Amazon activists were killed at a rate of one a week in 2011.[6] The toll includes landless rubber tappers and peasants trying to eke a living from the forest, Amerindian tribal members trying to protect their ancestral homes, Catholic priests trying to defend the forest and its traditional dwellers, scientists and prominent environmentalists such as José Claudio Ribeiro da Silva and his wife Maria do Espirito Santo, Brazilians who were ambushed and shot dead in Pará state, near the city of Marabá, after repeatedly reporting death threats against them by loggers and cattle ranchers. On too many occasions, Rosa has almost joined the forest's martyrs.

We're heading downriver to her latest venture, Serere Sanctuary, a 20,000-acre reserve, which she set up in 2008 to conserve the forest against poaching, logging and other threats. As our small craft speeds over rapids and dodges floating tree-trunks and rocks, Rosa feeds the tiny monkey a banana and milk from a syringe, and points out wildlife she spots. Fallen trees, partially submerged, provide sunbathing spots for turtles that pile on top of each other in long rows. In the next couple of weeks they will swim to the sandbars and lay their eggs. Capybaras pose stock-still, hoping for invisibility. Birds of every colour and size screech and call overhead. We see herons, macaws, toucans, fish eagles and vultures, small wading birds, tiny bright hummingbirds and kingfishers.

A clearing in the forest reveals a busy illegal timber port with logs of mahogany, cedar and other hardwoods stacked high and ready for barging out. Authorities periodically confiscate a few logs, but the bribes are too lucrative for officials to do anything effective to stop the trade. Further along, another clearing contains a camp of gold miners, churning up the deforested earth and flushing mercury and other toxins into the river. A boat laden with bags and bags of dead monkeys passes us, travelling in the opposite direction towards the markets in Rurrenabaque.

They are sights that Rosa must have seen hundreds of times before, but she is nevertheless visibly upset. 'I've spent decades living in this jungle and trying to work with the indigenous people here to protect it and we found a solution that was workable for everyone. This terrible destruction is just so unnecessary,' she says.

Rosa was perhaps born into her vocation. Her father died when she was just 7 months old, leaving her mother, Lucie, to scratch a living selling pharmaceuticals for commission among poor tin-mining communities. What Lucie saw during those years turned her into one of Bolivia's most celebrated human-rights activists, working tirelessly to protect enslaved indigenous people and their natural resources from exploitation. In pre-Columbian times, there were around 10 million people living in the Amazon rainforest; today fewer than 200,000 indigenous people remain in the entire forest. And yet, it has never been so threatened by humanity.

Rosa spent her childhood split between the Andean mining families (where her friends began working underground from age 8 and died on average at age 26 from silicosis), and living with the indigenous Tacana people of the Amazon rainforest. During the 1980s and 90s, she moved to the forest and began to help the disparate semi-nomadic Tacana communities to organise into political groups that could claim land titles to ensure their ancestral areas were safeguarded from exploitation by mining and other commercial interests that were making ever greater forays into the region. She describes those years as a terrifying time of continual coups and violent protests, with officials shooting at villagers from helicopters, and recalls an occasion when her childhood home was sprayed with gunfire.

It was out of this work that Rosa first conceived a plan for a large national park, around her home in the Amazon, to protect one of the world's most diverse areas of forestry. The range she chose included high Andean glaciers, cloud forest, dry forest, pampas and rainforest, hosting more than 1,000 species. It was also home to 1,700 indigenous people, and Rosa planned to create the park with their full support and participation. Part of her plan was to set up a sustainable tourism business that would provide livelihoods for the park's tribes and encourage the protection of its wildlife. Local people were sceptical – logging companies promised them schools, roads and money in return for mahogany and forest clearance. However, over time, it became apparent that the companies' promises were not being met – the loggers were stripping and burning the forest, shooting the wildlife, and their roads allowed incursion by others who also exploited the place.

For more than a year, Rosa kept up a campaign of persuasion, covering thousands of kilometres of arduous terrain, on mule, by foot or in a balsa-wood canoe (like ones the pre-Columbian Incas used), trying to win the trust and cooperation of rainforest communities for her Madidi project. It worked, and she began to train the forest people for eventual work with tourists. Rosa created the Eco Bolivia Foundation with her mother, and set out to convince the world that this precious part of the Amazon needed protection. Despite initial rejections by the Bolivian dictatorship, she managed to get the World Bank to support the park's creation and Madidi National Park opened in 1995: 4.7 million acres of pristine forest, home to tapirs and spectacled bears, jaguars and maned wolves, sloths, giant otters and over 1,000 of the world's 9,000 bird species.

Our boat pulls into a spot on the bank marked by steps cut into the mud and stone, and we climb up into bamboo forest that quickly turns to rainforest. We walk for two kilometres, past giant trees whose buttress roots dwarf us, strangler figs knotted into plaits and twists, and a palm tree that 'walks' around the forest dropping new roots in the direction of light channels. Beside our path, the forest is alive with noise and movement: the scurrying of small mammals, the crashing of monkeys in the canopy and countless shrieks and alarms of the many birds calling to each other through the dense foliage, and yet the creatures responsible for the commotion are hard to spot.

Charles Darwin described it well, when he first entered the Amazon in 1832: 'It is hard to say what set of objects is the most striking: the general luxuriance of the vegetation bears the victory, the elegance of the grasses, the novelty of the parasitical plants, the beauty of the flowers, the glossy green of the foliage, all tend to this end. A most paradoxical mixture of sound and silence pervades the shady parts of the wood: the noise from the insects is so loud that in the evening it can be heard even in a vessel anchored several hundred yards from the shore: yet within the forest a universal stillness appears to reign. To a person fond of natural history such a day as this brings with it pleasure more acute than he may ever again experience.'

Eventually we reach a small clearing in the forest housing a two-storey wooden construction with walls made of mosquito nets and hammocks strung out overlooking a large lake. A blue and yellow macaw screeches a welcome from its vantage point high in a cocoa tree. The lake ripples with fish and stingrays, reflecting the swoop of low-flying kingfishers.

I am shown to my cabin, a spacious, raised wooden platform with a roof, open to the forest on all sides, save for the mosquito netting. There is no power here, so I rely on moonlight, candles and my torch after dark. I am serenaded at dusk and dawn by the low rumble of howler monkeys, the multitude of birds and insects and the unidentifiable noises that permeate through the forest. Over the next few days, Rosa and I take walks through the jungle, disturbing tarantulas, which scurry back into their white sock nests, finding countless paths of busy leafcutter ants transporting their cargo like tiny green sails. Above us, we see tribes of playful monkeys, including howler, capuchins and spider monkeys. We spend a while with the bizarre frilly-headed, horned Serere bird (the Hoatzin or 'stink bird'), whose chicks are like pterodactyls, featuring claws on their wings that help them climb up muddy banks. They have an ugly barking song, and are one of the few creatures to maintain a healthy population – their terrible smell means no one hunts them for food.

The flora is equally fascinating: vines that when sliced spurt drinking water, nuts sweet as coconut and rich in protein, medicinal plants that heal wounds, cure arthritis and sooth insect bites, trees with bark like paper and others that taste of garlic. Coatis and possums climb the trees and wild pigs and peccaries rustle in the undergrowth. The animals are impossible to see most of the time, but we get enough rewarding glimpses to keep us keenly looking. As we walk, Rosa passes me snacks from the forest floor, including sweet palm nuts and bitter cocoa. Some 80% of foods originate in this tropical rainforest, including fruits like avocados, coconuts, figs, oranges, bananas and tomatoes; vegetables including corn, potatoes, rice, winter squash and yams; spices like black pepper, cayenne, chocolate, cinnamon, cloves, ginger, sugar cane, turmeric, coffee and vanilla and nuts including Brazil and cashew. And these forests have so far provided at least a quarter of active ingredients for Western pharmaceuticals, even though only a tiny percentage of plants here have been investigated by scientists.

Despite this extraordinary biodiversity, the soils in the Amazon are mostly extremely poor – indigenous people who have farmed here have had to design elaborate systems of banks and charcoal burial to enrich them enough for crops. Recently, scientists discovered that the rainforest is dependent on dust blown across the Atlantic from a specific, tiny area of the Sahara Desert to restock its soil with nutrients and minerals. Satellite images show that the Bodélé depression, near Lake Chad, which is one of the largest dust producers in the world, is the source of more than half the material that fertilises the rainforest.[7] If the Bodélé was not there, the Amazon might be a wet 'desert'.

Rainforests, with their hot humid climates, are the richest terrestrial ecosystems on the planet, home to more than half of the world's plants and animals – a total biomass that exceeds, by 50%, the biomass of all other terrestrial ecosystems combined. Many of the animals that exist uniquely in rainforests, like orang-utans, can only survive in undisturbed primary forest, or need large tracts of unbroken forest. This makes many species incompatible with humanity's tentacle-like incursions into intact forest, the piecemeal carving up of patches for cropland and sprawl of pollution, noise and deforestation that follows road-building. It's why the protection of such a large swathe of forest in Madidi was so important.

As we walk, Rosa tells me more of her extraordinary story. The creation of Madidi National Park won her international acclaim, but also many enemies. The commercial and political interests in the park's resources grew dangerously strong.

By 2001, Rosa had set up a lodge with cabins in the park and was ready to receive her first guests who would explore guided trails that led through carefully prepared communities. The site included a fully equipped conservation centre where up to 300 indigenous men and women could participate in workshops on monitoring protected areas, infrastructure management, and foreign-language tuition. But there was a problem. In 2000, she was featured for her efforts in a large *National Geographic* magazine article, which had garnered international support from donors and volunteers wanting to work with her. Ruiz was critical of the government's involvement in forest destruction and was becoming a thorn in the side of former dictator Gonzalo Sánchez de Lozada (granted US asylum by George W. Bush, but wanted on charges of genocide in Bolivia) and his cronies, who were making good money out of logging concessions, trafficking animals from the park and mining schemes. A sustained effort by the government and park authorities to discredit her had some small success, but still she clung on, continuing to work from her home. In 2003, she received death threats and an assassin was hired to kill her. The following year, her eco lodge and the twenty-six forest cabins were burned down with three of her staff and two children inside, who narrowly escaped being murdered – and all within plain view of park officials. The military turned up with machine guns and police, successfully evicting her from her property and the park, destroying her buildings and equipment, and stealing her personal belong-ings, she says.

'A project, my dream, that took me thirty years to put together came down in a few hours. They took everything, even my books, smashed my solar panels, destroyed the macaw-watching telescope – it was heart-breaking,' she says. 'And even though five people were nearly burned alive, nothing was done to catch the perpetrators and punish them.'

That year, revolution began in La Paz, the people ousted the dictatorship and installed a new leader. Then, in 2006, the country's first indigenous president, Evo Morales, was elected. Rosa was invited to the justice ministry for meetings to try to reinstate and compensate her. But soon after, during a routine swim across her lake, she was attacked by a 4.5-metre-long black caiman that nearly killed her. She spent three years in hospital having surgery to save her life and then to restore some function in her right leg.

We pause on a handily placed fallen tree, checking for large ants before sitting down so Rosa can catch her breath and rest her leg. The trouble with travelling through rainforest is that it involves leaving the world we have so perfectly ordered around us. We become, on entering, just another piece of flesh wandering around for anything to take bites out of, inhabit or incubate their eggs in. I am soon bitten to swollen lumps by mosquitoes, blackfly, sandflies, horseflies . . . Malaria is not a big problem here, but botfly, leishmaniasis and other horrible troubles are. I spray DEET but still they bite.

Above us, a large spider monkey approaches, swinging mournfully in the trees. When Rosa bought this land, it was a trashed zone of logged forest, where most of the wildlife had been shot or frightened. Some animals have now repopulated the area, and others, such as the spider monkey above us, are rescued animals she has released into the reserve. Like the creatures of the world's oceans and savannahs, forest-dwellers are facing an unprecedented threat – globally, the trade in wildlife is now the biggest illegal economy after arms and drugs, and here in the Amazon, the trades are all connected. Deforestation has accelerated since 2007 as farmers slash and burn forest to grow coca to supply the cocaine industry. Once a coca field is exhausted, farmers burn larger areas in rapid succession. Cocaine used to be produced in neighbouring countries, apart from a few remote factories where workers would tread the leaves by foot, but new machinery now enables production even in small apartments in La Paz, ramping up supply and demand for the lucrative export drug. The industry is linked with the traffic in wild animals: deforesting burns and kills animals leaving survivors vulnerable. The animals can be quickly hunted and piled into trucks – on average, just one in every ten animals caught will survive to be sold. There are said to be three powerful families in La Paz who control the bulk of the country's wild-animal trade. They employ hunters and truck drivers and export to countries around the world. The region's guerrilla groups, which formerly fought for political ideals, now use their weaponry to control their drug and wildlife fiefdoms.

Bolivia's new laws protecting indigenous culture are indadvertedly helping deplete the country's wildlife. Communities are allowed to hunt bushmeat for food and chop forest trees for their own use, but not to sell or trade in forest resources. This would have been fine a few decades ago, but with the increase in guns, in roads connecting remote villages to markets, and a fast-disappearing rainforest, a systematic slaughter of everything from monkeys to jaguars to tapirs is occurring, as well as the wholesale destruction of any mahogany still standing. Little of this wildlife is actually eaten by the community. Instead, the adults are skinned (their skins fetch a good price), their flesh sold as bushmeat, and surviving babies are sold as pets. A spider monkey goes for as little as $100 in the market in the El Alto district of La Paz, although the villager who captures it will only get a fraction of that. Rosa acquires these in various states. Often, the tips of a monkey's fingers will have been shot off while it was clinging to its mother; many of the rescue animals have bullets still lodged in their bodies or are missing anatomy.

Spider-monkey species are among the twenty-five most endangered primates globally. At a time when so many species are threatened with extinction across the world from the plains to the rivers and oceans – it's worth remembering that Earth's many and varied lifeforms are not simply our neighbours on this planet, but our relatives. And none are more closely related to us than the great apes. These intelligent forest-dwellers that share more than 97% of DNA with us are now in a perilous state, all threatened with extinction because of primate traders, bushmeat hunting, deforestation, war, encroachment on forests, climate change and diseases like ebola. Conservationists are fighting back in various ways. I visited the Susa troop of gorillas, high in the Virunga Mountains of Rwanda, that Dian Fossey famously helped preserve, before she was murdered by poachers. There are fewer than 400 mountain gorillas left in the wild, marooned on a fragment of their former habitat now that agriculture has stolen the lowlands; in 2010, researchers announced they had underestimated the threats posed to mountain gorillas, and they could become extinct in less than ten years.[8] So each troop is now followed at all times by armed guards to protect them from poachers. Although Fossey was against any gorilla tourism, small groups of tourists are allowed to visit some gorilla troops for a maximum of one hour per day and the income generated is helping to conserve them. In 2012, Ugandan gorilla numbers had rebounded by 10%.[9]

In the forests of Indonesian Borneo, Dutch conservationist Willie Smits is trying a different approach to help save the habitat of the orang-utan. Decades of logging, burning for palm oil plantation (the fires there are an annual hazard), and poaching have reduced Borneo's rainforest to around one-third of its 1970 extent – more than half of the world's tropical timber comes from the island. Rather than follow the orang-utans around with guns, Smits is buying up degraded forest, aiming to restore and protect it (with the help of local people), so that the apes return there voluntarily. So far, he has bought up 2,000 hectares of burned degraded land and is slowly replanting it with more than 1,000 native species and introduced

From top: Fishing widows on the bank of Lake Turkana; one of the many cattle carcasses I passed during the drought in the far north of Kenya.

Clockwise from bottom left: Luna fixes the gutter on the fog net above his Lima shanty-suburb; experimental fog net designs; the ubiquitous jerrycan transports water across Africa.

From top: A 'teenage' bull elephant; a Hadzabe wearing a baboon-fur headdress; a lion cub with a trophy of the Anthropocene savannah.

Clockwise from top left: A Maasai with a mobile phone in Kenya; Mark Gardner on Santa Cruz, Galapagos Islands; a caiman cruises the Pantanal, Brazil.

Facing: Rosa Maria Ruiz in her Serere reserve with a friendly scarlet macaw.

posite page, clockwise from top: Bushmeat for sale in the heart of the Amazon; Rosa feeds a
y spider monkey she rescued from traders; mountain gorilla mother and baby in the Virunga
hlands of Rwanda.

ve from left: Miner at Cerro Rico in Bolivia with a bag of coca leaves; heaped piles of lithium
minerals at the *salar*, Bolivia.

Clockwise from top: the clustered housing of Rocinha favela; Villa Hermosa slum on the outskirts of Cartagena; the Zen Rainman in his water-harvesting home in Bangalore.

rehabilitated animals (including orang-utans), in an enterprise that profits the indigenous people. Surveillance systems around the forest protect it from loggers and poachers. Smits's success shows that restoring biodiversity can be a profitable enterprise for local people and that with their cooperation a forest can recover from even the worst deforestation. It's a thesis Rosa wholeheartedly agrees with – a dream she was tantalisingly close to achieving at Madidi.

The spider monkey above us reaches a plaintive arm down, begging for some of the banana we're eating, revealing its past as a dependent pet. Rosa ignores it. 'Humans have done such damage to the forest and its animals, even within my lifetime,' she says. 'Logging, slash-and-burn agriculture, mining and colonisers with commercial interests from outside all threaten this special place. New roads are being built here and oil exploration continues with so many petroleum concessions,' Rosa says. 'When I was a child, I spoke Tacana here – now, it's a forgotten language; everyone speaks Spanish. And people who used to chop trees and hunt in the forest only for their own needs now use machines and guns and sell the forest's resources as far away as China.'

Now, Rosa says, she wants to move on from the past. She cannot stop the logging and hunting of wild animals in Madidi, but she's having some success stopping it on her own smaller Serere reserve. President Morales is accelerating the land titling of indigenous groups in the area, so Rosa is shifting her focus to working with these communities to help provide them with sustainable livelihoods, hoping that ecotourism may help the Tacana people bordering her Serere land in the way she had hoped that Madidi might have done for many more people who, lacking alternatives, continue to hunt, log or mine for gold. 'My dream continues to be for a "Madidi Mosaic", a tapestry of protected areas, where the lands in this region are owned and looked after by the indigenous people who live here, and where they benefit from the natural resources here in a sustainable way,' Rosa tells me, as we head back to the lodge.

Madidi now faces perhaps its greatest threat: President Morales has reintroduced a shelved plan for a hydrodam on the Bala narrows at the Andean headwaters of the Beni River that would drown more than a million acres of rainforest in Madidi and the neighbouring Pilon Lajas Biosphere Reserve. It's a plan that Eco Bolivia successfully fought off a decade ago. Even if opponents to the dam are successful again, and that seems doubtful, there are still more than sixty other large hydrodams planned for the Brazilian Amazon alone, including the highly controversial $15 billion Belo Monte Dam – the world's third biggest – which is due to open in 2015 in Pará. It will displace more than 20,000 people, including indigenous tribes, and flood vast areas of forest to produce electricity mainly for industry and manufacturing.

As Bolivia's fearless environmental advocate wipes down the baby spider monkey, cleaning it up and feeding it again, Rosa reflects on how Serere was when she bought the land. 'We had to remove twenty tonnes of garbage from the forest, we reforested, planting more fruit trees to raise the animal-carrying capacity here. And in just a few years, we already have a healthy top predator – jaguar – population here. Conservation is hard work and requires dedication and vigilance. But it's worth it.' The baby monkey points up to a nearby tree as two brightly coloured scarlet macaws fly in. 'This is what is so important to protect,' Rosa says, gesturing to the noisy foliage. 'This is my home and our country's heritage.'

This small part of the mighty Amazon rainforest is in good hands, but what about the rest of the 5.5 million square kilometres? In the Anthropocene we have to find a workable way to meet our desire for minerals, timber and other forestry resources, as well as our need to preserve a standing forest and its biodiversity.

I head west, and upriver, into the Madre de Dios region of Peru, the front line in the war over how humans use this ancient forest. Manu National Park, a protected area for four decades, is home to dazzling biodiversity, including at least 1,800 different bird species (more than any other protected area on Earth), thirteen primate species, 400 ant species and various indigenous human tribes, including hundreds of 'uncontacted' people, who choose to live a hunter-gatherer existence without the cultural and disease-carrying influence of more recent forest colonisers. (In 1984, large numbers of the Yora tribe died of flu after contact with gold miners.) In the Anthropocene, the latest to exploit uncontacted people are tour guides who take tourists on gawping missions to photograph tribes including Manu's Mascho-Piro and the Indian Andaman Islanders.

'This is one of the very few parts of the Amazon that was protected before it had been impacted by hunting and logging – it's one of the last places that you can see mahogany trees standing tall. The area from the cloud forest down is pristine and it's still fantastic for birdwatching,' says British-born ornithologist Barry Walker of the Manu Wildlife Centre, who's lived in the area and studied its birds for more than twenty years. Barry takes me into Manu via the cloud forest of waterfalls, mosses, lichens and a plethora of orchids. It is home to a 'lek' (regular bird gathering) of Peru's national bird, the bizarre, charismatic cock-of-the-rock, or tunqui, a bright amber bird whose eyes seem to be planted halfway up its neck. They dance impressively for us – well, actually, for the females, as it's mating season – and call out with an ugly chicken-like noise. Barry, a large, bitter-drinking Mancunian with ruddy cheeks and an easy-going nature, shows infectious enthusiasm for even the dullest bird, and time passes swiftly in his company.

Before long, we have reached the heat, humidity and intense green foliage of the rainforest proper. We are in the Amazon Basin now, on the banks of the Madre de Dios River, far beyond the Andes. The area was protected in the 1970s by the son of Polish zoologist Jan Kalinowski. In the late 1800s, Jan escaped imprisonment in Russia (charged for the suspicious behaviour of creeping around the Russian wilderness like a spy, while recording animal behaviour) by acquiring for the Russian tsar the fantastic prize of a giant (dead and stuffed) Siberian polar bear. On achieving his freedom, Kalinowski fled to Peru, married a local woman from Cusco and sired eighteen children, one of whom created the Manu protected reserve.

The future does not look good for this incredible forest, its people and animals. Hunt Oil are exploring for oil and gas at the edge of the park, and it's a serious threat, Barry says. Roughly 70% of Peru's portion of the Amazon is given over to oil concessions and seventeen hydrodams are planned here. Thousands of people have flocked to the area from as far away as China, and there are regular clashes between indigenous forest-dwellers and the new fortune-seekers. A few weeks before my visit, a British missionary working here was expelled from the country for helping indigenous people to stand up to oil incursions.

Barry was drawn here by a different forest attraction. A childhood spent collecting eggs fuelled his great love of birds, and he was determined to see for himself the birds of the Amazon. He arrived in Peru for the first time in the late 1970s, where he was captivated by its abundant birdlife. 'Many Amazon species were completely unknown, it was so little-studied,' he says. 'The habit of early ornithology was simply to shoot a bird, take it back to the city and give it a name. So no one knew whether the bird lived on the ground or in trees, what its call was or mating behaviour. It was hugely exciting.' He joined celebrated Amazon ornithologist Ted Parker, of Louisiana University, in birding trips deep into the forest where they spent years recording birdsong on an analogue tape recorder, tracking its singer and taking notes of the animal for later identification in the Lima museum of stuffed birds. Together, they recorded hundreds of birdsongs and described unknown behaviours of more.

We head downriver from the upper Rio Madre de Dios to the greater Manu River in a canoe that requires heaving over the riverbed every now and again because the deepening drought has so reduced the water level. Along the way, we spy cormorants, colourful yellow 'oropendola' birds that weave hanging nests, turtles and herons. We overnight in a series of forest lodges and tents, and rise early on the last day to go and see a high bank of clay visited by hundreds of squawking macaws and parrots. It's an impressive sight: a blast of colour and sound, radiant with scarlet macaws. These monogamous birds arrive in pairs or in threes (son or daughter in tow) to the lick, where they hang on to the wall and seem to manage to eat and shout at the same time. It's an important social and mating spot for the birds – those chaperoned youngsters will likely find a partner for life at the clay wall, and this is breeding season. During their long bird-spying trips, Barry and his fellow researchers at Manu Wildlife Centre were the first to report other wildlife making use of clay licks, including tapirs and black spider monkeys. Now, researchers are discovering geophagy in more and more primates; indeed it may be that it is the rule rather than the exception to eat dirt. On the riverbank we surprise a group of russet Bolivian howler monkeys, lounging on the bank, eating the clay with audible scoffing. This is the peak season for clay-eating, Barry explains, because in midwinter (well into the dry season), tasty, easily digestible fruits and berries are hard to find. The animals are forced to eat more of the toxin-loaded leaves and fibrous fruits that are difficult to digest. Clay minerals seem to help that process. In birds, the grit may provide extra help in the gizzard; in mammals, it's likely that the minerals and alkaloids help make the plant nutrients more bioavailable. Tapirs, the indigenous South American horse, also travel miles to visit a clay lick, and we overnight in a hide to watch them approach it. Will these remarkable creatures still come if the area is invaded by oil companies?

The pleasurable days I've spent in this unique part of the Amazon rain-forest are unlikely ever to be repeated. The future of this intricate ecosystem with its incredible wildlife is exceptionally bleak, in the face of the ultimate threat: a road. The 'jungle highway', a 5,400-kilometre interoceanic asphalt road, linking the Atlantic to the Pacific, opened in 2011, and is fast changing the area. The highway, which was built largely to export Brazilian-grown soybean and other crops to Asian markets via Peru's ports, cuts through the heart of the rainforest at some of its most sensitive and pristine points, including two of the largest protected areas, and very close to here, in Madre de Dios. Like the road through the Serengeti, the Interoceanic Highway promises development opportunities for poor people, but threatens entire ecosystems. Road-building is a defining characteristic of the Anthropocene, and roads are planned across the world's rainforests. In the Brazilian Amazon alone, 7,500 kilometres of new paved roads are under construction. After indigenous Bolivians marched hundreds of kilometres from the rainforest to La Paz demanding an end to a planned road through their territory, President Morales cancelled the project in 2011. But the protest was followed by another, albeit smaller one, a year later by road proponents who see the new connection as essential to connect their poor remote villages to economic development and marketplaces, and whether the project will stay shelved remains uncertain.

Road creation is the primary driver of deforestation. Whether it is to provide access for mines and dams or to link towns and villages, a road enables loggers, animal poachers and traffickers, and small-time miners to enter virgin territory. Following in their wake are the farmers, who deforest for cropland, and drug growers and processors. Scientists have shown that 95% of all deforestation takes place within twenty-five kilometres of a road and, on average, every road carved into the Amazon – and there have been 50,000 kilometres built in the past three years – is followed by a fifty-metre-wide halo of deforestation.[10] It is because of this that they recommend establishing a railway or using the river networks to support large planned hydrodams and mines in the forest. The Camisea gas project at Las Malvinas in Peru, for example, in a sensitive area of the Amazon, has no road connections and operates like an industrial island where everything leaves or arrives by boat, air or underground pipeline, which hugely limits its impact on the forest. Other ways of mitigating the disturbance include narrower roads that retain tree canopy above, or rope bridges which reduce roadkill deaths by allowing animals to cross safely. Ground-dwelling creatures can benefit from underpasses – an elephant underpass built in 2011 under a busy road in Kenya has been used by hundreds of elephants and helped reunite separated herds, preventing conflict with farmers and road users.

In Madre de Dios, the impact of the new paved highway is achingly clear. There is an escalating boom in illegal gold mining in the area, with the arrival of tens of thousands of peasants – a fivefold increase since 2006 – fuelling endemic lawlessness and corruption. The lower Madre de Dios River (which flows into Bolivia, becoming the Beni River, and then into Brazil, where it joins the Amazon) is now a clamour of mining activity, growing slum camps with malaria, prostitution and inflated rates of HIV, and a notable absence of wildlife. Human noise is a disturbance even to plant growth, let alone other species, research shows, because it disrupts the behaviour of pollinators. Driving along the newly paved road, I pass vast deforested pits, where workers wade thigh-deep through pools of coffee-coloured mud, dragging enormous hosepipes (powered by continu-ally operated diesel generators) – one of these monstrous boas is used to flood the area and loosen the soil, the other to suck and filter the muddy material. Mercury is then added to the trapped sand and mixed to form lumps of an amalgam of mercury and gold. The mercury is then burned off, leaving the gold. More than 200 people a day are flocking to the area's goldmines, and there are now so many deforested cavities in this area, they are starting to join up into a vast mud-slicked scar. As each area is depleted of gold, the miners move on to the next. More than 50,000 tonnes of mercury is poured into the Madre de Dios River annually as impoverished twelve-hours-a-day workers process the earth for gold (the workers keep 25%). Locals now avoid eating fish for fear of being poisoned. Many ecolo-gists actually favour the use of cyanide over mercury because it can be neutralised after use.

Several trucks pass by, piled high with illegally logged hardwoods, heading from the jungle's interior to the Pacific ports, while, in the other direction, the near-ceaseless traffic imports mining equipment and workers to the forest. As we approach the town, what's left of the rainforest degrades into pasture, although as yet devoid of cattle. It takes us just a few hours to get from the pristine jungle to the frontier town of Puerto Maldonado, a dusty outpost peopled with staggering drunks and strip clubs, hardware stores, traffic and the man-made world. For such a remote outpost, it is, like Rurrenabaque, surprisingly cosmopolitan. The new road has brought some small development – there are traffic lights now – along with many new social and environmental problems. With climate change hitting Andean farmers particularly hard, tens of thousands of people are willing to endure grim conditions in the quest for crumbs of precious metal – and, for them, the forest is a necessary sacrifice.

And yet, in the warming Anthropocene humanity relies more than ever on the Amazon rainforest. The photosynthetic activity of so much greenery produces more than 20% of the world's oxygen and absorbs about 1.5 billion tonnes a year of humanity's carbon dioxide emissions, which is why the Amazon is often described as the lungs of the planet. In the face of global warming, this task is now so essential that polluting countries are paying rainforest nations to keep their trees standing and not deforest. Ecuador, home to one of the most biodiverse regions on Earth, the Yasuni Amazon, even proposed leaving a $7.2 billion oil find in the ground and the forest undisturbed, on condition that rich countries paid the country half its worth. However, in 2013, with little more than $300 million in the donations jar, the president announced he had no choice but to go ahead with drilling – 40% of Yasuni is already being exploited by oil companies. The decision could cost us far more than $3.6 billion.

This enormously valuable, large-scale photosynthesis is a fragile process that depends on the health and 'quality' of the forest – how many large trees there are, rather than grasses and vines – and also on the climate and water availability. Deforestation significantly reduces the amount of carbon that can be absorbed, and so does drought. The severe droughts of 2005 and 2010 actually reversed the Amazon's activity, so that it became a net producer of carbon dioxide, rather than a consumer.[11] As the trees died off, their photosynthetic activity ceased; meanwhile, decomposition and respiration continued, both of which use oxygen and release carbon dioxide. In 2005, the Amazon released as much as 1.6 billion tonnes of carbon dioxide. The combined emissions caused by the two droughts cancelled out the net amount of carbon absorbed by the forest over the previous ten years.

The more frequent and severe droughts in the region had been predicted by climatologists because of humanity's global warming, but the scale surprised them. Forests, particularly in tropical zones, produce their own microclimate with clouds and rain – organic aerosols released by plants and fungi act as nuclei for water droplets to form around. This self-perpetuating water cycle can be used to great effect, as I saw in the Peruvian desert, where communities are harvesting fog in nets to irrigate a forest that will eventually do the job for them. Forests have such a transformative effect on local climate, soils and ecology, that they are being planted around the world to combat desertification and restore fertility to agricultural land – with the added advantage of locking carbon away to mitigate global warming.

This transformative power of forests was used by nineteenth-century scientists in one of the most remarkable geoengineering experiments ever carried out. After defeating Napoleon Bonaparte at the Battle of Waterloo, the British navy stationed a detachment on Ascension, a tiny, remote volcanic island in the middle of the Atlantic Ocean, in order to keep an eye on the exiled emperor, who was imprisoned on the nearby island of St Helena. Ascension, however, was an almost entirely barren rock without fresh water or vegetation – what little rain fell on the island soon ran off or evaporated away. Fortunately, some of the greatest scientific minds happened to be sailing around the world at that time. In 1836, geologist and naturalist Charles Darwin stopped at Ascension on his *Beagle* voyage, and was followed a few years later by his friend, botanist Joseph Hooker. Back in London, Darwin and Hooker formulated an elaborate plan to create water on Ascension by making a forest. The trees would capture the cloud vapour and create a humid fog, soils would accumulate storing more water, and before long there would be a functioning ecosystem. Working with the botanical gardens at Kew, of which Hooker's father was director, they approached the Royal Navy with the plan and, from 1850, ships were regularly directed to bring trees and seedlings from Kew to the Atlantic rock. The plants sent were a motley selection originating in Africa, South America, Europe and Asia, as the botanists experimented enthusiastically. And it worked. Within a couple of decades, the volcanic peaks were a forest of eucalyptus, Norfolk Island pine, bamboo and banana. It was a designer forest fit for the Anthropocene. Darwin's cloud forest, called Green Mountain by the locals, captures moisture from the sea mist, forming a steamy oasis that continues to flourish. Streams of water were created by the forest, and the naval base was even able to tend vegetable gardens in the rich soils. The Ascension example shows how forests deliver microclimates that support ecosystems and sustain human populations by providing water, and could be scaled up and repeated in locations across the globe.

On a far greater scale, the world's rainforests perform a similar task for millions of people. The Amazon Basin is one of the wettest places on Earth. Rain falls about 200 days each year, and the river carries by far the largest volume of water of any river in the world. Some 28 billion gallons per minute flow into the sea, more than ten times the flow of the Mississippi, and equivalent to one-fifth of the total continental run-off of all rivers on Earth. The discharge is so great that it noticeably dilutes the salinity of the Atlantic's waters for more than a hundred miles offshore. However, the rainforest not only produces the rain, it also depends on it for survival, and if the climate shifts, this beautiful, self-sustaining cycle breaks down. Based on humanity's carbon dioxide emissions, the prediction by climatologists is that the lushest region on Earth could be mostly replaced by desert during this century: as trees die, the forest emits more carbon, warming the earth more, killing more trees, and so on. One group of researchers predicts a drop in broadleaf tree cover from about 80% of the Amazon region in 2000 to around 28% in 2100. 'When the forest fraction begins to drop (from about 2040 onwards) grasses initially expand to occupy some of the vacant lands,' the researchers say from their model. 'However, the relentless warming and drying make conditions unfavourable even for this plant type, and the Amazon ends as predominantly baresoil [desert] by 2100.'[12] Another group goes even further, predicting: 'By the end of the twenty-first century, the mean broadleaf tree coverage of Amazonia has reduced from over 80% to less than 10%.' And they conclude: 'In approximately half of this area, the trees have been replaced by grass leading to a savannah-like landscape. Elsewhere, even grasses cannot be supported and the conditions become essentially desert-like.'[13]

I try to imagine the humming, buzzing, dripping oasis of deep forest replaced by lifeless silent desert, and fail. To think in my lifetime that people will point to a dry scrubland and talk of there once being a vast rainforest here with boa constrictors, uncontacted tribes and monkeys that swung from tall trees, is not just incredibly sad, but astonishing. The pace and scale of change between the Holocene and Anthropocene is perhaps nowhere so stark as here.

With humans so dependent on the Amazon's function as a carbon dioxide sponge, and the mighty rainforest on the verge of flipping into desert in the Anthropocene, governments are taking baby steps to address the problem. Tropical deforestation accounts for up to 20% of the carbon dioxide emitted by humans each year and so nations have proposed a forest-protection mechanism known as REDD+ (Reducing Emissions from Deforestation and Forest Degradation) in which developed nations can meet their international emissions-reduction obligations in part by paying tropical countries to preserve their forests. Essentially, rich countries are paying to keep carbon in trees and out of the atmosphere.

First, though, tropical nations have to calculate how much carbon is actually in the forest, a notoriously difficult task that depends on the types of plants and their density. A project in the forest near Puerto Maldonado hopes to solve that. Greg Asner, from the Carnegie Institution for Science in Washington DC, has developed an automated system, which involves flying over the canopy at an altitude of some 2,000 metres, while firing two powerful lasers that blast 400,000 pulses per minute at the ground in sweeping movements. The Light Detection and Ranging system uses a sensor to continuously record the laser signals as they bounce off leaves, branches and other objects, while a spectrometer identifies individual plant species – and their health – through details like the amount of lignin or cellulose in the plant, and the concentration of chlorophyll, nutrients and water in leaves. From this, he can map the entire forest at resolutions down to ten centimetres, meaning he can see individual trees, and also undergrowth shrubs, hanging vines, ferns and bromeliads in tree crowns. The technique allows researchers to calculate the amount of carbon stored in the vegetation, which depends on its biomass. Old-growth hardwood forest, for example, contains much more biomass per hectare than those dominated by bamboo or cropland. The system then extrapolates to produce an estimate for the entire forest. In 2009, Asner mapped 4.3 million hectares of the Madre de Dios region. The satellite images reveal deforestation extending from the dusty town in a series of long lines that branch out from the roads towards the area I visited with Barry. They are clear-cuts, where forest has been chopped down for farming or mining. Asner's analysis tool also reveals countless spots of small-scale logging far into the pristine forest, as well as illegal gold-mining operations, cocaine farms and small-scale agriculture. Now he's scaling up, mapping the entire Peruvian Amazon and then neighbouring Colombia. The colourful three-dimensional images produced by the system can track how much carbon is locked up in a forest over time, and be used for REDD+ payments.

In the Anthropocene, the ancient rainforests are changing more rapidly than at any time previously, but for the first time, humanity's all-seeing eye can observe these changes and at the scale of an individual tree being chopped by one man in Manu. It means we have the ability to record our own destruction – and to stop it. Asner revealed that twice as much forest has been affected by human activities than previous estimates, increasing the greenhouse gas emissions from the Amazon by up to 25%.[14] For the individuals who carve away at the great tropical forests, their piecemeal destruction of a few acres for timber, crops or ranching are livelihood improvements that make negligible impact on the enormity and permanence of a forest that extends further than they will usually have travelled in their lives. But the accumulation of millions of such forest-eaters, in addition to the large-scale commercial enterprises, are steadily eroding vast areas that have been forested for millennia. Asner found that the accumulation of small-scale logging operations have a footprint that is twenty times larger than the more obvious wholesale deforestation.[15] Even so-called sustainable logging practices – such as selective logging, in which one or two lucrative hardwoods are chopped down, leaving the rest intact – are enormously detrimental to the forest, he found. On average, for every selectively logged tree, another thirty are severely damaged, increasing fire risk, worsening soils and impacting biodiversity. He calculates that selective logging doubles the area of Amazon already burned or clear-cut for cattle ranching, farming and other development, to around 30,000 square kilometres per year. Deforestation is often carried out by some of the poorest people on the planet – a hardwood tree like mahogany can go for hundreds of dollars, and soy or cattle can produce an income for decades. Schemes like REDD+ that put a value on trees often result in governments or agencies protecting forests from the people who depend on them, which can create resentment. Generally, schemes that foster community ownership over forest resources seem to be more successful. But however well managed such initiatives are, tropical forests cannot support the vast populations that live in forest cities, the globalised market in forest resources and the assault of climate change. As countries develop, more people want more resources from the forest. Those richer countries often start protecting and restoring their own forests but they increase the demand on forests else-where. More than half of timber shipped globally is destined for China, which is funding road-building deep into tropical forests and undermining international attempts at sustainable forestry.[16] The global rise in meat-eating alone is fuelling massive demand for Brazilian soybean for cattle feed – already 17% of the Amazon has been destroyed for soya, cattle or timber.[17]

Unless humanity rapidly substitutes new materials for wood, meat, palm oil, sugar cane and so on, tropical rainforests and the unique animals that live within them will disappear within a generation. Sustainable plantation forests are one alternative to rainforest logging, with fast-growing trees like pines and eucalyptus being favoured. A biotech company, FuteraGene, has created a eucalyptus variety that is genetically modified to grow 40% faster than normal types, reaching over twenty-seven metres tall in just five years. The company, owned by Suzano, a massive Brazilian timber consortium, plans to grow vast plantations of these GM trees for timber and biofuels, and is applying to the Brazilian government for permission to grow the trees commercially there from 2015. Such plantations might successfully redirect a proportion of the logging, if they are grown outside native forest rather than in place of it. It is clear that in the coming decades, we will have to grow more biological factories like these, but as wildlife habitats, monocultures of eucalyptus, oil palm, rubber and the like are green deserts and no match for the extraordinary biodiversity that exists in other ecosystems.

Meanwhile, researchers are trying to learn more about what we can expect the natural rainforests of the future Anthropocene to look like. I head further north to the Brazilian Amazon, near the grand jungle city of Manaus, where experimenters aim to discover whether the Amazon will reach a tipping point and change rapidly to desert, or gradually worsen, perhaps giving us time to act.

Around 25 million people now live in Manaus and the other booming Amazonian cities, fuelling logging, agricultural expansion, industrial pollution and other damaging impacts. The area around Manaus, for example, is the only habitat of the endangered pied tamarin, a small primate species that's rapidly running out of home. 'We know next to nothing about the Amazon,' says Adalberto Luis Val, the director of INPA (the National Institute for Amazonian Research) in Manaus. 'Two years ago, we visited a place not far from here and discovered a new species of monkey. And we are finding countless new species of insects and spiders,' he says. Taxonomists are in a race against the Anthropocene changes to find and name new species. Researchers believe the Amazon is accumulating an 'extinction debt', in which species populations are rapidly declining but it will take a few more generations until they disappear completely.[18] Even if they do not become extinct, they may change dramatically under the pressures of climate change, Adalberto says. The trouble is, our knowledge of what changes will happen is based on computer simulations, on the palaeo record and on the few recorded occurrences of climate change already affecting species (usually birds). We can't, in other words, see into the future.

However, Adalberto is about to bring the future to us. In the first experiment of its kind – called ADAPTA – he is taking a peek at the Amazon of the coming decades, taking the three different warming predictions as his guide. Adalberto is a fish biologist, so he's beginning the experiment with fish, including the oscar, a type of cichlid. Despite its dull appearance, the oscar is rather special. It is able to shut down its metabolism, going into a type of hibernation, without a drop in temperature. It needs to do this because at night, the 'black' river (dark with decaying vegetation) in which it lives becomes massively depleted in oxygen and acidic (around pH 3.5 because of the dissolved carbon dioxide – photosynthesis doesn't work at night). This is an unusual situation for a creature to find itself in. Many 'black river' fish breathe air at the surface, so don't face the problem. The only other complex animals that deal with oxygen deficiencies live at high altitude and adapt with an enlarged heart. But the oscar copes with almost no oxygen in an entirely different way. First, it shuts down aerobic metabolism, switching to anaerobic, just as we do. And it reorganises its blood flow to different tissues, dumping build-up of lactic acid in those that can tolerate it, such as white muscle. Most fish would die if they didn't flush out their lactic acid. And then, at some point, a biological trigger that the researchers are still looking for, causes the oscar to flip from anaerobic metabolism to a kind of hibernation for the night. 'It looks dead, it plays dead, it pretty much is dead until the next day,' says Maria Tereza Fernandez Piedade, who's another oscar researcher. ADAPTA will look at ten Amazonian river fish species, including the oscar, a host of aquatic insects, microorganisms and flora, which are to be contained in four identical microcosms, three under the raised carbon dioxide, temperature and humidity constraints of different global warming scenarios, and one as a control. The microcosms will be left for years, before results are analysed.

Adalberto allowed me to take a peek at these possible Amazons of the future through sealed double doors in a 'clean room' shiny with equipment. There is an advanced high-throughput sequencer to look at genetic alternations caused by the climate change, and a complicated computer-controlled air system all backed up by an independent generator. I ask him what he expects to happen. 'Some fish may die, others will probably over-express or down-regulate certain genes linked to temperature, carbon dioxide and pH adaption,' Adalberto says. 'We don't know for certain, and that's the whole point.' The last time there was a massive increase in atmospheric carbon dioxide, it was accompanied by an explosion in numbers of fish species. But that occurred over a far longer timescale than the current rise in carbon dioxide. 'The Amazonian fish are adapted to high temperatures, but are extremely sensitive to fluctuations in those temperatures,' he says.

The Amazon is much more than the sum of its parts and, although I'd never say it to Adalberto, fish might not be the most crucial element of the forest. They are an indicator, though. And this is just the beginning – the experiment is already being enhanced to include a bigger ecosystem, including mosquitoes (some of which will carry the malarial and dengue parasites), mammals and trees. A similar experiment under way in Australia is exposing eucalyptus trees to 40% higher carbon dioxide concentrations (550 ppm) and warmer temperatures to see how they will respond. Initially there appears to be an increase in photosynthesis, but this is soon offset by the temperature increase, which stunts their growth. So far, forests are mostly being harmed by climate change, either because of increasing fire frequency or drought, or because warmer temperatures allow proliferation of pests and disease. Californian redwoods, however, seem to be benefiting from climate change, partly because it's reducing coastal fog, which let's more sun through to their leaves.[19] In some places, forests are springing up where there were none before. The Arctic tundra, a region usually covered in permafrost, only supported grasses and small shrubs during the Holocene, but melting soils have enabled green forests to creep northwards. Palm trees have been predicted for Antarctica, after researchers discovered similar trees existed there 55 million years ago, during a period of high concentrations of atmospheric carbon dioxide.[20]

Overall, however, the predictions for forest cover during the Anthropocene are dire – just when we need more rather than fewer trees to soak up humanity's greenhouse gases. Indeed the situation is so critical that some scientists are saying that instead of planting more trees, we need to improve on the tree design altogether, to find a more efficient way of sucking up carbon dioxide from the air. Nature's method, photosynthesis, was developed some 3 billion years ago, when a type of bacteria first evolved the ability to directly eat the sun, using the energy to make sugars out of the carbon dioxide in the air. It proved a hugely useful adaptation and the bacteria multiplied swiftly and prolifically, filling the air with their poisonous waste product, oxygen. Those organisms that weren't killed by the oxygen soon evolved and profited from it, giving rise to the life on Earth today. Plants relied on the products produced by these bacteria until, at some point 2.6 billion years ago, a plant incorporated the bacterium into its cell, where it carried out the photosynthesis as a part of the plant. Almost all life – including human – relies on the photosynthetic ability of plants to absorb carbon, the essential building block of life, from the air, using the sun for fuel. And the Holocene climate our human civilisation evolved into depends on forests to modulate the levels of greenhouse gases in the atmosphere.

I travel to New Jersey to meet a man who believes he can better this process with a forest of artificial trees.

Klaus Lackner, director of the Lenfest Center for Sustainable Energy at Columbia University, is an enthusiastic German-born scientist, who, like James Lovelock, came to the field of Earth systems management after a stint at NASA. (Perhaps there is something about studying outer Space that makes people realise the preciousness of our home planet.) Klaus is looking at ways to modulate the global temperature by removing carbon dioxide directly from the air just like a leaf does. If it works, it would be one of the few ways of geoengineering the planet with multiple benefits, beyond simply cooling the atmosphere – oceans would become less acidic, for example.

In recent years there have been attempts to remove the carbon dioxide from its source in power plants using scrubber devices fitted to the chimneys. The carbon dioxide can then be cooled and pumped for storage in deep underground rock chambers, replacing the fluid in saline aquifers, for example. Another storage option is to use the gas to replace crude oil deposits, helping drilling companies to pump out oil from hard-to-reach places, in a process known as advanced oil recovery. Removing greenhouse gas pollution from power plants – called carbon capture and storage – is a useful way of preventing additional carbon dioxide from entering the atmosphere as we continue to burn fossil fuels. But what about the gas that is already out there? In a power-plant chimney, carbon dioxide is present at concentrations of as much as 12% of a relatively small amount of exhaust air. Removing the gas takes a lot of energy, so it is expensive, but possible. The problem with removing carbon dioxide from the atmosphere is that it is in such a low concentration – just 0.04% of all the gas by volume (or 400 ppm) – and so enormous amounts of air must be processed to remove enough to make a difference, and it would require long-term commitment.[21] As a result, most scientists baulk at the idea.

But Klaus has come up with a technique that could solve the problem. He has designed an artificial tree that soaks up carbon dioxide from the air using 'leaves' that are 1,000 times more efficient than natural leaves. 'We don't need to expose the leaves to sunlight for photosynthesis like a real tree does,' Lackner explains. 'So our leaves can be much more closely spaced and overlapped – even configured in a honeycomb formation to make them more efficient.'

The leaves look like sheets of papery plastic and are coated in a resin that contains sodium carbonate, which pulls carbon dioxide out of the air and stores it as a bicarbonate (baking soda) on the leaf. To remove the carbon dioxide, the leaves are rinsed in water vapour and can dry naturally in the wind, soaking up more carbon dioxide.

Klaus calculates that his tree can remove one tonne of carbon dioxide a day. Ten million of these trees could remove 3.6 billion tonnes of carbon dioxide a year – equivalent to about 10% of mankind's annual carbon dioxide emissions. 'Our total emissions could be removed with 100 million trees,' he says, 'whereas we would need 1,000 times that in real trees to have the same effect.' Not only are natural trees struggling to grow in the Anthropocene, planting so many would mean taking up valuable land that could be used to grow food. If the artificial trees were mass-produced they would each cost around the same as a car, he says, pointing out that 70 million cars are produced each year. And each would fit on a truck to be positioned at sites around the world. 'The great thing about the atmosphere is it's a good mixer, so carbon dioxide produced in an American city could be removed in the Arabian desert,' he says.

The carbon dioxide from the process could be cooled and stored, he says. Many scientists are concerned that even if we did remove all our carbon dioxide, there isn't enough space on Earth to store it securely in saline aquifers or oil wells. But geologists are coming up with alternatives. For example, peridotite, which is a mixture of serpentine and olivine rock, is a great absorber of carbon dioxide, sealing the gas as stable magnesium carbonate mineral. In Oman alone, there is a mountain that contains some 30,000 cubic kilometres of peridotite. Another option could be basalt rock cliffs which contain holes – solidified gas bubbles from the basalt's formation from volcanic lava flows millions of years ago. Pumping carbon dioxide into these ancient bubbles causes it to react to form stable limestone – calcium carbonate. These carbon dioxide absorption processes occur naturally, but on geological timescales. To speed up the reaction, scientists are experimenting with dissolving the gas in water first and then injecting it into the rocks under high pressures.

However, Klaus thinks the gas is too useful to petrify. His idea is to use the carbon dioxide to make liquid fuels for transport vehicles. Carbon dioxide can react with water to produce carbon monoxide and hydrogen – a combination known as syngas because it can be readily turned into hydrocarbon fuels such as methanol or diesel. The process requires an energy input, but this could be provided by renewable sources, such as wind energy, Klaus suggests.

We have the technology to suck carbon dioxide out of the air – and keep it out – but whether it is economically viable is a different question. Klaus says his trees would do the job for around $200 per tonne of removed carbon dioxide. At that price, it doesn't make financial sense even for oil companies who would pay in the region of $100 per tonne to use the gas in enhanced oil recovery. Ultimately, we have to decide whether the cost of the technology is socially worth the price – and this will continue to be the decider as we innovate more ways to carry out the tasks that the natural world performs for us for free.

Currently, photosynthesis, whether by rainforests, artificial trees or the algal forests of the oceans, represents the only way to return our carbon dioxide levels back to Holocene norms. Further into the Anthropocene, it will become easier and cheaper to remove carbon dioxide at source, and more necessary. We may well see enormous forests of artificial trees cleaning our air. But, as I leave Klaus's lab, my mind carries me back to the huge Amazonian trees that sing with insects and birdlife. Rosa's forest is so much more than an array of carbon-sucking trees – it is life itself. How fickle will our care for forests be, once we discover artificial replacements for these services?

9

ROCKS

*A*lmost all the known matter in the universe, 99.99%, is hydrogen and helium, which makes Earth's material abundance an extraordinary aberration. Even more extraordinary is how our beautiful aberration evolved its complexity, because when Earth formed 4.6 billion years ago, it was a churning mixture of only around seventy different minerals. As the planet cooled, the heavier of these – elements like iron – were drawn by gravity into the core.

Around 3.9 billion years ago, our spinning planet entered a violent era known as the Late Heavy Bombardment, during which time a barrage of asteroids smashed into Earth, pummelling gigantic craters in the surface. These asteroidal invasions brought valuable cargo – a treasure trove of precious metals, including gold, platinum and tungsten, which were embedded in the planet's mantle. Earth's continual volcanic upheavals and the abrasive wear of the oceans more than doubled the mineral count, as different elements were brought together and reacted under a range of pressures and *temperatures. Other planets in our solar system peaked at this point, but Earth went on to accrue thousands of different minerals. For that, we have Life to thank.*

The shifting and crashing of Earth's tectonic plates – the floating granite crust above the denser basalt – offered up the ingredients for life: the raw materials (including carbon, nitrogen, phosphorus and sulphur), the energy, and the continual cycling of these materials that helped ensure life's perpetuity. Life's marvellous trick was photosynthesis. The ground beneath our feet, this unique Earth, is a product of the dance between biology, chemistry and geology: life has created this rocky planet as surely as the sun did. The planet's internal engine provides the surface with a tiny 0.1 watts per square metre; however, four times as much energy is trapped in chemical form from sunlight by photosynthesis. This chemical energy helped to drive geological processes, which boosted life on Earth, in a virtuous cycle. Photosynthesis filled the atmosphere with oxygen, giving rise to thousands of new minerals, such as metal oxides. And the organic matter itself became geology – the newly abundant clays helped lubricate tectonics, the movement of vast granite plates, increasing continental movement. Clays accumulated into sedimentary layers and were cycled down into magma and then regurgitated as granite. The rocks of this planet have been almost entirely modified by life.

As living matter died, it became interred in the planet – the forests petrified into coal, the dead ocean matter liquidised into petroleum or became trapped as natural gas. On the surface of Earth, glaciers and rivers eroded the rocks to carve valleys and canyons. Volcanoes and earthquakes continued to issue minerals and gases from the unsettled belly of the Earth, changing the climate and creating or swallowing land mass in unpredictable fits.

The Holocene is a relatively stable geological time – the continents have moved less than a kilometre in 10,000 years – but humans have been a new geological agent. We have carved and collected rocks for tools and building materials, burrowed into the ground for metals and fuels. We have prized the useful metals, like iron and tin, but attached more value to the useless:

decorative inert metals like gold or crystals like diamond, whose worth has been the source of empires and of wars.

In the Anthropocene, our Earth-moving capabilities have become truly planetary in scale: we now move more rocky materials through mining and other extractions than do glaciers and rivers combined. The Anthropocene has been built by disembowelling the Earth. Everything we manufacture – and the energy we use to do so – is based on the minerals we mine.

The Industrial Revolution was fuelled by subterranean fossils, and the great social changes that followed created an escalating lust for materials gleaned from Earth's rocks. Since the mid-twentieth century, mass produc-tion, free-market capitalism, consumerism, the technology revolution and globalisation have all played a part in dramatically increasing the demand for mined minerals. In many cases, this demand is outstripping the Earth's supply and some metals are becoming increasingly inaccessible or rare. Many of the metals we use for twenty-first-century gadgets, such as smartphones, are the rarest on Earth, leading some to suggest we mine asteroids or even other planets for future supplies. In the Anthropocene, the scale of human-ity's plundering of Earth's mineral resources has never been greater, but we are now exploring new ways of producing energy and materials from abundant rather than rare resources, and new ways of living within the availability of our planet's rocks.

There is a small cluster of low, unheated stone hovels clinging to the moun-tainside and between them is the low dark entrance, stained black with llama blood, remnant of a sacrifice to the Tio (the Devil) made a couple of weeks ago. Miners are extremely superstitious creatures, the reality of their predicament making them desperate for hope of supernatural help. In the sunny outside, they are fervent Catholics; once in the underworld, it is the Tio who holds them in his hands.

It is here that most of the climate-change refugees from Bolivia's villages end up – making a pact with the Devil that few will survive. I've come to see what happens to these people once drought kills their crops and forces them from their country farms and homes – and it's not pretty.

At over 4,000 metres, dilapidated, poverty-stricken Potosí is one of the world's highest cities but it is overshadowed by the rainbow-coloured Cerro Rico (Rich Hill), which looms above the citizens – an imposing reminder of the cause of the city's splendour and horror. Potosí was founded in 1545 following the discovery of silver in the hill, the veins of which proved to be the world's most lucrative, bankrolling the Spanish Empire for more than two centuries. The extraordinary quantities of this precious metal – it is said that at one time you could have built a silver bridge from here to Spain and still have enough silver to carry across it – led to the city of Potosí becoming the richest, biggest and most populous in the Western world, with more than 200,000 inhabitants in the seventeenth century. The city had its own mint, whose mark – the letters PTSI superimposed on each other – is thought to be the origin of the dollar sign. In all, nearly 70,000 tonnes of silver were shipped from Cerro Rico during colonial times.

It came at a terrible price: 8 million people died working in the industry during the 350 years of Spanish occupation. The first to be 'used up' were the indigenous slaves, who expired in such numbers that the Spanish had to import tens of thousands of African slaves and, in 1572, invoked a military law requiring all slaves to work twelve-hour shifts, remaining underground for four months at a time. Average life expectancy of a miner was just six months.

By the 1800s, the silver was depleted and its global price diminished, sending the city into a decline that it is only recently recovering from, thanks to the demand for tin, lead and zinc. But Cerro Rico still draws silver miners. They get a worse deal than Faust, exchanging not just their souls but their bodies too. Miners here die within ten years of entering the mine; on average, before the age of 35. They die from silicosis, accidents and poisonings from the various noxious chemicals they are exposed to, including cyanide, mercury and carbon monoxide. For their efforts, those working in the sought-after cooperatives can earn an average of 1,500 bolivianos a week ($225); those working alone usually earn a great deal less. It all depends on the quantity and quality of the minerals they produce. At the time of writing (January 2014) silver is selling at $20 per ounce, but once the intermediaries have been paid, the miner gets a small percentage of this. If a miner manages to join a cooperative (he must pay $7,000 for the privilege), then 16% of his earnings (evenly shared by all members) goes into the cooperative. After tax, this money is used to ensure that miners have a pension and health care when they inevitably get silicosis or suffer an accident, and that their widows get a stipend.

Juan Mamani Choque worked the mines for three years, carrying 50 kg sacks of rock from the deep (sixty metres below) drilling depths to the next level up (twenty-five metres below) where the trolleys could move it. Juan weighed just 45 kg at the time, working in the same way as had his father, dead from silicosis. At the age of 25, Juan slipped and fell down a shaft, injuring his back. During his recuperation, his wife talked him out of returning to the mines, and he went back to school, eventually graduating from university as a language teacher.

Juan takes me to the miners' market, where for $2 I buy a stick of nitroglycerine, a stick of ammonium nitrate and a fuse long enough for a four-minute escape. We take a bus up to the Candelaria Bajo mine, one of the oldest of the 700 mines in Cerro Rico, dating back at least 350 years, and walk to the blood-splattered entrance.

Like the millions who have gone before me, I take a last look at the sun and then enter the underworld. It is immediately dark and dusty, the air infused with a peculiar smell – a caustic combination of the many chemicals here. I stumble along behind Juan, crouching under the low rocks, trying to forget that ten years ago geologists predicted that the mountain, riddled with tunnels and crevices, would collapse within eight to ten years. I bash my head frequently, alternatively grateful for my hard hat and cursing it for falling over my eyes and obscuring my view. Suddenly Juan shouts for me to get to the side, off the tracks. We struggle to mount a rock just as a series of steel trolleys comes speeding down the tracks towards us, pushed and pulled by ghostly men, wide-eyed from chewing coca leaves.

We continue on, heading deeper into the mine. The air becomes impossible to breathe even through my scarf, and it becomes harder to see with every step. Soon we are reduced to crawling on hands and knees through tunnels tight enough to panic in, and still we descend. Sliding down a rabbit hole – although no rabbit would live this far down – we reach a lower level. We are a couple of kilometres inside the mountain now and it is stiflingly hot, lung-searingly difficult to breathe and incredibly exhausting – and we are not even working. Further along still, we descend through a passage so tight that I must slide on my belly, and reach two men shovelling rocks on to trolleys. I offer them some of my water and, after two more trolleys have delivered their loads, one pauses for a few minutes to talk to me.

Damaso Condori is 40 years old and he's been here just five years, driven from his country village by worsening droughts to seek his fortune in Cerro Rico. He started to cough (the first sign of silicosis) a few months ago and he is worried for his family back in the village, he says. But for now he must keep working here because there is no other way to pay for his five children's education. At 4,200 metres above sea level, in temperatures above forty degrees, Damaso and his workmate are each shifting forty tonnes of rock in a (short) eight-hour shift. The air is so thin and the dust so thick that I have a pounding headache and streaming eyes simply from standing here. Damaso returns to his shovelling as another quartet of men arrive with their trolley-loads, and we turn to head back up. It's a long and painful journey, hauling ourselves up the shafts we came down, and I am humbled, remembering that Juan did this same journey many times a day carrying 50 kg on his back. What if your son said he was going to work in the mines, I ask him. 'I would put dynamite up his arse,' Juan says. 'No way.'

Precious silver was once used mainly in coins and jewellery. Later, it was discovered that when the metal was dissolved in nitric acid, a transparent, photosensitive crystal – silver nitrate – emerged, which proved key to the development of photography. It was so important that in the 1970s there were fears that silver would run out and it would mean the end of photography. Then digital photography was invented. Now, silver's industrial uses – which outstrip its decorative market – include electronics, catalysis and medical applications. Silver has the highest electrical (and thermal) conductivity of any metal, so it is used in a range of electronics. The medicinal properties of 'silver bullets' have been known at least since Hippocrates, the Ancient Greek 'father of medicine', and rely on the toxicity of silver and its compounds to bacteria and fungi, making silver compounds ideal for use in antiseptics and wound dressings. Silver is also used in heart valves and catheters, and researchers are now investigating its potential in the fight against cancer.

But silver is becoming scarcer. According to a 2010 US Geological Survey report, there are global known reserves of 400,000 tonnes of silver, and annual production is 21,000 tonnes, meaning reserves will last until 2029. In reality, though, as the price of silver rises, exploration becomes more cost-effective and new deposits are usually found – although it may not last beyond this century. One solution is to recycle the silver already in circulation. A tonne of ore usually contains less than three grams of silver; whereas a tonne of discarded mobile phones (6,000 handsets) can contain 3.5 kg of silver, so it makes more sense to 'mine' old handsets – electronic waste now contains precious-metal deposits that are on average forty to fifty times richer than ores mined in the ground. But as the metal becomes scarcer and its price goes up, replacements will have to be found for many of its uses. Silver could be replaced by aluminium, for example, or by complex metal oxides, which are good conductors. If the price of gold dips below silver, which may happen, it could replace silver in some applications.

Even with potential alternatives, while the price is high and there is still some silver left in the ground, desperate people will continue to burrow for it. And as minerals become scarcer, more resources (land, energy and water) are needed to extract them. Mining copper now it is scarcer requires on average ten times the ore it once did to collect the same volume of metal, for example. Demand has pushed up the price for a range of useful metals from gold to rare earth metals, and as extraction becomes more lucrative it is fuelling the destruction of rainforests and rivers, and financing violence and human-rights abuses across the poor world. And humanity is using ever more of everything – by 2050, global use of fossil fuels, minerals, ores and biomass is expected to treble, soaring to 140 billion tonnes a year.[1] The price of metals and other commodities has helped generate wealth in some countries, although whether this trickles down to the poorest depends on governance. Botswana, for example, has used its diamond mines – which are part government-owned – to fund education, health care and other public projects. The Democratic Republic of Congo, on the other hand, has not.

On the way up through the mine, Juan and I detour through another indescribably awful passage to visit the Tio – every mine has its own Tio. We find the ghoulish figure sitting with hideous features and an unfeasibly large penis, surrounded by what at first appears to be a pile of junk. I soon realise that the cigarettes, coca leaves, bottles of whiskey and rum, food and soft-porn playing cards are all gifts for Tio. The miners come here regularly and appease him with these offerings, hoping he'll be distracted long enough not to kill them in an accident.

After what feels like an entire day, but is in fact only an hour and a half, we emerge blinking into the sunlight. I gasp in the cold fresh air, grateful to be alive, to have cheated Tio. The dust in the mine has taken my voice, I will have to shower for twenty minutes to clean off the grime and my clothes stink of the underworld, but for me, it was a brief foray. I simply cannot imagine having to work in those conditions for twelve to fourteen hours, sometimes doing a double night-shift, and for years. Children work in these mines from the age of 9. There are no middle-aged men here; only widows shovelling rocks outside.

At Potosí's hospital for miners, I find a pitiful scene: men in their 30s, looking ancient, lie listlessly in beds with drips in their arms and enormous oxygen tanks behind them. This is the fallout of our love of silver, the result of climate change forcing villagers from their farms into the mines, the result of government policy that allows people to work and live in such conditions.

Simon Arcibia, bed five, was born in Potosí and worked in the mines from the age of 17. He's been in the clinic, which is paid for by his coopera-tive, for a month with silicosis and he tells me that he's getting better. Really, I say, delighted. 'Yes, usually people die, but I won't. I'll be okay. I have a son, you see.' I ask him if his son works in the mine. 'No he's studying. I want him to be a doctor. It's expensive, but the whole cooperative is club-bing together, and I am working more shifts.'

We chat a while and then, as I turn to leave, he says: 'I think the government should close the mines. They killed my father and now they are killing me.'

For tens of thousands of years humans have chipped away beneath the soils of this vast rocky planet, plundering the hard surface for tools, building materials and sparkling jewels. We're not the only species to recognise the value in strong, inert materials – chimps use stone tools and Neanderthals mined flint in Europe – but we're the only creatures to delve beyond stones to the metals and other minerals within. Whole civilisations have been based on our extractions – certain metals became so important that they defined cultural periods: the Iron Age and the Bronze Age, for example. Empires extended far across the seas in search of precious metals, gems and fuels. The Romans invaded Britain for its metals, using hydraulic mining to harvest gold from deep in the rocks. Spain's sudden improvement in fortune, 400 years ago, is largely attributable to silver scraped from these Potosí hills.

Some of this dug-up treasure – iron, for instance – was created billions of years ago inside a star; others, such as coal, were more recently formed from living forests that were buried in bogs and compressed by sediments some 300 million years ago. Some of these mined materials, like iron, are common in the Earth's crust, and others are rare, but none is replaceable or renewable within a time frame of thousands of years.

In the Stone Age, technologies were primitive and minimal, and were fashioned entirely from renewable materials. As cities developed and entire industries emerged to feed new civilisations, our impact on the planet grew. The Industrial Revolution saw a sharp increase in mining. Steam power, fuelled by coal, rapidly increased our capacity to delve for materials (including coal), transport and process it. In 1700, more than 80% of the world's coal was mined in Britain, the nation that drove this movement in the extraction, fabrication and distribution of new materials across the globe. Over the eighteenth and nineteenth centuries, as technology improved, speculators explored new territory across Africa, Australia and the United States, looking for gold and other high-value metals and gems. But it is the past sixty years that has seen a truly global revolution in the scale of our mining and extractions. Nothing compares to the post-war Great Acceleration, a rapid increase in human activity driven by population expansion, globalisation, technological and communications improvement, improved farming methods and medical advances. The hike in use of mined materials during the Great Acceleration can be seen in everything from carbon dioxide release, to water use, to the number of cars, to ozone depletion, to deforestation, to GDP, to consumption. As our insatiable lust for new materials grows, we are churning up more and more of our planet's structure and spreading it around the surface.

Humanity is now the biggest terrestrial shifter on the planet – humans move more than three times as much sediment and rocky materials as all the world's rivers, glaciers, wind and rain combined. Every year, rivers move about 13 billion tonnes of sediment.[2] To put that in perspective, 1 billion tonnes is equivalent to two Great Walls of China. Humans mine around 8 billion tonnes of coal each year – by 2030, it will be 13 billion tonnes of coal. And we produce 13 billion tonnes of aggregate, 2 billion tonnes of iron ore each year, and billions of tonnes of other materials.[3] There are 568,000 abandoned mines in the US alone, and many millions more throughout the world. We are sculpting the Earth's surface and taking material away at rates not seen before in geological time. There are millions of holes and burrows through the rock – kilometres long – from our drilling expeditions. Throughout the world – even in some of the most inhospitable places such as the freezing Arctic Ocean, the burning Atacama Desert and the deepest Atlantic – humans are plunging far below the planet's surface to reach oil and gas, diamonds and copper, uranium and rare metals that were unknown decades earlier but are now an indispensable part of the conveniences of modern life. New finds promise untold wealth to the territories they lie beneath – and to the speculating companies that dig and drill – so nations from Uganda to Brazil are eagerly pursuing all explorations. The world has watched some of the poorest nations dramatically rise to prosperity after substantial finds – Saudi Arabia, with its vast oil and gas wells, is just one example.

However, such wealth has not come without its problems. Mining is a dirty, dangerous business. The world watched in horror as thirty-three miners were trapped for months below ground in Chile in 2010. They were rescued, but a few months later, twenty-nine mine workers in New Zealand were less fortunate. These are just a couple of incidents that occurred recently in some of the world's best-run mines operated by the biggest companies. However, each year, thousands of small-time miners, many operating in illegal shafts under terrible conditions, perish without publicity

– in China alone, around fifty miners die each week.[4] Mines are often plunged into land owned by poor local people with few rights, who have received little or no profit or compensation for their polluted fields, air, waterways and wildlands. Nigeria's Niger Delta is a ruined slick of oil spills and leaks, which continues to poison the health of some of the world's poorest people, ruins their drinking water and farmland, and kills fish. The ongoing environmental catastrophe there shows no signs of abating while one of the world's richest oil companies pollutes with near-impunity and the poor continue to illegally tap the pipes, spilling and spreading the noxious liquid. Across the globe, from the Amazon to the Siberia, oil formed millions of years ago from the burial of algae and plankton has been spilt and spread around the terrestrial surface with ecologically devastating consequences.

Currently, the world uses over 85 million barrels of petroleum every day, and as its price rises companies are becoming ever more ambitious in their extractions, even trying to harvest crude oil from tar sands. Although tar sands can hold large volumes of bituminous oil, extracting it is a filthy, energy-intensive process that leaves the landscape devastated in its wake. Nevertheless, for countries such as Canada, the allure of domestically available supplies of such an essential commodity is strong and could well outweigh the substantial environmental and public-health concerns. Mining involves huge amounts of water and energy, often in places where both are scarce. Because of this, mines are often accompanied by on-site power stations, hydrodams, processing facilities or refineries, highways and other infrastructure, and waste facilities from large treatment ponds to noxious piles of sludge. The methods used to separate out desired metals from the rest of the rock are also often poisonous – whole river systems have been contaminated with mercury from gold panning, for example. The actual extraction of the materials is a small part of far bigger environmental change, often in fragile areas, such as rainforests or the Arctic.

We have carved the tops from mountains, created new mountains of slag and rock, even made islands where none existed before. The palm-shaped island off the coast of Dubai was formed from some 400 million tonnes of geological material moved from one place to another – the kind of terrestrial shift that would only otherwise result from a volcanic eruption or tectonic jolt. In the Anthropocene, humanity is incrementally reshaping the planet with earth-moving machinery. The rock beneath our feet has been drilled and plundered, and injected with waste material such as toxic metals and sewage, which can leak out and contaminate groundwater supplies.

The Industrial Revolutionaries of the nineteenth century left a legacy of pollution in Britain: there are parts of the Peak District where the peat is more acidic than lemon juice because of acid rain from factory chimneys; rivers still contaminated with the toxic tailings of lead and tin mining; and coastal zones ravaged by colliery dumps. As places get richer, however, they begin outsourcing the dirty, polluting parts of making stuff to poorer countries. Most of China's pollution – filthy emissions, discoloured rivers, poisoned soils and ravaged wildlands – is associated with the recent fabrication of goods for export to Britain and other places.

My foray into the mine at Potosí leaves me with a practical problem – my camera has been affected by the dust and grime and the shutter no longer works. In the capital, La Paz, I seek out a camera repairer. Come back in a couple of hours, he tells me, and I feel grateful that the problem arose in the developing world. So many times during my journey, there's been something that's needed fixing, from rips in my backpack to memory cards that have lost data. From India to Ethiopia, I have had no trouble finding someone who can repair what is broken or find an ingenious way of side-stepping the issue. In rich countries, such items would be thrown away and replaced with new, but the developing world is full of menders, make-doers, inspired users of others' scrap. I've seen a bicycle in Nairobi made from bits of car, a colander and a leather belt; aerials made from all manner of implements; and houses constructed out of boat sails, rice sacks and plastic drinks bottles. However, on returning to the camera workshop, I am in for a shock. This sort of camera cannot be repaired, the man tells me. It's 'too new'. Thirty years ago, just a handful of camera models would have been around each year, service and repair manuals and spare parts would have been available for each one, as well as a thriving repair industry. But things have changed. Like the majority of consumer electronics, my camera has not been designed to be easily reparable, and the manufacturer no longer releases repair manuals.

My camera doesn't come with parts designed to be replaced – the rechargeable battery can be changed when it wears out, but at a similar expense to buying a new camera. In some gadgets the batteries are soldered to the internal circuitry, forcing owners to replace the entire product when the battery dies. As gadgets become progressively cheaper, consumers become ever more willing to 'upgrade'. Since the mobile-phone handset market reached saturation in the UK and US in the early 2000s, phone companies compete for business by automatically upgrading customers' phones every year, replacing functioning phones simply for fashion reasons or for minor technological additions that could easily be achieved through software upgrades to existing handsets.

Around 50 million tonnes of electronic waste, including mobile phones, is now generated annually, containing engineered parts assembled from hundreds of different materials.[5] The international trade in e-waste is highly lucrative but much of it is currently recycled by people, including children, working in conditions where they are exposed to neurotoxic heavy metals, dioxins and plastics that are burned to get at the precious metals within. At its best, used electronic equipment is cleansed of data, refurbished and repaired. At its worst, it is shredded and disposed of, wasting the mined materials and other resources that went into manufacturing it, and causing replacements to be made. Manufacturing electronic equipment now uses

320 tonnes of gold and 7,500 tonnes of silver annually, and yet less than 15% of this hoard is recovered.[6]

The idea that something that works fine should be replaced is now so ingrained in our culture that few question it. But it is a fairly recent concept, brought about by a revolution in the advertising and manufacturing industries, which thrived on twentieth-century changes, including urbanisation, mass production, public broadcast media and globalisation, which has turned the entire world into a factory and also a marketplace for the stuff humanity produces. The clothes, steel, gadgets, toys, computers, mobile phones and cameras that are now so cheap that they can be tossed away and replaced at the smallest fault, are made in low-cost factories where the health and safety of workers is often neglected, where toxic by-products and waste materials are flushed into rivers or left to pollute the soil. Some 20% of China's arable soil is now polluted with heavy metals.

The aim of most manufacturing companies is to optimise the path and frequency from mined materials to product to customer. The earliest example of manufacturers manipulating people into frequently replacing a product may be the so-called 'light-bulb conspiracy', in which a group of companies set up a secret cartel to prevent companies from selling light bulbs with a life span longer than 1,000 hours, even though 100,000-hour bulbs could be produced.[7] Manufacturers of nylon stockings and tights also famously instructed their engineers – who boasted that their nylons were strong enough to pull a car – to make less strong fibres that would fail or ladder. In that way, women would need to replace them more frequently.

This way of selling more products by designing in a set lifespan is known as planned obsolescence, and many politicians and economists believe it to be an economic necessity. The idea was born in the US during the 1930s depression as a way to get the economy moving again and provide employment. By the 1950s, a sophisticated advertising industry and easy credit were persuading people to shop till they dropped. Consumerism was born. The fashion industry is predicated on planned obsolescence, and other industries are following this high-turnover model and bringing out products that have cosmetic gimmicks or seasonal appeal but which will soon appear dated. Meanwhile, many products are designed to fail or be difficult to repair or upgrade. Television sets have heat-sensitive condensers deliberately fitted on to the circuit board next to a hot transistor. Printers contain chips that are programmed to stop working after a preset number of pages.

Fresh-faced, energetic and innovative, Californian computer wizard Kyle Wiens is the sophisticated electronics tinkerer that the industry emperors like Microsoft and Apple dream of employing. But Kyle is a guerrilla geek – he slipped through their net and crossed over to the consumer side. Instead of dreaming up intricate ways of fitting more and more components into ever-slimmer gadgets, Kyle is taking them apart. It started in 2003 when he was 18 and studying engineering at California Polytechnic State University in San Francisco. 'My iBook stopped working. It was out of warranty and I couldn't find a service manual for it, so as my friend [fellow student Luke Soules] and I took it apart in our dorm room to fix it, we created our own manual.' They then printed copies of their iBook manual to distribute among fellow students.

Theirs was not the only device without a service manual – Kyle reels off a list of global manufacturers that make everything from washing machines to mobile phones. In fact, he says, it's extremely unusual for products to have service manuals nowadays. The turning point was when VCRs came out in the 1980s, Kyle says, and his eyes flick skywards as he thinks back. I realise, with a shock, that Kyle wasn't born until 1985 – he's talking about a make-do-and-mend culture that he has never known. Until then, the majority of electricals were made in Europe or America, near to the customers and service technicians who repaired faulty goods with product manuals and parts supplied by the manufacturers. When

VCRs came out, most were manufactured in Japan and elsewhere in Asia. 'As the product manufacture went to Asia, the companies stopped producing service manuals and thousands of repair-shop technicians went out of business. When their product broke, customers threw it out and bought the latest model.' And people who bought reconditioned second-hand computers or TVs would also have to buy new.

As a result, most of the big-name companies from Apple to Toshiba now fiercely guard their service manuals, and use copyright law to prevent people from making them public online. 'The reason they don't issue the manuals is to stifle the resale market, kill the repair business and to strengthen the planned obsolescence of their products,' Kyle says. And it also enables companies to significantly mark up different models. For example, the difference between a 16 GB iPad and a 64 GB one is a chip that costs an extra $7, he says, whereas the price difference for consumers is $200. If consumers could replace the chip, or change the batteries – which are not replaceable and designed to fail in twelve to twenty-four months – then the companies would lose a lucrative high-end market.

Kyle and Luke started taking apart computers and other equipment, getting around the copyright protection by creating their own service manuals from scratch, and posting them online for free. The response was immediate and beyond all their expectations. 'In the first weekend, we got over 10,000 hits,' Kyle says. The pair called their site iFixit.com and over the past decade, they have expanded to hundreds more manuals. They encourage home hackers to create their own manuals for the site and they also sell the tools and spare parts users need to fix their own electricals. 'We're having an ecological impact through reducing waste and landfill, and conserving the mined resources – it takes around 3,000 lb of raw material to make a 5lb laptop,' Kyle says. And they are also contributing to a cultural shift in perception, he says. 'When someone has repaired their broken device, it's a life-changing experience. That person will become a different consumer, they will choose new products based on how long-lasting they are, and how easy they are to take apart and repair.'

The market may force their hand anyway. While over the twentieth century there was a steady fall in the price of commodities, since 2000 that price has been climbing. 'We're moving towards a resource-limited economy, which means manufacturers are going to have to head towards remanu-facturing and repairs as the cost of rare earth metals and other resources slash profits,' Kyle says. Rare earth metals are so called because these seventeen elements occur in such tiny amounts in deposits, and in combin-ations with other metals including radioactive uranium and thorium, that they are difficult, expensive and resource-intensive to extract and purify. More than 95% of the world's supply of rare earths is produced by China, but global demand for the electronics essentials has risen so sharply over the past few years that in 2012 China began restricting export, sparking protests from other nations. The shortage is also affecting the renewable energy sector, because everything from wind turbines to efficient fluorescent light bulbs to solar panels use rare earth metals. (And it's not just exotic metals that are becoming scarcer and worth more – lead, copper and others are now in such demand that thieves are stealing church roofs, manhole covers and even railway cables.)

Take indium, for example. Television screens, computer monitors, photovoltaic cells in solar panels, and phone touchscreens are all laced with indium oxide, because it is transparent and it also conducts electricity. It is also one of the rarest metals on Earth. As a result, its price has rocketed in recent years – it went from $60 per kilo in 2003 to $1,000 per kilo in just three years – giving rise to a whole new indium smuggling industry, primarily out of China. Analysts are warning that global supplies of indium could run out as soon as 2017. Such a prospect is not as alarming as running out of food or water, but over the past few decades digital displays have become so enmeshed in our lives that they are integral to our social inter-actions and livelihoods, from rural East Africa to the offices of Wall Street.

Turkana fisherwomen trading their wares via SMS to clients based hundreds of kilometres away is an opportunity that depends on indium.

Indium was discovered 150 years ago and is harvested as a by-product of zinc mining – it exists almost entirely in trace amounts (sometimes as little as 1 ppm) inside deposits of other ores such as zinc and lead. Because it's not mined in its own right, it means that greater demand for it won't necessarily result in a greater output of the metal. Most of the indium on Earth is just single atoms stuck inside rock that can never be utilised.

One solution is to increase recycling efforts – because of its value, the indium recycling market is already bigger than primary production. The problem is that the metal – commonly manufactured in its useful oxide form, as indium tin oxide (ITO) – is used in such tiny amounts that recovering it from electronic products is expensive. A single screen typically contains less than 0.5 g of ITO.

With no let-up on demand for displays – there were more than 1.5 billion mobile-phone handsets sold in 2011 alone – scientists are working on alternatives to indium. Finding an optically transparent conductor that's as efficient and light as ITO is challenging but there are some candidates. So-called non-stochiometric tin oxides, which use the far more abundant aluminium, are one option that could be incorporated fairly easily into current manufacturing set-ups. They don't perform as well as ITO and they have other drawbacks, including the fact that tin is itself running out, with reserves estimated to last another twenty to forty years.

In the Anthropocene, designers need to build sustainability into their products right from the start rather than substituting materials at a later stage as they run out. This means going for resource-safe technologies, basing our innovations on readily available materials that are economically and environmentally benign. Carbon nanotubes, which are so tiny as to be transparent, are also excellent conductors, and offer perhaps the most sustainable solution to the indium problem. Carbon is one of the most abundant elements on Earth, and sheets of carbon nanotubes, called graphene, have several performance advantages over ITO.

The invention of 3D printing, which will enable people to manufacture things as and when they need them, may spell an end to unnecessary products – such as novelty items – being made and marketed, and an end to excess packaging. Or, it may lead to an increase in the production of poorly made, one-use disposable items. Currently, only a limited range of plastic items can be made in this way, but as more materials and techniques become available, it would be wise to design concurrent recycling methods. A more interesting revolution will be one in which materials and products have biologically mimicking life-change response designed into them. For example, when we use a muscle repeatedly, it gets stronger rather than wearing out, just as when sunlight hits a tree from a specific direction, the plant responds by preferentially growing in that direction to optimise its photosynthesis. Materials scientists are looking at designing structural materials that will evolve in service. Aircraft wings, for example, which currently develop metal fatigue because of constant vibration, would instead grow stronger, remaining safe for much longer. Structural materials could be multifunctional, also harvesting energy; ships' hulls could be anti-microbiological, eliminating fouling and improving transport efficiency.

Humanity currently uses 30% more natural resources a year than the planet can replenish.[8] The idea that scarcity of natural resources poses limits to growth seems logical – indeed, the end of unsustainable capitalism has been predicted for centuries now – except that every time humanity is said to be on the brink of collapse, we find a way around the problem. The Green Revolution that tripled agricultural yields in the 1970s saw off one predicted crash. There was a huge fear in the nineteenth century that New York City and London would drown in their transport pollution (horse manure), or that they would run out of transport fuel (horse feed). The invention of the motor car put paid to those concerns within a decade, but gave rise to other worries, like rubber and oil shortages. The new explosion in fracking, whereby engineers tap previously inaccessible horizontal drill sites deep underground, has increased global supplies of natural gas from decades to centuries.

Such game-changing discoveries are hard if not impossible to predict. So should we rely on them to get us out of other resource shortages? Like every other life form on this planet, humans are limited to whatever is contained on Earth and its atmosphere, with only solar energy being limitless. Humans are masters of conversion and adaptation. We will always find a solution to mineral shortages, but the problem is that the price for extracting scarce materials is usually paid for in environmental terms, whether it is the huge energy and water costs – and pollution – involved, or the destruction of precious landscapes. Humanity now uses roughly half of global natural production. Every other species that achieves this level of 'success', from bacteria to rats, reaches a tipping point at which its population size outstrips its ability to feed itself. Humanity is reliant on food, but also on external energy and other materials. The crash could be far more severe for us.

Arguably, the simplest way to make resources go further is to improve efficiency, which means slashing energy consumption, cutting waste and shifting to a low-carbon lifestyle. The savings can be immense from practices as simple as insulating buildings effectively so that human body heat alone is sufficient to maintain temperature. The International Energy Agency forecasts that by 2035, US oil consumption, which peaked in 2005, could decline to 1960s levels with strong efficiency measures, and is expected to be as much as a third less than in 2011.[9] But, as with most other resources, China's and India's shares will dominate the global agenda. In China, oil consumption is expected to be up as much as two-thirds from the 2011 level, and India's is predicted to more than double. Japan has already shown the world that efficiency is possible. In 2011, despite a very hot sticky summer, during which forty-three out of fifty-four nuclear power stations were out of use, Japan did not suffer national blackouts. Instead, the nation pulled together to ramp up energy efficiency, practising *setsuden* (energy saving), based on a strong sense of social responsibility – and all in a matter of weeks. Efficiency needs to go further than energy – all resources need to be used more judiciously, and the lifetime of materials (and of products) needs to be prolonged far further. Researching her film *The Story of Stuff*, Annie Leonard discovered that of the materials flowing through the consumer economy, only 1% remain in use six months after sale.

In nature, resources are cycled – for example, water taken up by plants is eaten by animals which breathe out carbon dioxide, returning the gas to the atmosphere for the plants. Nothing is lost. But the scale with which we humans are using resources is too rapid for natural systems to manage. Microorganisms, for example, can degrade our waste, but not at a rate that matches our production. Likewise, trees and plants can take in our industrially produced carbon dioxide, but not fast enough. The answer to our resource predicament lies in a synthetic version of this natural cycling, so-called closed-loop or cradle-to-cradle or industrial symbiosis manufacture, in which products are created using renewable energy, all the materials used in the process are reused and the final product is also fully recycled to minimise original resource use. It means no waste, no pollution, and far greater efficiencies. The US carpet manufacturer Interface, for example, which is the world's biggest producer of floor coverings, is aiming for zero carbon emissions through closed-loop manufacture. The company's founder Ray Anderson has promised to eliminate any negative impacts they have on the environment by 2020. In a circular economy – in contrast to the linear manufacturing route: mining materials, fabricating, selling, throwing them away – products would be more easily disassembled, so that the resources can be recovered and used to make new products, keeping them in circulation. A report by McKinsey into the idea found that the benefits to Europe's economy alone could be $630 billion, based on cycling just 15% of materials in

48% of manufacturing and just one cycle. The town of Kalundborg in Denmark has been practising this closed-loop resource use since the 1960s – almost everything there is recycled and reused in what the townspeople describe as an 'industrial ecosystem', saving 272,000 tonnes of carbon dioxide a year.

However, if we are to restrict humanity's resource use to what is easily available in Earth's rocks, we have to tackle the demand for new stuff as well as its production. Although there are societies around the world that continue to live a Stone Age existence, including the Hadzabe bushmen of Tanzania, most people are following the the Great Acceleration trajectory of development set by industrialised nations. Peasants scraping for subsistence in the dirt of India do not toil under romanticised notions. They too would like to experience the same acceleration in life expectancy, clean water provision, toilets, electricity, Internet access, a car, air conditioning, a refrigerator, washing machine . . . But, as global citizens sharing this planet, we cannot afford for billions more people to eat and live as the average North American now does at the expense of the environment. Climate change, ecosystem services degradation and limitations on natural resources threaten the lives of millions.

The scale of our globalised buying, selling and moving of stuff has never been more clearly brought home to me than on the banks of the Panama Canal, as I watched an awesome ship glide magnificently into one of the locks. Onboard were thousands of containers, the humble steel box, invented six decades ago, that revolutionised global trade and made stuff cheap. The efficiency of container ships for transportation of goods, loading and unloading, vastly lowered the cost of international trade – transport cost is now less than 1% of the price of most goods – meaning that it became cheaper to manufacture goods where labour costs were low. All around, enormous digging machinery was busy widening and deepening the canal to accept even larger ships, capable of transporting 13,000 containers – one such megaship is built every three days to carry cheap consumer goods from factory to purchaser, with six of the world's ten biggest ports now in China. As China gets richer, it will push the dirty industry elsewhere – this is already happening internally, with the wealthier cities, including Shanghai and Beijing, forcing polluting industries west into the countryside or to smaller cities. Following a spate of unwelcome publicity, some of the larger Western-owned companies are beginning to ensure better factory conditions for workers in China, and improving environmental performance. Local people are taking unprecedented action against polluting industry, in some cases forcing them out of towns and cities, and the Chinese government is starting to take notice. China may well be the first country to clean up its industry while still developing.

One of the main challenges in the Anthropocene, it's worth repeating, is that many poor countries want to develop their economies with however much pollution that entails, and then clean up when they are rich – after all, that's what rich countries did, the argument goes. And there's a chance that – with the increased pace of development and superior environmental clean-up techniques – we could ride out the next few decades without devastating consequences. The odds don't look good, though: rich cities still suffer from smog, much of our biodiversity has been irrevocably lost, and we're still pumping out greenhouse gases. Some point out that there is more than enough environmental space on this planet for everyone to live sustainably, but the developed world needs to scale back its consumption and allow for a more equitable sharing of resources.

As I lug my backpack from one hostel to another, taking buses and trains and occasionally humping it uphill on foot, it becomes leaner and lighter. Out of necessity, nagged by an aching back, I whittle down my few possessions into fewer and fewer essentials. On a journey of more than two years, it's surprising how little I actually need. Later, when I return to London and recover my boxes out of storage, I'm astonished and a little bewildered by how much stuff I have. When will I wear so many clothes;

read so many books; what am I going to do with all these collected knick-knacks?

There are encouraging signs that our runaway gorging on new stuff is beginning to wane, in the West at least – that we've reached 'peak stuff', that people are eating less environmentally costly meat and that Americans are copying their European counterparts in adopting more fuel-efficient cars and driving less. Environment analyst Chris Goodall calculates that the UK reached peak consumption (of everything from food to fuel to furniture) in 2001 and has declined since. Because the economy rose significantly during that time, Goodall believes that Britain may have achieved the 'green' growth dream of economists, by decoupling resource consumption from economic growth. The growth of Internet sites such as Freecycle and eBay, for example, enables people to keep products circulating through more than one owner – and out of landfill. Nevertheless, these are small steps in economies still based around worshipping consumerism. Citizens are encouraged to buy stuff from childhood onwards, very little of which they actually need, and consumerism is the main economic strategy for governments trying to climb out of recession. Meanwhile, the consumption epidemic has spread throughout the world, and is particularly obvious in Communist China. Shanghai, whose per capita energy use is already higher than London's, is a city of shopping malls in which the nouveau riche demonstrate their success and status by buying lots and lots of unnecessary stuff. We will reach a global 'peak stuff' eventually, in part because the rate of population growth is declining, so there will be fewer new consumers born to outweigh those who are dying. But this may not come soon enough to avert environmental crises.

One way of prompting people to be more thoughtful about their purchases would be to levy an environmental services fee. The 'polluter pays' principle already covers dirty factories in many countries, for example, who must pay to clean up their damage to soil, water or air, and to safely dispose of toxic waste. If this was extended to include carbon dioxide pollution, fresh water and energy use by manufacturers, the price of a product would more accurately reflect its environmental cost. The charge would encourage more sustainable production and also help pay for remedial efforts, including carbon capture and storage, waste-water clean-up and recycling.

Socially educating people not to seek happiness through excessive consumption of stuff is a huge and probably doomed undertaking, however many times it is shown that happiness does not that way lie. There is now a deeply rooted cultural belief that shopping is necessary and a patriotic undertaking for national economic prosperity, and that buying more things makes a person more desirable and improves their status. Entire industries exist to perpetuate this philosophy, even though most people know, deep down, that the majority of the stuff they buy is unnecessary and wasteful of money and resources. Purveyors of green lifestyles and sustainable fashion products have had some little success at convincing small sectors of society to buy products that are less environmentally damaging. Some consumer-targeted campaigns that have focused on ethical issues have been very successful at persuading people not to buy fur or to buy fish caught in a way that doesn't harm dolphins. Studies have shown, however, that people who take small steps to act sustainably by using a carpooling scheme, for example, then 'balance' their behaviour by being less likely to recycle.[10]

In general, driving consumer change through economic incentives or legal restrictions works better than fighting entrenched cultural practices. Feed-in tariffs – which allow people to sell their home-made energy to the grid at a higher rate than the market norm, providing grants and cheap payback schemes for rooftop solar panels, or simply subsidising grid energy produced from renewables to make it cheaper than that produced from fossil fuels – have been shown to be highly effective ways of stimulating renewable energy uptake. In countries where plastic bags are paid for or banned, there is an immediate drop in their use. Legal restrictions on the energy efficiency and emissions of new houses, cars and white goods have also spurred change both in consumer perception of what is culturally normal, and in technological investment and progress towards more sustainable product design. If materials and built products were scaled on their ease of recyclability, for example, and taxes levied on low-performing items, the market would respond: more sustainable products would be made, and there would be concurrent evolution in efficient recycling technology.

But so far, governments have shied away from the more radical steps needed to overhaul consumer choice and spur a low-carbon future in which resources are used according to their availability in the Earth's crust, the energy needed to extract them, and the environmental costs arising from their use. Governments massively subsidise the fossil-fuel industries, for example, artificially lowering the price of energy for consumers. Part of the reason they do this is because energy is a necessity and poor people need to be able to afford it. But by keeping prices artificially low, householders are encouraged to see energy as cheap and therefore not value it – they are less likely to insulate their homes or cut waste in other ways. One possible solution is to introduce tariffs that give a certain quota of cheap energy, which then rises in price after that quota is used. If energy use is clearly metered on electrical equipment, people would be encouraged to keep within their 'cheap-energy budget'. Carbon taxing or 'pricing' would raise significant amounts of revenue for governments in an equitable way, while at the same time stimulating a market for low-carbon design and manufacture. Governments can also play a role in creating markets and fostering industries that produce low-carbon sustainable materials. For example, when public buildings such as schools, hospitals and town halls, or transport vehicles and infrastructure are built, contracts should only go to those using the latest sustainable technology. More than half of industrial carbon dioxide emissions come from producing steel, cement, plastic, paper and aluminium, and efficiencies in production can be made with proven technologies, according to the IEA, cutting energy consumption by 34% in steel, 38% for paper and 40% for cement.

If the Anthropocene is to continue to be a liveable, pleasant and healthy environment for billions of people, then humanity needs to curb the use of its most prolifically mined resources: fossil fuels. Currently, 86% of global energy use comes from fossil fuels, and energy consumption is expected to grow 50% by 2050.[11] The filthiest of these is coal – the coal power stations built during the first decade of the twenty-first century alone will emit more carbon dioxide over the next twenty-five years than the total since the Industrial Revolution. China, which now consumes more coal than the entire planet did in 1980, is valiantly attempting to decarbonise and clean its environment. Nevertheless, three-quarters of the 1,200 new coal-fired stations in planning around the world are in China and India. Attempts at big intergovernmental agreements on climate action may eventually prove fruitful, but having watched and attended them myself, I'm doubtful whether the economic, diplomatic and social complexities involved with so many very different nations tackling complicated issues will allow for the kinds of practical solutions needed to reduce carbon emissions. Instead, clusters of nations, such as the European Union, and cohorts of businesses are forging ahead with their own targets to reduce reliance on fossil fuels. Denmark, for example, has determined to be entirely fossil-fuel independent by 2050. The small Danish island of Samsø already generates all its own electricity (100% renewable, mainly wind) for the 4,300 inhabitants, and has enough left over to send 40% to the mainland.

Decarbonising is hard. Access to energy is an essential part of development, and around the world 1.3 billion people lack electricity, meaning water cannot be pumped and children cannot study in the dark (which means after 6 p.m. in the tropics). Petroleum gives people essential mobility – physically, socially and economically – allowing them to transport goods, trade at greater distance, and practise more efficient agriculture. Oil is also the basis for that essential, yet problematic, material of the Anthropocene, plastic, although a remarkable Japanese inventor has achieved the alchemy of turning plastic back into oil and gasoline in a neat desktop machine, making a valuable product from our biggest waste problem. Akinori Ito's machine is already an Internet sensation, racking up millions of views on YouTube.[12] In the video, Ito stuffs handfuls of plastic waste (including carrier bags and styrofoam cups) into the device and flips a switch. As the machine heats up, a stream of golden liquid dribbles out. Ito pours the oil on to a steel tray, flicks a lighter, and tall, smokeless flames jump up. The process, called pyrolysis (heating in the absence of oxygen), uses less than one kilowatt-hour to convert a kilogram of plastic into a litre of oil, and essentially replicates the process that occurred underground to turn dead algae and plants into crude oil. At the moment it's energy intensive, but as the technology improves and crude oil becomes more expensive, this type of recycling will become more widespread.

There are replacements for all the essential roles that oil, gas and coal play in our lives, including bioplastics derived from plants, but for developed countries, it means turning back on 150 years of investment in the fossil-fuel industry, infrastructure of grids, power stations and generators, and manufacturing from cars to pharmaceuticals – not to mention upsetting the powerful fossil-fuel lobby. For developing countries, it is easier to adopt the energy systems already established in the West than pioneer new methods, many of which have unresolved issues, such as how to store and distribute energy for when the supply fails and demand is high. Into the Anthropocene, countries will need to cooperate better on efficient energy generation, distribution and storage. Grids will transcend national borders, becoming regional networks that, like a European–North African supergrid, supply users with solar power from the desert in the south, with wind and wave energy from coastal zones, with nuclear and hydropower from river-based plants, and with geothermal power from the mountains. Fossil-fuel generation will then be concentrated in those plants that are cleanest and most efficient, while others are phased out.

The most promising source of low-carbon energy is nuclear. Developing countries are pursuing nuclear power enthusiastically, and skills learned by engineers in China are already being deployed in the West, which has neglected the industry for decades. Hundreds of nuclear power stations are in planning or being constructed around the world, reducing the proportion (and perhaps even the amount) of dirty electricity made from coal. Estimated global reserves of uranium should last around 200 years at current rates of consumption – far longer if someone figures out a way to efficiently harvest the vast quantities of uranium in seawater. Mining the radioactive mineral is dirty, and nuclear waste poses additional problems, but the great amount of energy that can be released with impressively few greenhouse gas emissions – less than 5% of the carbon emissions of a coal power station over the lifecycle of the plants – makes nuclear energy a particularly attractive option. Aside from capital costs, the biggest problem facing nuclear expansion, though, is public perception of the environmental and health risks involved with radioactive materials. The actual numbers of people injured or killed by the nuclear power industry since its inception is many orders of magnitude lower than for any of the fossil-fuel industries. Particulates in air pollution (mostly derived from burning coal and biomatter for energy) cause an average 447 deaths a day (1.3 million a year).[13] Pollution from fossil fuels causes twenty-nine deaths per hundred gigawatts of energy generated, versus 0.73 for nuclear and 0.27 for hydropower.[14] However, atomic reactions are still powerfully associated with the horrors of the 1945 bombings of Hiroshima and Nagasaki, which killed as many as 250,000 people. There have been relatively few disasters at nuclear power stations, compared with industrial (or urban) gas, oil and coal explosions. Nevertheless, when they do occur, they generate far more public fear and panic.

The disaster at Fukushima in 2011, which was the second biggest industrial nuclear incident after Chernobyl, occurred when an earthquake and tsunami destroyed the plant. Just four people died at the plant – none from radiation. Many people, though, died because of fear of radiation: vulnerable hospital patients and elderly people, who were evacuated rapidly and died en route. The death toll from the earthquake and tsunami, however, was around 20,000. Environmental repercussions, including radioactive contamination of fish, had significantly reduced eighteen months after the disaster, but were still evident. However, on a global scale, it was sadly no worse than the ongoing chemical, heavy metal poisons and acid waste that pollute soils, waterways, coasts and food sources elsewhere. Nevertheless, the way the nuclear industry operates, with a lack of transparency and public accountability, is largely to blame for the ongoing fear, and perhaps even the extent of the disaster itself.

Nuclear power will certainly play a part in Anthropocene electricity generation, to provide essential energy to those who currently don't have access. The options being explored include integral fast reactors, which use conventional nuclear waste (plutonium) as a fuel, and thorium nuclear reactors. Thorium is more abundant than uranium and the reactors can be used to consume existing plutonium in nuclear-waste stockpiles. Thorium reactors produce less waste, and the runaway chain reactions that can lead to nuclear disasters with uranium cannot occur. India, China, Russia, France and the US are now all pursuing the technology.

There is another way of using the enormous power of atomic energy without the risks of a nuclear fission reaction: fusion, the holy grail of energy sources. I went to France to look at the International Thermonuclear Experimental Reactor (ITER), humankind's most ambitious project: to replicate the sun here on Earth.

In the dusty highlands of Provence, workers have excavated a vast rectangular pit seventeen metres down into the unforgiving rocks. When construction is complete, the pit will host a machine seventy-three metres high that will attempt to create boundless energy by fusing together two hydrogen nuclei, in much the same way as stars like our sun do. The result is an atom of helium plus a highly energetic neutron particle. Physicists aim to capture the energy from the emitted neutrons and use it to drive steam turbines to produce electricity. They have designed a doughnut-shaped reaction chamber, called a tokamak, which deploys a powerful magnetic field to suspend and compress the high-temperature hydrogen plasma for fusion. Once the reaction is initiated, the heat produced by the atomic fusion will contribute to keep the core hot. But unlike a fission reaction that takes place in nuclear power stations and atomic bombs, the fusion reaction is not self-perpetuating. It requires a constant input of material or else it quickly fizzles out, so it is far safer.

The concept was discussed and argued over for several decades before finally being agreed in 2007 as a cooperation between thirty-four countries representing more than half of the world's population. Since then, the budget has trebled, the scale of the reactor halved and the completion date pushed back, but progress is being made. ITER will make the first equipment tests in 2020, and the first fusion tests are planned for 2028. The physicists hope to prove that they can produce ten times as much energy as the experiment requires: use fifty megawatts (in heating the plasma and cooling the reactor) to get 500 megawatts out. Larger tokamaks should, theoretically, be able to deliver an even greater output-to-input power ratio, in the gigawatts.

It is a big gamble. So far, the world's best and biggest tokamak, the JET experiment in the UK, hasn't even managed to break even, energy-wise. (Although, in 2013, the US National Ignition Facility in California managed to produce more energy than was put in, in a fusion experiment lasting a fraction of a second) But, if ITER is successful, the first demonstration fusion plants will be built in about 2040, capable of actually using and storing the energy generated for electricity production. If we do manage to replicate the sun on Earth, the consequences will be spectacular. An era of genuinely cheap energy – both environmentally and financially – would have far-reaching implications for everything from poverty reduction to conflict easement.

In the meantime, though, countries around the world are taking tentative steps towards improving the way they harvest energy from mined resources. None more so than China, which is already the biggest polluter on the planet (producing one-quarter of humanity's greenhouse gas emissions) and one of the least efficient energy users. With a population of more than 1.3 billion, what China does has global consequences. The newly opened GreenGen energy plant in Tianjin, south of Beijing, may be the world's best hope for a workable solution to our energy troubles. The integrated gasification combined cycle (IGCC) power station produces electricity by gasifying coal, removing the hydrogen and using the generated syngas to drive turbines. The sky-staining sulphurous pollutants, nitrous emissions and other impurities are almost entirely scrubbed by the IGCC process, and the efficient method produces just one-tenth of the usual carbon dioxide emissions per kilowatt-hour, and the greenhouse gas is streamed out in an almost pure form, making it easy to capture and store. It is more expensive to produce electricity this way – eighty cents per kilowatt-hour rather than fifty – although the company intends to slash costs through further efficiency innovations. When I visited, the 3,000 tonnes of carbon dioxide produced by the facility was being sold at a profit to beverage companies to make fizzy drinks, but GreenGen's owner, the state electricity giant Huaneng Group, planned to sequester the gas – as much as 3 million tonnes a year – in offshore oil wells in the Bohai Sea to aid oil recovery there. Further south, at Shidongkou, outside Shanghai, Huaneng has opened the world's largest post-combustion carbon dioxide capture facility, capturing 100,000 tonnes of carbon dioxide, ostensibly to sell to the beverage industry. This quantity of gas far exceeds the requirements of the soft-drinks industry, however – for now, the surplus carbon dioxide is simply released into the atmosphere.

In the dry steppes of Inner Mongolia and the Gobi, in China's desert north, home to many of the nation's coal seams, scientists are experimenting with storing carbon dioxide in deep saline aquifers. These vast, stable spots should be able to accommodate hundreds of thousands of tonnes of carbon dioxide, but that is still only a fraction of the several billion tonnes the nation produces each year. China's coal consumption has increased threefold since 2000 – much of it low-grade dirty fuel, and there are worrying signs it may be using the black stuff for transport too. Ordos, in Inner Mongolia, has the world's biggest plant converting coal to liquid diesel. While the facility looks clean and non-polluting – like the IGCC plant – it is the most environmentally damaging of all. Cracking the hydrocarbons in coal to produce diesel generates up to twice as much carbon dioxide as using oil, and the process requires enormous volumes of water – 6.5 tonnes per tonne of liquid fuel. Such quantities would be tricky to manage near a river system, but in the dry desert, they are disastrous. Water is being pumped from an aquifer more than seventy kilometres away. The world simply cannot afford for China to run its cars on liquid fuel converted from coal.

So what are the alternatives? China has been forging ahead with electrification – its high-speed trains zip across the country at hundreds of kilometres an hour – and electricity could be produced from a range of low-carbon sources including nuclear, wind and solar. China is already investing heavily in wind, carpeting areas of the Gobi Desert in turbines, and aims to send the power generated from the remote north on high-voltage cross-country transmission lines to cities. Giant solar farms are also springing up across the country from the Gobi to sites closer to the end users, including a 20-megawatt farm in Xuzhou, Jiangsu province. China is the biggest producer of solar PV panels and a large part of the reason for the technology's rapid fall in price. Impressive though the investment in solar and wind is, power generated by renewables simply supplements that generated from coal, rather than replacing it. China needs to be bolder – its ambitious plans for a carbon trading system, being trialled in cities and provinces across the nation, are a strong reason for optimism. Putting a price on carbon will make sequestering carbon dioxide financially attractive for IGCC plants, and could well out-price coal-to-liquid-fuel technology. But perhaps most importantly, it will bind forever the economics of development and pollution in the rush to commercialise new energy technologies.

Humanity depends on China getting it right. But also on the thousands of small steps made by individuals around the world moving away from fossil fuels to new energy sources. In India I met farmers turning silage into fuel, in Britain and the US I talked to scientists trying to extract carbon dioxide from the air in different ways to turn into hydrocarbons for transport fuel, and in Kenya I met a woman who was selling the sun by charging neighbours to charge their batteries and phones from her solar panels. One of the cuter ideas I came across was in a village in Peru, where a couple of semi-retired agronomy professors were generating biogas from guinea-pig poo and using it to light their homes, cook their food, grow their vegetables and power their televisions.

Into the Anthropocene, humanity's quest for new low-carbon alternatives to fossil fuels will mean electrification of heating, lighting and transport, with vehicle fuel tanks replaced with batteries. The material for this lightweight high-energy density power source – lithium – will also need to be mined, and one of the few places in the world with supplies is Bolivia. More than 500 years since it was home to one of the biggest, richest cities in the world, mining is pushing the impoverished nation back on to the world stage.

I arrive at the mudbrick town of Uyuni in the south of Bolivia during one of the frequent blackouts. A screaming typhoon is lashing through the town, the air is opaque with dust and flung stones, the town is boarded up but windows are still smashing. Uyuni is a desolate, windswept outpost at the best of times. Its exposed location was chosen deliberately when the town was founded in 1889 as a pit stop for trains between Argentina, Chile and the Bolivian mines north of here: a sheltered site at the foot of a mountain would leave the silver-laden train carriages easy prey for robbers. The town is 3,670 metres above sea level and the extreme cold reportedly prevented epidemics breaking out among Uyuni's first citizens – railroad workers and army personnel at the military base – rendering the town's new cemetery embarrassingly barren for a few years. This was apparently such a problem for the townspeople that they 'borrowed' corpses from the nearest mining settlement of Pulacayo until they could produce their own.

I've come here to visit the nearby salt flats, the *salar*. I locate a driver and we head out of town on a road that soon disappears into desert. We drive through icy streams and over frozen ground, casting a sympathetic glance at the few flamingos that haven't escaped to warmer lakes. Golden vicuña, the wild ancestors of domesticated alpaca, leap in alarm as we break the horizon with a shiny steel can. We pass the world's second biggest mine (the continent's biggest) at San Cristobel, which turns over 100,000 tonnes of dirt a day, exporting mainly silver and zinc. New houses huddle around a seventeenth-century Jesuit church, which was moved here brick by brick when the villagers relocated from a higher site.

The landscapes are big and brutal: wind-sculpted rocks stand incongruously against earth and sky; volcanoes crown the horizon in fantastic colours, their ancient lava petrified where it spilled in foamy release; and the minerals of this planet are laid bare and bright, naked in the absence of vegetation. Every few kilometres the landscape transforms utterly. Near the Chilean and Argentine borders, we pass salt flats of borax, being mined for export, the wind casting veils of white dust over their shimmering expanses. The lakes are glassy with ice and spiked with crystals of algae and salts on a terracotta backdrop of mountains and volcanoes, the biggest of which, Licancabur, is so alien that NASA tested its Mars Rover in its crater. In that same crater, 5,916 metres up, were found the mummified corpses of sacrificed young girls, sent there by the Incas to freeze to death.

After two days and nights, we enter the surreal landscape of the world's biggest salt flat, visible by Neil Armstrong from the moon. At 10,500 square kilometres, it's the size of a small country, or a sea – which it was once. Many years ago, the Atlantic Ocean flowed from present-day Argentina as far west as La Paz in a huge channel. Land eventually sealed off this liquid ingression from the ocean, creating a vast interior lake. The clash of tectonic plates that created the Andes forced the lake higher, where, over time, it evaporated producing this incredible expanse of brilliant white salt. The sea of white, hemmed by the azure sky, is unlike anything I've experienced and my eyes are confused by the enormity of it. Perspectives are distorted and it becomes impossible to gauge how near or far, big or small something is. In the distance, I see a small island in the *salar* and, on approaching, discover it to be alive with more than 6,000 cacti. This is Isla Incahuasi, an unreal place of gold and green in the deathly white *salar*, and a place of refuge for the Incas who once crossed these flats with their llama trains. At the heart of the island, which is an atoll of fossilised algae and coral, stands a small ceremonial centre and a gorge thirty-five metres deep which once held precious fresh water.

The *salar* is tessellated with polygons, cracks in the five-metre-deep salt layer down to an underground lake beneath. Capillary action drives lake water up through the cracks where the salt crystallises at the surface forming ridges. In many places, the tessellations are disturbed by large 'pools' of salt. These are the 'eyes' of the salar, where water is nearer the surface, and lethal patches for those who drive over them. Crossing the *salar*, we reach the salt-mining community of Colchani. Hunched men and boys with pickaxes and shovels carve up the salt by hand, piling the white ground into conical heaps, where it evaporates in the sun and is transported out by truck. It's back-breaking work in the freezing wind and blinding sun. Boys paint their faces with charcoal to dull the reflection and help spare their sight, but the old men are unpainted, burnt as bark, their eyes milky with cataracts. More than 20,000 tonnes of table salt is produced annually from this *puerto seco* (dry port) of Colchani, but the white stuff is more than a seasoning. The Uyuni *salar* holds more than half of the world's supply of lithium. This soft, silvery paste of a metal is the lightest solid element and, until recently, had few uses beyond tranquillisers in psychiatric medicine to treat mania. But since the 1990s, lithium has been used in a range of long-lasting light batteries that transformed consumer electronics. It is used in mobile phones, laptops and cameras, and now lithium is taking on the twentieth century's dominant resource: aiming to replace petroleum in the cars of the Anthropocene.

Lithium is essential for the efficient, light and quick-to-recharge batteries of electric cars. If Bolivia could somehow tap into the lithium production market, the impoverished country would finally be rich – the reserves are thought to be worth $1 trillion. At least, that's what President Evo Morales is hoping. More than 100 million tonnes of lithium is dissolved in the lake under the salt flat, but getting it out and industrialising production is not easy, and Morales is forestalling foreign investors. The possibility of so much of this essential material being available at one handy site is an exciting prospect, and everyone from Japan to China to Finland to France wants in on the act. But history has not been kind to Bolivia. The country, one of the continent's richest in terms of natural resources, is home to many of the poorest people, largely because of the way that foreigners have exploited it for everything from gold to silver to zinc to gas. Morales is so far refusing to let this precious find out of Bolivian state control. 'The *salar* will not become another Cerro Rico,' he has vowed.

He has set up a pilot project to extract lithium. The milky brine is poured into shallow pools on the *salar* and left for days to evaporate off, leaving behind the valuable minerals: potassium, manganese and borax, and the lithium salts. These are then refined by filtering and evaporating in a series of ponds, until pure lithium carbonate can be harvested. It is very different from the horrendous mining conditions I saw in Potosí. Sceptics say that Bolivia lacks the know-how and infrastructure to extract lithium efficiently, certainly in the quantities needed – around 500,000 tonnes a year to supply a small market, car manufacturers estimate. But the cards are in Bolivia's hands – the *salar*'s lithium reserves are indispensable to electric-car manufacturers and storage batteries for a host of other renewable power sources in the green energy revolution. Morales plans to industrialise the nation with the wealth generated, intending to produce not just the lithium, but also the batteries and eventually the cars. It's an ambitious plan for a nation whose roads are almost undriveable.

In July 2012, the government agreed its first lithium investment, a fifty-fifty agreement with South Korea's POSCO iron and steel consortium to build a pilot factory for the *salar* that will produce lithium and battery components. The deal will allow Bolivia to learn manufacturing expertise from South Korea, which is the largest producer of lithium batteries, and gives South Korea unique access to the lucrative fields. Three months later, Morales' government signed a Law of Mother Earth, giving nature the same rights as humans, with penalties for violating rights. Under the law, mining, oil and gas, and other extractive industries that underpin the nation's economy, are in a unique and uncertain position. Lithium extraction, however, has a low environmental toll and so progress has been unaffected.

As we drive back from the *salar*, it feels as if Mother Earth is reasserting her power. The winds take a more violent twist and by the time we enter Uyuni, the town has disappeared in a blinding blanket of swirling dust. It takes some days until the storm abates. When I re-emerge from my hostel cell to inspect the aftermath, I see the few trees are ensnared by garbage and a thick layer of dust coats everything, rendering the entire town in brown monotone. Windows have been smashed, adobe walls have crumbled back to the ground from which they were made, and corrugated roofs picked off by the gales lie crumpled on the streets. In the main square, people are rescuing flamingos blown there with broken wings. There is still no power, no water, no diesel.

It's as if the town and its materials are disappearing from this hostile desert, and I am strongly reminded of those ruined other places – the ghosts of Pompeii, Hampi and Tikal. In the Anthropocene, our domination of the planet is so thorough that it is easy to believe that humans will last forever. But, like every other species, our existence is transient. In many places, we will leave strong signs in the geology of the far future – in the fossil record – that we were here, and of our effect. Yet in the windswept desert of Uyuni, so far from rivers and oceans, our buildings will crumble back to dust and all evidence of a human society scratching out a living here will be lost. If the *salar*'s resources are to be developed – if humans are to dominate nature in this remote and alien environment – more robust and lasting infrastructure will need to be built. Roads will need to be laid, energy piped in and sturdy buildings constructed. People will have to stamp their mark – humanise this environment – in the way that they have done from the Amazon to the Sahara: a city will be built.

10

CITIES

*C*ities are the most artificial environments on Earth. They are entirely synthetic human creations, microcosms of the planet fashioned for our species. In them are represented the high views of mountains, the protection of dry caves, the fresh water of lakes and rivers, the food, materials and fuels from the natural world. Cities have fundamentally changed the planetary landscape, and they have also changed us, their inventors.

Humans evolved as hunter-gatherers roaming a savannah landscape and it took until the start of the Holocene (more than 100,000 years) for the first villages to appear, with the advent of farming. It took millennia more for the early cities to emerge, though. As village populations expanded, agriculture became much more efficient and people were able to generate a surplus of food from multiple annual harvests. Villages began trading and growing, drawing people to markets for food, cloth, pottery and other materials, and *simply to meet. In time, municipal buildings and governing systems grew up to serve these honeypots of humanity: the city was born.*

The first cities we know of were built by the Sumerians around 6,000 years ago in the fertile river valleys of Mesopotamia. It was here, in legendary cities such as Ur, Babylon and Nineveh, that the first written languages, irrigated agriculture, piped water and sewerage were developed, and the first standing armies of professional soldiers emerged. Civilisation, that great loaded descriptor of human society, originated with cities because once food could be generated by relatively few workers, a proportion of society was free to pursue occupations other than acquiring enough daily calories – they could be teachers or engineers. The division of labour that was possible in these first cities allowed for the flourishing of ideas – the arts, sciences, political and social philosophies, and innovations in how to live, all of which are unique to humans. The city allowed these ideas to spread quickly among the urban population and to other cities as people travelled to exchange goods and ideas. Archimedes was as much a product of the city state of Syracuse in Sicily, as the city was of him. Apart from food – and possibly the wheel – what invention, what creation, what of any consequence has ever emerged from the countryside? The story of our remarkable species, is born of cities. When humans became urban, they also became urbane.

This compulsion to arrange our lives and nests into closely packed populations, living in houses organised however chaotically around social buildings like a market, town hall, centre of worship or political seat of governance, is unique to humans and widespread among us. The mastery over our environment that allowed large populations to live together in a small area – diverted and trapped water, acquisition and transport of distant resources, efficient large-scale farming – are all developments that led us to dominate on a planetary scale. But it took the improvement in the last century of sanitation and medicine for city populations to truly explode. Clean water and antibiotics dramatically slashed the death toll and, for the first time, large concentrations of humans could live in relative safety.

Cities have grown from the once vast Nineveh – home to 120,000 people in 650 BC – to the megacities of the Anthropocene with more than 10 million inhabitants. The Anthropocene is the urban age. Our species is undergoing the biggest migration in human history – already more than half of us live in cities; by 2050, around 7 billion of us will. We have become Homo urbanus – a different creature, a faster-thinking, more reactive, more genetically diverse human. Human history is increasingly urban history.

A million-person city will be built every ten days over the next eighty years.[1] There are currently around thirty megacities on the planet, and by 2050 they are expected to merge into dozens of megaregions, like Hong Kong–Shenzhen–Guangzhou in China, where more than 100 million people will live in a seemingly endless city. The Tokyo metropolis, Japan's national capital region, already hosts 36.7 million people with a population density more than double that of Bangladesh, and the largest metropolitan economy in the world. Cities are becoming more powerful than nation states.

The majority of urbanisation in the coming decades will consist of poor people in Africa and Asia migrating from rural areas for paid work. Almost all of them will live in slums at densities as great as 2,500 people per hectare, sharing as many toilets as the average American family home. Globally, the urban construction boom is causing flooding, erosion and loss in water and soil quality elsewhere as sand is mined and rivers are dredged to provide building materials for the new cities. But we are still in transition at the start of the Anthropocene and cities are sophisticated organisms: the bigger they grow, the faster they operate and innovate. They are the embodiment of human civilisation and personify our species' best and worst characteristics. The cities that emerge out of the natural landscape may be more brilliant than anything known today. They may be less demanding of planetary resources, more savannah-like in their parks and roof gardens and more self-sufficient, recycling energy, materials, water and air. They will undoubtedly generate new ideas, culture and technologies, propelling us beyond anything imaginable today.

Cities are the most artificial environment on Earth, and the one in which humans feel most at home. Cities can even be considered the planetary citizens of the Anthropocene.

The residents have placed stepping stones in the fetid stream of raw sewage that marks a street between rows of houses, but in some places, the green liquid is too high and the stones are submerged. Most of the local women are removing their shoes and treading barefoot into the mire. Having watched me gingerly tiptoe from stone to stone, a small elderly man in gumboots offers me a piggyback ride across the sunken section. I'm tempted, but swallowing my revulsion, I plunge forth, my rubber sandals alternately sticking and sliding in the slime. Recognisable chunks of excrement float around me.

This swamp of human inhabitance in northern Colombia is called Villa Hermosa – it translates as Beautiful Villa, but if there was ever an attractive mansion here, it's long gone. These houses are shelters of plastic bags and plywood, with leaky roofs of tin or plastic sheets, whose sanitary facilities amount to two flimsy curtains for privacy in the bog outside. Children paddle and play in the stinking water, which is giving them dysentery and cholera and breeding the mosquitoes that lead to still more deaths from malaria, yellow fever and dengue. Just one house has access to running water and, in entrepreneurial fashion, the household supplies the rest of the 10,000-strong community for a fee.

It's been described by public-health officials as resembling an African refugee camp, but Villa Hermosa is just minutes from the country's most popular tourist destination; it's one of several shanty suburbs grasping at the silken skirts of Cartagena. A short distance from here, international passengers disembark from shiny cruise ships to sip lattes in the city's bougainvillea-lined colonial squares. While the nation has prospered over the past decade with a drop in poverty and crime – homicides have fallen from 29,000 a year to 16,000 – Colombia continues to have one of the greatest disparities between rich and poor.[2] In Cartagena, a city of 1 million people, more than 600,000 are poor and tens of thousands live in conditions of squalor, malnourished and without access to clean water and sanitation. Many arrived over the past decade, fleeing climate-change impacts and violence elsewhere in the country. They are farmers unable to farm, villagers who have escaped guerrilla groups like FARC and paramilitary horrors.

I am visiting the village with Jaidith Tawil Dominguez, who works for J. Gonzalez Foundation, one of the NGOs trying to help the children of Villa Hermosa. The charity has a nursery and centre for young children, and is creating a recreation park in a place where there are no facilities for kids. The people here are mainly the old, young children and women who have lost their husbands, brothers and sons to the violence. There are no teenagers but many orphans. More than 70% of the population consists of displaced people who cannot return because their former homes are still not secure.

Freddie Uribe, the community leader, fled his village ten years ago after seeing his brother and friends killed. He cannot return, and even if the violence ended, he wouldn't. 'What could I go back to?' he asks. 'My house, if it's still standing, will be lived in by someone else. I have a wife and family here now. My life is here,' he says, wading through the swamp. Freddie works as a carpenter now, doing business within the community.

Violence has followed the migrants here. The neighbouring communities are all ruled by drug gangs and are rife with prostitution. But without a school in Villa Hermosa, children as young as 6 have to walk through these dangerous neighbourhoods every day – unless it rains, when the community is effectively sealed off. The rains have been excessive this year, causing widespread flooding. It is an El Niño year, but Jaidith says the floods have been getting worse over the past few years and she blames climate change. There is no drainage in the slum settlement and the water mingles with the sewage and sits for weeks.

Jaidith takes me to the nursery-school project, where fifty kids are looked after for free from 9 a.m. to 4 p.m., including breakfast and lunch. It allows their parent or carer to work in Cartagena. The building is constructed from a floor of wooden pallets and walled with planks, with holes for windows to allow light into the bare interior. Inside, tens of infants lie in rows on the floor with their eyes shut – it's nap time. I point to the child nearest me, a girl who looks about 3 years old with the pale skin bumps I've seen in Indian leprosy sufferers. What's wrong with her face, I ask. I'm told it is a symptom of poor sanitation – the dirty water – and also that she is 5 years old, but malnourished. There are cases of leprosy here, but other bacterial and fungal infections are far more common and more likely, the nursery carer says. Jaidith shows me the vat of soup that will be poured into bowls and distributed to each child at lunchtime. For most it will be their only proper meal of the day.

I've seen similar levels of extreme deprivation in most cities and towns in the developing world. In Khulna, for example, a city in southern Bangladesh where shrimp farming had brought considerable wealth to local landowners, 40,000 flood-displaced migrants were joining the city's slums each year. I'd met some of these fortune-seekers: ten people sharing one-room hovels, many forced from their villages by river erosion, by increasing flooding due to sea-level rise, by loss of farmland due to increased salinity, and by crippling poverty. Everyone had come hoping for a better life for themselves and opportunities for their children. Although Khulna was 150 kilometres from the Bay of Bengal, rising sea level was already flooding the slum – during the highest tides, water was at least knee high. People sought refuge on the road, which was slightly higher, living under tarpaulin or palm leaves. The Khulna slum-dwellers also had strange skin complaints, some of which were from arsenic in the water, but also due to their unsanitary living conditions and malnutrition.

Often the poverty is most pronounced in urban centres that are experiencing the most dramatic improvements in prosperity, such as

Nairobi and Mumbai. Mumbai, which is on track to become the world's biggest city, is home to around 9 million slum-dwellers – more than half the population – many of whom live in the shadow of some of the world's most expensive new apartments. By 2025, the World Bank estimates that 22.5 million people will be living in Mumbai slums. Even in the heart of Europe slums exist, such as Canada Real Galiana on Madrid's borders, where poverty is rampant and infrastructure as poor as in any developing-world shanty town. Nowhere on Earth is wealth disparity so obvious as in a city.

As I make my way out of Villa Hermosa to return to the splendour of Cartagena with its bright lights, bars and restaurants, a small boy grabs my arm. Grinning, he points to my camera and to himself, asking me to take his photo. I oblige and the boy stands proudly in nothing but a threadbare pair of filthy shorts, his arms, back and forehead covered in lumps. The shutter snaps and the child rushes round to look at his image on the screen. Delighted, he stands hand in hand with a friend to wave off this visitor from another world who glimpsed his for an hour or so.

These children are living in a uniquely urban time in Earth's history. In the 1,000 years before 1800, just 2% of the world's people lived in urban areas; by 2050, around three-quarters of the 10 billion people on Earth will do, and the evolution of places like Villa Hermosa will define the future Anthropocene environment.

I feel a particular thrill when I'm in the throbbing heart of one of humanity's great cities. The press of people all weaving past each other with barely a bump, the four-deep race of cars that stop sharply at a red signal and then thrust forward at green in a pulse that persists day and night. The bars that buzz with chatter, music, people and warmth, the order of planned buildings, and streets that lead to squares and markets.

The denser the city, the more productive, efficient and powerful it becomes. If the population of a city is doubled, average wages go up by 15%, as do other measures of productivity, like patents per capita. Economic output of a city of 10 million people will be 15–20% higher than that of two cities of 5 million people. Incomes are on average five times higher in urbanised areas in countries with a largely rural population. And at the same time, resource use and carbon emissions plummet by 15% for every doubling in density, because of more efficient use of infrastructure and better use of public transportation. These population-density effects, discovered in 2007 by theoretical physicists, help explain the pull of the city and offer insight into how humanity works.[3] And they even seem to apply to hunter-gatherer societies: a group becomes 15% more efficient at extracting resources like food from the land with each doubling in population. Or, as population doubles, it means 15% more resources and time can be applied to something other than food generation. Complex societies owe their evolution to high-density living.[4] Cities are accelerating humanity's domination of the planet, bringing people into ever greater contact. Transport links and communication between cities, from superhighways to express trains and planes, allow businesses to operate across cities around the world, shrinking the human world and making the global local. The great homogenisation of the Anthropocene includes human culture and lifestyle as much as any effect on the natural ecosystem. And cities are the biggest expression of that. They truly are universal. I feel at home in cities precisely because they all offer essentially the same experience. Some are more violent or more sleepy or richer, but the urban environment is at its heart the same. There is not the vast diversity of landscape and experience that exists across the natural world.

The urban revolution of the Anthropocene could prove the solution to many of the environmental and social problems of our age, allowing humans to inhabit the planet in vast numbers but in the most sustainable way. Or, it could finally prove our species' undoing, the apocalyptic version of the dystopian megacity so often portrayed in science fiction.

Urbanisation in many of the places I visited in the developing world is mired in squalor, the accumulation of rubbish, and polluted waters and land. This is the result of fast, unplanned migration to what are often 'illegal' sites, meaning that even if municipalities wanted to provide infrastructure such as waste collection and sewerage they would struggle. However, even once established for decades, sustainability in many urban areas in poor nations is far worse than in rural areas. This is because as people get richer in cities, they use more energy, waste more water and other resources and eat more than poor rural villagers. The opposite is true in rich countries where rural people are often wealthier than city-dwellers. As most of the urban growth in coming decades will be in poor countries, particularly in Asia and Latin America (Africa's urbanisation is stark and rapid only in a few countries, although this could change), it is associated with a general rise in global resource use as people improve their standard of living. The challenge then is to create the most sustainable cities – ones where people lead dignified lives without generating excessive waste.

The evidence so far seems to point to this happening – some 230 million people have moved out of slum housing since 2000, for example.[5] But whether the city of the Anthropocene will be environmentally sustainable depends on how places like Villa Hermosa and Khulna evolve. Will they follow the inefficient North American model – a suburban sprawl of highway-linked satellite towns – or the closely packed high-rises of Hong Kong and Singapore, for example? Residents of dense apartment blocks produce half as much carbon dioxide as those living in suburban family homes.[6] Seoul is one example of how a city can transform in a couple of decades from a slum in which one-third lived in low-rise squatter settlements to a shining functioning city of metro-linked skyscrapers in which the majority of the 25 million people live in pleasant surroundings.

Most cities were never designed or planned, they grew up – sometimes over thousands of years – in an ad hoc pattern. Occasionally, sections would be entirely rebuilt according to architects' plans, but these opportunities were usually the result of disasters, like earthquakes or bombings, or because of grand schemes, such as slum clearance, industrial development or large-scale municipal constructions, like a new highway or transport system. Many planning schemes aimed at social improvements actually made the situation worse, because while the new infrastructure may have led to better health for residents, many factors of city living were ignored or worsened, such as the existing social structures and networks, the way residents used their streets and public spaces, and practicalities including ease of access, shopping and integration with the rest of the city. Some planners deliberately tried to break down the social strength of a city to impose top-down authority, such as post-Revolution Parisian architects who cleared the intricate medieval alleyways and passages, and replaced them with broad boulevards that could be easily controlled by military and police forces.

Villa Hermosa, the slums of Khulna, and most of the current new city expansions have not been planned by architects and town planners. They are informal – often illegal – settlements that cluster and grow up around an existing town or industry. Already, there are over 1 billion slum-dwellers in informal housing – by 2030, one in four people will be an urban squatter, and by 2050, one in three will be a squatter, the United Nations predicts. Most of these slums start off as a collection of rudimentary shelters without access to water, sanitation, electricity or paved streets.

How do people end up there? In the Peruvian capital, millions of impoverished migrants squat in cheap, three-ply dwellings that pepper the sand dunes around the old colonial city. These Quechua-speaking country peasants live in squalor, trying to scrape a living shining shoes for rich city people or selling snacks at the windows of cars waiting at traffic lights. But what I had assumed to be the accumulation of a million arbitrary decisions to move from the countryside to the big smoke, turns out to be a highly organised, billion-dollar business in internal people trafficking, controlled, in the main, by one incredibly powerful individual, whose money ensures political sway. German Cardinas León reportedly locates land on the outskirts of the city, ascertains that the owner is overseas, and then collects people to invade the land – for a fee.

German is reputedly intolerant of the curious – instead, I talk to people who have dealt with him; people who have been trafficked by him. There are many, but the stories are all similar, so here is Abel Cruz's tale: 'I am a farmer from Echarte village near Cusco. It's a very poor area. I had some pigs and grew cocoa and vegetables, but there was drought and our productivity went down. One day in 2003, a man came into the village and asked if we wanted a better life in Lima. He said we would live in a nice house and find good jobs and a good school for our two little boys. I thought about it with my wife for a few months, because we didn't want to leave our families and our home. But the drought got worse. We paid the man 1,500 soles and packed up a few belongings and left the countryside. We were told to meet the man at 5 a.m. with the money and a few sheets of plywood or *esteras* [bamboo sheeting]. When we met him, there were lots of other families just like us waiting there. We were taken to a sand dune and told to fence off an area of the dune and start constructing our houses from the material we brought with us. He helped arrange for a private company to connect us to the electricity and cable TV for a fee. We have no water or sanitation. Our toilet is a silo, a hole we dig in the floor of the room and cover with plywood. Every couple of years, it gets full so we make another hole. It takes ten years for the whole of the floor space to be full of shit, and I don't know what we'll do then.'

Despite the obvious hardship, the poverty, garbage, the stench and lack even of pathways, which send an elderly woman scrabbling with her groceries up sand and rocks to her hovel, the draw of the city is stronger than ever. Already, 75% of Latin Americans live in cities, by 2050 92% of South Americans will do, driven there by environmental degradation of their rural lands, conflict and lack of employment. And hope. However grim it looks here for these shanty-town dwellers, in many ways it is far better than what they have left. Employment opportunities, income levels and access to services from markets to health care are all better than in rural areas. These migrants are trading their ancestral homes and land, community and a traditional way of life for the uncertainty of squatting and the hope of improved livelihoods. A person's prospects have always been better in the gleaming metropolis, where salaries are higher and there are multiple opportunities to make a buck. But until the twentieth century, although you were likely to be richer, you were also more likely to die – a phenomenon known as the 'graveyard effect'.

Densely populated cities were perfect hosts for diseases such as smallpox, bubonic plague and measles, and in the past caused huge epidemics and regularly killed off substantial proportions of the population. The plague wiped out half of London's population in 1348, for example. Such was the death rate in cities that populations were only maintained through the regular immigration of countryfolk. Life expectancy in 1861 of a boy born in Liverpool was twenty-six years, compared to fifty-six for one born in the rural village of Okehampton in Devon.[7] Medical and sanitary advances have now made cities safer places to live, and they contain a greater number of good quality hospitals, so people are more likely to survive if they do become sick or injured. Even though slums continue to be excellent breeding grounds for new diseases such as HIV, SARS and influenzas like H5N1, life expectancy in most cities is now higher than in rural areas. In developed-world cities, such as New York, where life expectancy is at least one year greater than in its surroundings, much of this is attributable to the reduction in traffic accidents – young New Yorkers are far more likely to take the subway home after drinking in a bar than their rural counterparts who risk death by drink driving, for example.[8] However, in slums in the developing world that lack clean water and sewerage, like Villa Hermosa and Khulna, life expectancies are less than half those in healthy cities, with high child mortality.[9]

Although it may look like utter destitution, slums are rich seedlings for the vibrant cities of the future. Many of today's established capitals, including London and New York, launched from similar embryonic beginnings. A few minutes in many of the shanty towns of Africa, Asia or Latin America reveals a bustling informal labour market of street stalls, repair shops, barbers, ad hoc cinemas, and all manner of enterprise. Already, more than half of the global workforce is informal and unregistered; by 2020, two-thirds will be.[10] Although the informal economy doesn't generate tax revenues, it does provide employment for the hundreds of millions of people whose job needs could never be met otherwise. And even if the street-food vendors, waste-pickers and stallholders selling mobile-phone credits do not directly pay the taxman, their indirect contributions to the economy and society are often immense. Over the next couple of decades around half of global economic growth is expected to come from developing-world cities – much of that from unorthodox slum enterprise.[11] Slums are already part of the global economy – the markets and stalls of Latin America's shanty towns sell clothes and electronics manufactured in China, for example. Cities connect poor people with rich markets.

Community schemes including childcare, cooperative selling and bulk buying, microloans and share agreements for expensive infrastructure projects often blossom in the absence of top-down state provision. Transformative innovations, like the M-Pesa mobile-phone banking system (see Chapter 1) first took off in the informal markets among street traders of Kenyan shanty towns. When slum communities organise themselves, as many have begun to do either individually or as part of the wider, Slum-Dwellers International network which originated in India, they can achieve infrastructure improvements for a fraction of the cost of state provision. Residents of the Orangi slum in Karachi, Pakistan, for example, built their own sewerage system in the 1980s, slashing infant mortality. In Delhi, a bank run by and for street children since 2001, called the Children's Development Khazana ('treasure chest'), allows its 1,000 young customers to safely store and save their meagre earnings in twelve branches across the city, accumulating 5% interest.

In a Durban slum, I met a group of waste-pickers, who were campaigning for greater recognition of the work they do and some protection for their livelihoods as South Africa mechanises and develops. Waste-picking is one of the world's lowest-status jobs but is the only way that millions of people have of acquiring food for survival, and these shadowy gutter figures are a feature of every city in the developing world. Their work improves urban sanitation and recycling rates, solving waste-management issues. And the collected organic waste is often composted and used to nurture urban gardens, producing food and money. Around 800 million urban farmers harvest 15% of the world's food supply – a share that would grow if governments promoted, rather than discouraged, the practice.[12]

However, as economies become richer, corporations are given contracts to perform waste-management services in the most cost-effective way: incineration or landfill. No respectable country wants to see bare-footed rag-wearing tramps bin-diving in its newly shining cities, so these companies are starting to own the space and occupations of the most destitute. As a result, conflict is breaking out as corporations seal off landfill sites and security patrols ban people from collecting waste from shops and residences. In response, waste-pickers have been banding together in cooperatives, with the help of NGOs, and have even formed a global alliance to defend their livelihoods. In poor countries, as it should be everywhere, recycling is a matter of common sense – there is almost nothing that cannot be reused or have some further value eked out of it, whether that be feeding it to a cow or having its materials disassembled and sold on by waste-pickers. Waste-pickers recycle more than 95% of rubbish, reducing greenhouse gas emissions, resource depletion, and energy use.

The new cooperatives are making some progress. Durban has an estimated 15,000 waste-pickers and the few I spoke to said they had never had it so good. They now receive a better rate per kilo for the cardboard they collect, a trolley so that they don't have to haul it on their head, and overalls. They are now a recognised part of the local economy with functioning relationships with businesses.

Cooperation, whether to achieve wider policy change and improvement in working conditions or build essential infrastructure, is a slum's biggest strength and essential to nurture as cities 'upgrade' their poorest areas. Societies that evolve in such close and dependent proximity often achieve remarkable feats – I've seen neighbours pull together to fix each other's houses after storms, women breastfeeding other women's infants while their mothers spend the day at market for both of them, and neighbours all watching television at one person's house. Just as the economic prospects for urban dwellers are better, so are the social ones. Women, for example, often have greater freedom in a city, where they can earn money and start their own businesses – something unthinkable in many traditional and repressive rural villages. Minority groups also often find more tolerance in cities.

That's not to glorify life in these communities, which is usually hard, disease-prone and crime-ridden, where residents are prey to exploitative 'landlords' and left to fend for themselves at the margins of prosperous society. But the sheer size of this Anthropocene urban migration, the sophisticated tools like smartphones that even poor people have at their disposal, and the globalisation of culture and commerce means that citizens of these newest urban additions are able to influence the humanity superorganism like never before.

Some cities make this easier than others. While the slums of Lima, Villa Hermosa and Khulna represent some of the worst examples of urban living, enlightened planners are looking not to flatten slums altogether and move people into alien housing structures, or relocate slum-dwellers to dysfunctional satellite areas. Instead, they are beginning to understand the social wealth of these existing communities, and realising that the best way to capitalise on that is to incorporate these dynamic, lively parts of the city into the established whole, by providing the tools for growth, integration and citizen strength. That means improving networks, infrastructure, communication and transport links between different social and geographical parts of the whole city. The result of successful intervention can be incredible, as I discovered in Colombia and Brazil.

A day's bus ride south from Caribbean Cartagena lies Medellín, Colombia's second largest city, perched in a mountain valley 1,500 metres above sea level. Medellín is not a city for sightseers, it is that very rare thing in the developing world: a modern, clean, functioning metropolis of 3.5 million people with a smart, efficient metro system, pelican crossings and safe, metered taxis. The local shopping mall is a gleaming edifice to twenty-first-century consumerism, where the security guards glide around on Segways, the price tags are electronic LCDs, fountain and light shows decorate the centre, and bands entertain shoppers. This very liveable city is unusual for South America, but for Medellín, it's nothing short of a miracle. In less than a decade, the city has been transformed from the world's murder capital of notorious slums to one of its safest. It's an inspiring example of how no-go areas of violence and terror can be turned around using good local planning, intelligent governance and investment in infrastructure. Plus, the use of controversial police tactics to rid the city of its infamous drug gangs.

In the 1990s, Medellín was the headquarters of the international cocaine trade, lorded over by Pablo Escobar, a man so rich that he once offered to pay off Colombia's $10 billion national debt. Stories about Escobar's phenomenal wealth are many (some here regard him as a hero because of his donations to the poor) and include his payments to hitmen of $1,000 per cop killed, that he once burned $2 million in cash to keep warm while on the run, and that he spent $1,000 a week on elastic bands to wrap his stacks of cash, writing off 10% in notes nibbled by rats in the warehouses where he kept his bundles of $100 bills. Despite his church-building ways, Escobar was a ruthless and paranoid murderer. At the height of his reign of terror, there were more than 27,000 murders in a single year in Medellín. In December 1993, after a right-wing paramilitary operation financed by the US to rid the city of Escobar and his associates – more than 300 were killed – Colombian police finally shot dead the man himself. Escobar's successors were killed a few years later.

The city was then transformed through massive investment in infrastructure, including clever transport planning that links the poorest slums with the rich downtown areas, using an affordable metro system with cable cars that ride up to the most impoverished areas. New brick and concrete housing – sensitive to the needs and lifestyles of the community – has also been built to replace the dark six-to-a-room timber shacks. People have been elevated from terrible conditions to a brighter modernity with running water and electricity, but crucially within their own neighbourhoods. Canals and parks have been built, the streets are cleaned and drainage ditches maintained. Skyscrapers pierce the heavens at a faster rate than in any other city on the continent.

Medellín is no megacity, but smaller cities like this will bear the brunt of humanity's great urbanisation. Less economically important nationally and internationally, such second-tier cities often miss out on the attention and resources directed at improving megacities and tourist centres. Nevertheless, the example of Medellín shows that the vast migration from countryside to city needn't be a crisis but an opportunity.

That's something Brazil, one of the most economically unequal countries, has taken to heart. In the cities of Brazil, the urban poor live in slums called *favelas* (after the first such settlement, which was established on a hill called Morro da Favela), made up of squatters who build incredibly dense shanty towns of increasingly robust materials. Around one in five Rio de Janeiro citizens lives in a *favela*, and while Brazil accelerates into one of the world's largest economies, millions of *favelados* have been left behind in poverty with little opportunity for education, healthcare or sanitation. The largest, most prosperous and one of the oldest *favelas* in the city is Rocinha, and in comparison to slums I've visited in other parts of the world, this one is streets above – literally. Rocinha's position high on the hill is advantageous because there's no flood risk, the view is magnificent and its residents paying 300 reais a month look down on São Conrado, one of the most expensive bayside neighbourhoods where rents go for 5 million reais a month. This reversal of the usual social order of things give Rio's *favelas* a unique advantage over other disenfranchised communities.

The hill rises under a tapestry of buildings so tightly packed that it's impossible to see where one ends and the other begins. Residents can meet their day-to-day requirements within a couple of hundred metres' stroll, so from a transport perspective, slums are the most sustainable of Anthropocene neighbourhoods.

Getting up there, however, means a perilous ride on the back of a motorbike, or more than an hour's trudge up steeply winding, dangerous alleyways. I'm far removed from the efficient public-transport network that serves the city proper, where buses careen along the grid of streets and taxis jostle for road space above the rapid metro – one tourist area even has its own restored tram service. The provision of transport across the city is as divisive socially as it is geographically. But it is about to change. In recent years, the government has switched tack from its former hostility towards the *favelas*, to an attitude of acceptance and even inclusion. People in Rocinha stopped fearing that they would be evicted and the new sense of security led to a rapid regeneration programme in which residents formed *mutirões* (construction cooperatives) and replaced their timber and tarpaulin shacks for brick and concrete – an investment they were unwilling to risk before.

Leaving the main street, I am immediately ensconced in a labyrinth of winding alleyways and paths that lead up steep stairs through houses and out the other side. It's too narrow for vehicles here and I step gingerly over garbage and dog shit trying to keep my feet as dry as possible. Every passerby greets me with a smile or wave, music is everywhere, kids tag along for a while and cats stalk me. I duck under trailing powerlines and over water pipes – almost every household has clean running water, pumped from below using electricity. There is a semblance of a sewerage system, although the stench hints at problems there. Meanwhile, I pass a couple of guys high up unwinding a blue wire for Internet connectivity. The entire *favela* will be provided with free WiFi under the government's new development programme. I stop at a bakery to buy fresh pastries – they're more expensive up here than in the main city, but local people can buy everything on credit, which is the only way to do business in a *favela* where people have fragile incomes.

The *favelas* first started at the end of the nineteenth century when freed slaves made their homes on what land was unoccupied. They grew larger and more concentrated during the country's industrialisation, when the poor migrated from the countryside and were unable to pay Rio's very high rents. Now, much of the growth is internal – people are born and raise families in the *favelas*. Despite the enormous problems (for example, there are only four schools in Rocinha, and none for older children), the *favelas* are ideally located for residents working as maids or labourers in the richest neighbourhoods, the rents are very cheap and there are no taxes because few own their own land. And in many ways the community is strong and supportive here.

Undermining any semblance of normal life, however, are the drug gangs that control every aspect of life here. I pass a cute 3-year-old girl who cheerfully raises her fist in a hostile gang salute, and teenagers stand at the corners with walki-talkies and guns under their arms. They are members of the Amigos Dos Amigos gang that controls Rocinha's 250,000 people. They're recruited young – more than half of 15-year-olds here are already members, and they are usually either dead or in jail ten to fifteen years after joining. Their leader is Antonio Francisco Bonfim Lopes ('Nem'), who has effectively evaded arrest, even faking his own death certificate. Every *favela* is controlled by a gang and they make good business, especially here, where they process one to two tonnes each of cocaine and marijuana a month, reportedly turning over $150 million a month – tax-free, of course.[13] The gangs make life dangerous and noisy for ordinary people living here. Constant war with the police, who use helicopters day and night, adds to the general mayhem of bikes roaring up and down and fireworks and crackers exploding in warning of a police presence. Their reputation for violence and crime tarnishes all the residents, who are discriminated against in job applications, health care and other ways. In almost all situations, it is easier to join a gang than battle against mainstream society for a place in its ranks. But Rio, conscious it needs to smarten its image as host of the 2014 World Cup and the 2016 Olympics, is stepping up its war on the *favela* gangs. In 2011, the *favela* was permanently occupied by police under a new 'pacification' operation involving the police and army, aimed at driving drug traffickers out of the slum. So far, few people have been arrested but hundreds of kilos of drugs, guns, ammunition and counterfeit products have been seized.

What may prove far more effective at achieving a lasting peace and improving the livelihoods here, though, is a government pledge to invest more than $58.5 million to build a cable-car system and funicular up the steep hills of the *favela*. As in Medellín, the idea is to integrate these people with those of the established city, to remove the stigma of living in a *favela* by normalising transport links and connecting *favela* stations to the modern metro, suburban trains and large bus fleet that serves the 'formal' city. Instead of a dodgy ride uphill on the back of a teenager's motorbike, and a risky, time-consuming journey on foot through meandering alleyways, residents and visitors, including disabled people, will be able to access homes and services in this densely packed warren. The gondola system carries people over the labyrinth and connects them with the city proper, making the shanty towns more visible and less ignorable. The Teleférico do Alemão, a gondola system linking the Complexo do Alemão *favelas* to the rest of the city, is already open and at 3.5 kilometres, it's the longest cable-car line in the world. The journey time from the base to top has been cut from at least an hour by foot to just fifteen minutes. Each cabin has a solar panel installed, which powers the lighting, sound and video surveillance systems. And the stations host community services, such as job training, education, medical services and legal advice, all helping foster growth and inclusion. It's a radical alternative to the common approach of governments to slums: bulldozing or neglect.

Housing is also going to be improved, with architects being consulted on how best to upgrade overcrowded spaces while maintaining the *favela*'s integrity. Great success has been achieved in Heliopolis *favela* in São Paulo, where innovative architect Ruy Ohtake designed eleven brightly coloured circular blocks for residents, to be a 'dignified way of living' in which all the apartments look outwards towards the light. The apartments have been designed to be socially inclusive, residents pay a rate of half their salary – however little that may be – for their units, and they are accessible to disabled people. Around the blocks are recreation areas for kids, and community services including an education centre that hosts seven schools, beautifully designed by Ohtake to be light, open and cohesive. Such transformations are only possible when cities have strong independent governance with authority and finances to act. Budgets, planning and transport decisions are often still managed at a national level, leaving many cities around the world with little or no ability to tax citizens. It means that city councils and leaders have to beg national politicians for upgrades to sewerage, roads or changes to areas that once might have affected a few thousand people, but now affect millions.

Around the world, architects are rethinking the city in the age of high population, strained resource use and global environmental impacts. Nations will recede in importance in the Anthropocene, as people live, work and identify with new 'city states'.[14] A Londoner may have little in common with a farmer from Norfolk. She is ever more likely to share similar interests, concerns and lifestyle with someone from Shanghai,

New York or Cairo. Cities will become the individuals of the Anthropocene, changing lives as impressively as Mahabir Pun and Chewang Norphel, and indelibly shaping the planetary environment.

Nowhere is the pace of Anthropocene urbanisation being felt more keenly than in China, home to more than 1.3 billion people, more than half of whom now live in cities – an urban population that at 690 million is more than double the entire US population. By 2030, 75% of Chinese will live in cities; thirty years ago, less than 20% did. In some places, construction is outpacing urban migration with blocks of new apartment complexes, offices and shopping malls finished and unoccupied for years. Entire 'ghost' cities are springing up in the reversal of American ghost towns like Detroit. These are not places bereaved of former occupants, but rather cities built from scratch with broad highways, vast fountain-filled squares, shopping plazas, huge buildings and towering housing blocks, that stand empty, awaiting citizens that never materialised. Ordos new town, in Inner Mongolia, is the biggest of these peculiar Anthropocene stage-set cities. It has been empty since its conception in the early 2000s, but perhaps one day residents will move in, marvelling at the ready-made infrastructure that couldn't contrast more with the millennial timescale by which other cities have grown incrementally.

While the Chinese government is encouraging urbanisation, most people are moving voluntarily to improve their prospects – salaries are at least double in Chinese cities, and there are far more employment opportunities, especially for those whose rural situation has been made intolerable by pollution, drought, industrial encroachment or compulsory purchase for urban development. Urbanisation has been one of the most important factors in China's phenomenal economic growth – in 1980, China was poorer than Afghanistan – and the rapid industrialisation over the past decade, providing essential labour and new consumers. But it also comes saddled with tremendous challenges. Over the next two decades, the urban population will balloon by more than 300 million people, all of whom need housing, infrastructure, water, food, jobs – and the rising issues of pollution and social inequality must also be tempered. In a typically top-down style, the government is tackling all of these at once: the biggest polluters have been moved out of the biggest cities (into rural areas or other cities), slum populations (called 'urban villages') have dropped from 37% to 28% since 2000, and China is in the midst of a building frenzy.[15] But cheap and poor construction methods are making the process more filthy than it needs to be. Concrete is the second most used product on the planet after water, and more than half of it is produced in China, where making one tonne of cement emits more than one tonne of carbon dioxide. And many of the buildings being thrown up over the past few decades are of such shoddy construction that they have to be demolished and replaced within twenty years, using more materials and emitting more carbon.

In a few places, planners are designing entirely new cities for the Anthropocene, trying to avoid errors of the past and achieve a sustainable solution from the outset. A few minutes from the world's fifth biggest port at Tianjin is what's billed as the city of the future – a place so new that most of it is still being built. Tianjin Eco-city is a collaborative project between the Chinese and Singaporean governments that will house 350,000 people in a 'low-carbon, green and pleasant environment' by 2020. As I approach the city under an ever-present pall of filthy air, across a wasteland of contaminated soil and water, I am fairly sceptical. Other much-touted eco-cities, such as Dongtan, near Shanghai, have been scrapped over funding and political issues, and because they failed to consider the needs of ordinary residents in ostentatious master plans. The site chosen for this project was an industrial dumping ground for toxic waste, barren salt flats abutting one of the world's most polluted seas. This was deliberate, says Ho Tong Yen, head of the Sino-Singapore Tianjin Eco-city Development and Investment, charged with building the city. 'In the past, so-called eco-cities have been built in ecologically important areas or on useful arable land. We wanted to show that it's possible to clean up a polluted area and make it useful and liveable.'

The clean-up took the best part of three years, and included the development of a newly patented technology to remove the heavy metals from a central reservoir – soon to be a boating lake. The hard graft's paid off. I enter the part-complete city down an avenue lined with fragrant trees and solar panels. Among the newly planted saplings I spot five wind turbines and solar-powered street lighting. One-fifth of the electricity used here will be from solar, wind and ground-source heat pumps, which use the temperature difference in the ground for energy. As I tour a nearly finished school, I surprise another innovation into action: Dutch-owned Phillips is trying out its new sound- and motion-sensitive lights, which default to off unless the switch hears or feels someone approach. Other innovations include a planned pneumatic municipal waste-collection system, produced by Swedish Envac, which will eliminate the need for refuse trucks, and driverless cars are being trialled by General Motors. Buildings will have smart controls, automatically raising and lowering window blinds to regulate light and temperature, for example.

In 2012, the first sixty families moved into the city's residential buildings, all of which are designed to a minimum green buildings standard, including water-saving sanitary fittings, insulated walls and double-glazed windows, as well as a south-facing orientation to optimise passive heat. Such techniques may be standard in some countries, but in China they are rare. Also rare is the emphasis on liveability. Parks and green spaces are planned around the city, and reed beds have been created to attract birdlife and help clean the water. Lanes and alleyways have been strung through the usual grid layout of big blocks, meaning communities can develop, everywhere is walkable or cycle-able and people don't feel socially excluded from areas. Free recreation facilities will be provided within 500 metres of anywhere. A green spine called the 'eco valley' runs through the heart of the city with cycle routes and a tram. Residents will be encouraged to use regular low-carbon transport or walk, rather than driving, although cars won't be banned, Ho says. 'We don't want to create obstacles for people, but rather make it conducive to use alternatives.' Niche designs that have focused blindly on eco-technologies haven't worked, he says. 'This eco-city will be practical – it will work.'

It certainly feels like a more pleasant place to live than the traffic-choked polluted cities further inland, even at this incomplete stage. But whether it lives up to its green credentials will depend in part on the type of society it nurtures. Perhaps its most novel, yet important mission is social inclusivity. One-fifth of the housing will be subsidised for low-wage workers and their families. 'We want to avoid the idea that this is a haven for rich people or second-homers from Beijing,' says Ho. 'Being green isn't a luxury, it's an affordable necessity. This city should be a practical, replicable, scalable model for elsewhere in China and the world.'

Water provision is one of the bigger challenges – this is an arid area. The eco-city is planning a desalination plant, and a lot of effort is being put into conserving water and recycling it for irrigation and toilet flushing. Waste water will be sent to a plant for anaerobic biodigestion, and the methane emitted in the process used to produce energy. Nevertheless, supplementary water will need to be piped in to meet demand, stretching sources elsewhere. Many of China's new cities are being built in locations that are so dry they could barely support a village without vast hydroengineering. China is literally moving mountains – 700 in the case of the planned metropolis at Lanzhou – constructing rivers of concrete tunnels thousands of kilometres long to transport water to where its Anthropocene populations will be placed, and growing food hydroponically in the absence of soil to support them. How successful the planet's desert cities will be at providing their citizens with water over the coming decades is questionable. Most American cities don't recycle their water supplies – something Singaporeans, Namibians and others have been doing for decades. However, necessity is forcing citizens to confront their squeamishness and deal with the reality of drinking each other's processed urine. With modern purification facilities, there is no public-health reason to discard the vast volumes that cities currently waste.

Recycling, efficiency of resource use and decentralised generation will characterise the cities of the Anthropocene. Citizens are likely to generate their own supplies of water, electricity and fuels to supplement the increasingly expensive municipal provisions. Many buildings will harvest their own rainwater, or gather it in community schemes using fog nets or soak-away schemes. Householders will filter, store and reuse grey water from showers and sinks for toilet flushing and watering the garden. Water conservation will become commonplace and regulatory as it already is in arid countries like Australia. That means dual-flush mechanisms on toilets, efficient washing machines and dishwashers, aerator-flow taps and showers, hosepipe-use restrictions, and a metered supply so households can see immediately how much water they're using.

Electricity generation will also become decentralised. The large power stations that currently supply cities will continue to exist, but individuals and communities will increasingly generate their own, through solar thermal generation to heat water and buildings, and photovoltaic panels, for example. In the next decade, photovoltaic paint, blinds and glass that can generate electricity from sunlight will be used for public and private buildings. The urban environment will generate its own electricity and heat through passive measures, rather like crystal radios used to. It is also likely that pavements, streets, stairs and corridors in buildings will be fitted with piezoelectric generators that charge up through footfall – a school in England already has carpet tiles that are powered by passing children. Pipes carrying water could be embedded in dark asphalt in pavements, roofs or streets and use the solar energy generated during the day. Innovations in solar paint and roof tiles will help ensure that buildings of the Anthropocene become net energy producers rather than users. The technology is currently expensive compared with grid-produced electricity from fossil-fuel power stations, but innovations in the design and materials of panels, mass production and government incentives are driving down costs. And the relative inefficiency of solar panels, which convert less than 20% of the sunlight hitting them into electricity, is being improved continually. Carbon nanotubes, for example, may be used in conjunction with silicon, to harvest the infrared (heat) rays as well as the visible light. Rubbish produced by householders and office workers is also likely to be used onsite to generate combined heat and power (electricity) for residential and commercial blocks. In Paris and London, the heat from passing trains and from the bodies of waiting commuters in underground metro stations is being used to heat the apartment blocks above. Many cities have plans to become independent from the national grid in the next couple of decades.

Efficiency measures in the generation, transport and use of energy need to be vastly improved so that, for example, waste heat is recovered from boilers and stored or used elsewhere, leakage is reduced with insulation and improved materials, and automated 'smart' controls regulate energy use throughout buildings from switching off lights in unoccupied rooms to adjusting heating and cooling systems. Decentralised and renewable energy production, which is often erratic or (in the case of solar or tidal) depends on the time of day, require a smart grid that is far more flexible than the twentieth-century ones used in most cities today. A smart grid relies on sensors and feedback mechanisms to detect customer demand and adjust the load accordingly. For example, at peak demand – 7 a.m. when people switch their kettles on – standby generators or storage devices can be brought online, while some gadgets, such as refrigerators or air-conditioning systems can be automatically dialled down or switched off for a short time. Most countries are looking to adopt smart grids in the coming years, partly because they are far more efficient and so use less energy, but also because they allow for renewable energy integration – monitoring and sensing demand and supply in real time and balancing loads accordingly – and for individual householders to feed their excess electricity back to the grid. Grid upgrades are essential – even large cities like New York have been experiencing crippling blackouts – but expensive and disruptive, and resisted by energy-intensive industries such as cement works, which prefer baseload power plants.

Energy storage and distribution solutions are becoming increasingly urgent in the Anthropocene. Storing the energy locally that buildings produce for later use will eventually be far more efficient than feeding it into the grid and using the grid as the energy store. There are plenty of options but none is perfect. Batteries are improving every year, with the most likely grid-scale option being liquid metal batteries, which are still at the prototype stage. Using the energy to compress air or steam, which can then be released and converted to electricity is another option. Another is looking at hydrogen for storage, which can then be burned as a fuel when needed. Another option is to store the energy in the fast spin of a friction-free flywheel – the energy is then accessed by using the flywheel to turn a rotor in an electric generator, which slows the wheel.

Distributed energy and water systems are just some of the ways that citizens of the Anthropocene will interact with their urban environment. Rather than being passive users of municipal services, citizens – and the city itself – will generate real-time responses from providers, who can make continual adjustments to improve efficiency, minimise waste and generate a more intuitive personalised operation. So-called 'smart cities' communicate through sensors embedded in infrastructure, or information sent by individuals that is either automatically generated or deliberately sent. These networked cities – including New Songdo City currently being built in South Korea, and PlanIT Valley in Portugal – generate intelligent adjustments to everything from street lighting to mass transit routes and times, based on real-time feedbacks. Other cities are using intelligent sensors for regulating utilities, designing flood-defence systems, regulating traffic lights and flow, reducing emergency vehicle response times, speeding baggage flows through airports, locating parking spaces for drivers, optimising waste management, reducing peak-load demand on electric grids and even cutting crime rates. Masdar, a new city being built in the desert of Abu Dhabi, has many of these elements designed into it from the start. The entire city is on a raised platform so that the smart-metered services – from waste to water – can be monitored and accessed from underneath. Masdar plans to be carbon neutral and is powered by an enormous solar station and wind farms, with buildings that incorporate smart shading, solar panels and architecture to maximise cooling breezes. The city, which aims to be completed by 2020, is car-free with above- and below-ground driverless electric transport pods that operate like a personal rapid transit system.

The human network is key to next-generation smart cities, to reducing energy consumption and more effective use of the city in all ways. Crowd-sourcing and collaborative mapping are two such techniques that over the past few years have revolutionised information creation on the Internet without the need for massive infrastructure projects to plant sensors across cities. For example, data points on Google Maps are built by a large network of hundreds of millions of anonymous phone-users whose devices continually send GPS-based status-updates. In the case of its traffic app, this reveals how the traffic is flowing or stationary in the city. In Dublin, the crowd-sourced data app ParkYa finds the nearest free parking spots for drivers. In the aftermath of the 2010 Haiti earthquake, volunteers, citizens and government agencies were able to pinpoint aid flow through the island, where needs were most urgent and where disease was breaking out. The ability to track the world in real time, using GPS and other systems, is generating a global nervous system, which can be used for social better-ment. In a Kolkata slum, a gang of 12-year-old children called the Daredevils are dramatically improving health outcomes in their area, using cheap and simple technology that wasn't invented twenty years ago. Like so many slum neighbourhoods, the notorious 2 Nehru Colony doesn't officially exist, meaning it has no access to government services such as sanitation and electricity. The youngsters set out to literally put themselves on the map. They went door to door, taking photos with their mobile phones, registering residents and detailing each child born in the colony. They send the information by SMS text to a database that links the data to a map hand-drawn by the kids, which is overlaid to GPS coordinates. By registering their existence on Google Maps the group has doubled the rate of polio vaccination from 40% to 80%, decreased diarrhoea and malaria rates in the slum, and is lobbying for electricity.[16] Rio's slums are doing similar: *favela* children are mapping their neighbourhood from the air by flying kites that have glass bottles suspended from them with a camera inside. They identify problem areas, such as buildings in danger of collapse or landslides, sanitation or garbage issues, or hazardous powerlines, and take images with GPS-enabled smartphones, prompting intervention.

Rigging cities with sensors is just a first step, though. The designers and engineers planning smart cities a decade ago could not have predicted the way humanity would become an integral part of the network. Now that the technology exists for individuals to communicate instantly with companies, government departments, to broadcast to millions or to specific groups over the Internet, the city has gained an entirely new dimension. This 'virtual city' of communities formed online, using social networks like Twitter or Facebook, is incredibly powerful and not limited by the geographical contours of the real city. Like-minded individuals can find each other easily, gathering in online forums or through hashtags and comment streams in the same way as special-interest clubs and café movements coalesce in the real city. Virtual applications make it easier to sift through a crowd – the Grindr app, for example, allows gay people to find other users of the app in a public setting. Online clubs – like the shopping network Groupon – are attempting to personalise trade exchanges and perhaps develop a proxy for the relationship people might have with a neighbourhood store.

Those petitioning for social or political change can hold governments and companies accountable with an ease never before possible. Vast amounts of data are now published online and can be searched and filtered in minutes with simple algorithms. The virtual and real cities are closely enmeshed. Information-gathering and community-building can take place more easily online than in the vast cities of the Anthropocene, where members of a group may live far from each other or be unable to meet easily for momentum-building. But the discussions and real-world changes these online gatherings initiate move easily to government chambers, mainstream media outlets in television, radio and press, or on to the streets. The Arab revolutions across Northern Africa and the Middle East since 2010 were coordinated via the virtual city of Twitter, Facebook, SMS messaging and other apps, but they took place on the streets and squares of the real cities, uniting flesh-and-blood individuals who had gathered online using computers and smartphones. Starbucks was compelled by a Twitter campaign to pay millions of pounds of tax to the UK government after its perfectly legal offshore tax avoidance was revealed in 2012. The virtual city is as global as it is local. I can get hourly updates on air-pollution levels in my neighbourhood or buy a new battery for my phone from Korea. People from across the world can gather online to share ideas, pressure for change, innovate, spread their artistic talents or make friends. The virtual city provides a way of shrinking and filtering the real megacity, saving time and energy on real journeys across complicated spaces, of accessing multiple conversations with relative anonymity, and of individually helping steer humanity through collaborative creativity and problem solving. It enhances but doesn't replace the real city with its face-to-face social cues, physical exchanges and wealth of information humans use to make judgements about trustworthiness and other value-laden decisions.

The virtual city does have a more problematic side, however. Never has there been so much information about so much of our lives in such an accessible form. In the course of a day, the average person in a Western city is exposed to as much data as someone in the fifteenth century would encounter in their entire life.[17] The digital birth of a baby now precedes the analogue version by an average three months, as parents post sonogram images on Facebook and register their infant's domain name before the child is even born. Governments, groups, individuals and corporations can access data about us and use it for their own purposes. This erosion of individual privacy can be benign or malevolent, but it is already a part of life in the Anthropocene. Customer data collected by the US supermarket Target allows it to identify with a high degree of accuracy which shoppers have recently conceived and when their due date is.[18] The store uses this information to target such women for advertising of its pregnancy and baby products in a timely fashion, even if she has not yet told anyone else. Sinister? Maybe. What about police officers identifying householders as marijuana growers by analysing energy-use data? Or neighbours targeting individuals for cyber or physical bullying because of information they discover online? We're all generating data; in the Anthropocene, we will have to decide who owns our data and whether it can be shared.

Cities are already the most environmentally sustainable way of housing humankind, but as hubs of industry, transport, and domestic and office buildings, they use a large proportion of global energy and generate more than 80% of carbon emissions if food energy and other consumption is accounted for.[19] This geographic concentration of such a hefty proportion of emissions actually presents an opportunity for easier fixes – it's easier to improve sustainability in a small area than over a large region.

The more compact the city, the more energy efficient transportation and everything else is. Underground space, already so riddled with cables, pipes, sewerage, subway lines and basements, will be exploited more deeply. Some cities, such as Singapore, already have vast networks of malls underlying the streets, which descend several storeys. And Hong Kong is considering burrowing into its rock for housing and other developments. However, to fit 6 billion people into cities by 2050, they will have to grow upwards. Cities have always defined themselves by their skylines, whether the vast gold-plated Mayan temples of pre-Columbian America, the Dome of the Rock in Jerusalem, the mythical Tower of Babel, the steeples and spires of Europe or the Empire State Building in Manhattan. Now, the new cities of the Anthropocene are continuing this trend – more than 350 skyscrapers have been constructed since 2001, when Malaysia's twin Petronas Towers were completed as the world's highest buildings. By 2016, the Petronas won't even merit a spot in the top ten. The current tallest building, at 828 metres, is the Burj Khalifa in Dubai, a sky-piercing struc-ture of such immensity that it bears no relation to the human-scaled Bedouin huts that peppered this location just a few decades ago. Three-quarters of the world's tallest buildings are now in Asia and the Middle East, over a century after the first skyscraper opened, half a world away in Chicago. These icons of the urban age are as much vanity spires as a housing solution – the top one-third of most of the world's tallest are unoccupiable by design.[20] Nevertheless, skyscrapers connected by sky-bridges and corridors, served by elevated mass transit, with above-ground streets, markets, parks and gardens, will dominate the cities of the Anthropocene. Scaling up doesn't have to be as ostentatious as the Burj Khalifa, but it can be as innovative. Architects are using new materials, such as steel-reinforced concrete, and new techniques, such as computer modelling, to see how their buildings will function when complete – how people will move around, live and relax in these towering houses. The technology allows them to tweak their designs early, to improve escape routes or optimise apartment views and sun penetration, for example. Cleverly designed, complexes holding thousands of people needn't feel crowded or cramped. And, in contrast to privately owned vanity spires, they can be at the heart of the community – Bogotá is home to BD Batacá, the world's first crowd-funded skyscraper.

To improve the sustainability of skyscrapers, architects need to go beyond the not insignificant economy of scale achieved to include low-energy innovations and transform this traditional symbol of opulence and indulgence into a green beacon. The Bank of America Tower, which opened in 2009 in New York City, achieves just that. It is made largely from recycled materials, captures its own rainwater and conserves water in other ways, such as with waterless urinals. It filters incoming and outgoing exhaust air, produces two-thirds of its own energy using an onsite, gas-fuelled combined heat and power generator, and conserves energy with insulating glass and 'ice batteries' in the basement, which use off-peak energy at night to freeze water and allow it to melt during the day to cool the air in the building. In a city where 75% of carbon emissions come from energy used in buildings, this office tower has won a slew of awards and been given a platinum Leadership in Energy and Environmental Design rating. In the Anthropocene, features like these will have to become the norm rather than exceptions. Quick fixes, such as painting roofs white – and in some cases lightening streets – to lower urban temperatures, fitting low-energy lighting (20% of a building's energy is consumed through lighting), and improving insulation are likely to become mandatory through policy tweaks to building regulations and property sales and rental codes. We may see the day when all new buildings must be energetically self-sufficient through effective insulation and generation systems like ground-source heat pumps or combined heat and power production from waste. It doesn't have to be high-tech: clear plastic two-litre drinks bottles filled with water, and punched through a hole in the roof to refract sunlight, are bringing daylight into the poorest dark dwellings. Other innovative energy generators that are in development include lighting that relies on bioluminescent bacteria or algae. People will also have to curb the use of energy-guzzling electronics, or use renewable recharging methods – in Britain alone, electricity consumption from domestic appliances has doubled since the 1970s, even while products have become more efficient.

While suburbia was built around the automobile, the city of the Anthropocene needs to be built around efficient, low-emitting mass transit. In central areas, it makes sense to ban private cars altogether – something that is gradually being encouraged by the heavy congestion commuters face, expensive parking and by specific taxes for vehicle drivers, such as London's congestion charge. In Bogotá, for example, cars are banned from the city on Sundays, when the streets fill with pedestrians, cyclists and roller-bladers. Americans, who currently spend an average nine years of their lives sitting in their cars, may have now passed 'peak car', with car ownership and driving rates dropping.[21] Even in China, the world's biggest car market (where ownership has risen twentyfold since 2000), 'peak car' could come sooner than the two decades estimated to reach saturation: fuel prices are escalating, the cities' brand new six-lane highways are already clogged with traffic and car fumes, and with more of the population living in dense cities and 'meeting' on social networks, car ownership is increasingly becoming an inconvenience. Instead, citizens need a rapid, regular and reliable network of alternatives that allow for fast cross-city journeys, as well as strategic, more personalised options. Underground trains and carriages on suspended tracks above the city can whizz commuters around avoiding jams. Electric trams and buses at street level can provide hop-on-hop-off convenience. In another inspired move, Bogotá has transformed its public transport inexpensively by retrofitting its buses into an above-ground metro system: the pavements are raised at stops to become platforms, the bus doors open like metro ones and the ticket booths are similar. Engineers in the US are designing superfast maglev trains that hover friction-free above a rail through magnetic repulsion and could, in principle, eventually run under the Atlantic in vacuum tunnels, connecting European and American cities. In China, there are conceptual designs for an electric megabus that would carry hundreds of commuters on wheeled stilts, five metres above the ground, straddling gridlocked traffic on the road beneath.

Bicycles are excellent for personalised travel, and electric bikes – which I saw in popular use in Beijing – can cover longer distances or hills without exertion from the rider. The Paris rent-a-bike scheme, Vélib, which has since been copied in cities around the world, allows users to pick up a bike in one location and drop it off in another for a small fee. In Israel, electric-car users can use swap stations, run by Better Place, to exchange their depleted batteries for fresh ones. This cuts the cost of the car (the battery is one of the most expensive parts) and allows it to be upgradable as improved batteries enter the market, while keeping the value of the car. A comprehensive network of switch stations removes the problem of 'range anxiety' (when drivers fear their car will run out of juice). Once vehicles are fuelled by electricity, direct pollution (including ozone and oxides of nitrogen) disappears and power can be provided using low-carbon sources such as nuclear or renewables. Better Place is now rolling out to other countries. Electric cars can even be used as an alternative power source, discharging electricity during power cuts. But while electric cars are an emissions improvement on conventional gasoline burners, they and their batteries still represent a metal and mineral guzzler, requiring ever more rare earth metals, for example, which if not efficiently recycled, end up as toxic waste. Better electric public transport is by far the most sustainable solution.

Like everywhere else on Earth, cities of the Anthropocene will have to deal with climate change. As the atmosphere heats up, it holds more water, producing heavier downpours. In concrete paved urban areas, the rain cannot soak away and often leads to flooding – metro systems already expend vast amounts of energy pumping out water and this will only increase. Storm drains and soakaways more usually used for tropical cities are now being incorporated in European urban areas, and some cities are innovating with porous cements on pavements and green spaces. Heatwaves are another problem that is worsening. Cities already suffer from the 'urban heat island' effect, in which the energy from industry and buildings, and poor air throughflow, leaves built-up areas hotter than the surrounding landscapes. Delhi and Mumbai, for example, get 5–7°C hotter than surrounding rural areas during summer nights.[22] Global warming is making this a killer – people are especially susceptible to high night-time temperatures – and in cities, residents fearful of crime often don't open their windows. The US Natural Resources Defense Council calculates that climate change will treble heat-related deaths in the nation by 2100. Painting roofs and pavements white is a cheap way of lowering urban temperatures. However, as people in the developing world become richer, they are installing air-conditioning units – sales in India and China are growing at 20% a year – and the greenhouse gas emissions of the coolants used (called HCFCs, which replaced the ozone-layer-depleting CFCs) could contribute almost one-third of all global warming by 2050, researchers estimate. Simple passive innovations, like shaping the building to direct airflow, can result in significant energy savings – London's Foster-designed 'Gherkin' tower directs wind flow from the street up its spiral height, reducing by almost half the need for air conditioning.

One of the problems with cities, and construction generally, is that they are designed with little flexibility. They need to meet certain structural and safety standards, to withstand worst-case weather and seismic challenges, for example, but not necessarily to operate best during the majority of the time. For certain buildings, like lighthouses, this may be the only concern, but for most buildings and urban spaces, liveability under multiple weather conditions is also important. We adapt our clothing – why not our buildings? The Media-TIC building in Barcelona has been designed with an inflatable plastic ETFE skin that is regulated by a solar-powered automatic digital light sensor. As the sunlight changes throughout the day, different air chambers in the skin are expanded or contracted. The responsive skin allows visible light through, but filters UV rays. Compared to glass, ETFE film is 1% the weight, transmits more light and costs up to 70% less to install. It's also resilient (able to bear 400 times its own weight), self-cleaning due to its non-stick surface, and recyclable. Some buildings are incorporating light-reactive glass, in a similar way to that used in sunglasses, to reduce heat and glare during hotter parts of the day (reducing the need for air conditioning), while allowing the maximum light through during other parts of the day (reducing the need for electric lighting). Automatic light-responsive sunshades play a similar role in other buildings. Other sensors can be used to regulate energy and resource use inside buildings, operating lights when movement is detected (as I saw in Tianjin Eco-city), for example, or sensing when to divert people, water or heating. If a regularly trodden route of carpet starts wearing out, say, subtle lighting changes can encourage people to take a fresh route, or if activity is detected in only a portion of an office space, heating can be restricted to that area.

Architects are also looking at incorporating responsive, living materials into traditional buildings to make them more environmentally sustainable and adaptive. Many now have green roofs and walls, which filter air, keep the buildings cool, conserve water, invite biodiversity and look attractive. But what about making the structures themselves out of living materials? A Dutch team has created a self-repairing concrete, which is impregnated with bacteria that excrete limestone when exposed to water. Any cracks that appear in the concrete and allow water in will activate the bacteria, lying dormant in nutrient-rich capsules, and the limestone released fills the cracks. Other proposals include growing structural parts of buildings and street furniture around frames, using synthetic biology to prompt living cells to produce useful materials. Designers also imagine living, responsive buildings that act as an immersive interactive installation able to move and breathe alongside their occupants. Engineers will be using ideas from biology and incorporating self-repair, light and heat response, and other adaptive features into the materials and structures of future cities.

Sustainability is about more than resource efficiency, though. Cities and buildings also need to meet our human needs in a liveable space, including sociability, calm, recreation and ease of movement. There are plenty of examples of ghostscrapers, malls and entire districts, abandoned because they failed the liveability test. Achieving a city in which citizens feel ownership and pride in their surroundings means embracing the landscapes humans enjoy – and most of those include natural elements. Cities are the ultimate manifestation of our species' stamp on the planet and subjugation of nature. They are environments entirely shaped and created by humans to protect them from the natural world. But cities are also where humans most clearly define their relationship with nature, allowing select bits in and banishing others. The cities of the Anthropocene will incorporate the natural world in new and innovative ways. Parks and green spaces will be multiplied from ground level upwards, attracting birds and wildlife to sky-gardens, tens of floors up. In Singapore, for example, the iconic Marina Bay Sands hotel features a skypark on the fifty-sixth floor, with trees, leisure facilities including a pool, and an incredible view. It's an example of how specific elements found in the natural world, such as a mountaintop view, a lake and palm trees, have been cherry-picked and combined to provide an easy, entirely artificial landscape for the city. There is already a new field of urban ecology for scientists who study the city and biophysical interactions within it in a similar way to traditional ecosystem research.

Vertical gardens and farms are also being planted, although the energy involved in irrigating and maintaining them makes them impractical for food production on a larger scale. However, growing food in the urban environment on regular multistorey plots is likely to increase as hobby farmers, beekeepers and specialist growers take advantage of cleaner air, water and soils of Anthropocene cities, and vacant sites are used more effectively. In Berlin, rooftop fish farms have been started, with the waste going to feed agricultural plots in the city. Creative growers are already converting industrial spaces, street corners and rooftops to micro-wildernesses or manicured into formal gardens. A disused raised railway in New York City has become a popular park, self-styled 'guerrilla gardeners' are planting flowers and trees in plots among the tarmac and traffic of London's highways, and once-polluted industrial wastelands now chirp with birdsong, rivers swim with fish and populations of animals that have become rare in the countryside are thriving in urban niches. In fact, a surprising amount of wildlife now depends on the man-made environment. From the clouds of huge Sydney fruit bats to London's wily foxes, to skyscraper-nesting peregrine falcons, animals have made cities their home – in some cases, their natural habitats have disappeared. In other places, the mix of human-introduced plant and animal species, and those opportunists that migrate to the urban environment, are interacting to produce unique ecosystems that exist nowhere else. Seagulls, for example, now often live in cities hundreds of kilometres from the coast. As their traditional food – fish – becomes scarcer, they scavenge human rubbish. Whether humanity learns to value these new flourishings is still to be seen, but the survivors of these menagerie experiments in the human garden will produce a genetic legacy. In centuries to come, new species that could have not have formed under any natural circumstances will be testimony to our mixing. Already, urban moths have evolved changes in shade suiting their dull concrete habitat compared to the tree trunks they formerly lived among, songbirds have got louder to compete with traffic noise, and urban varieties of other creatures exhibit different physiology, behaviour or brain size.[23] It isn't just animals and plants that are genetically altering because of cities; humans are too. City-dwellers have measurable brain differences, some of which may help urbanites handle stress and crowding, although disorders like schizophrenia are twice as common in city populations.[24]

The great urban migration of the Anthropocene is effectively a global mix of humans, most of whom had been genetically isolated for generations in villages. As a result, humans are now more genetically similar than at any time in the last 100,000 years, according to biologist Steve Jones. The spread of genetic diversity can be traced back to the invention of the bicycle, he says, which encouraged the intermarriage of people between villages and towns. But the urbanisation occurring now is generating unprecedented mixing. Cosmopolitan cities like New York are home to people speaking more than 800 different languages – thought to be the highest language density in the world – and in London, less than half of the population is made of white Britons – down from 58% a decade ago. Meanwhile, languages around the world are declining at a faster rate than ever – one of the 7,000 global tongues dies every two weeks.

It is uncertain how permanent our megacities will be, though. Many of their residents, like migrants everywhere, talk of 'home' as somewhere else entirely. Even second- and third-generation urbanites often describe their 'home village' and their dreams of one day 'going back' to build a big house or a ranch. If prosperity affects these citizens of new cities the way it has in Europe and America, there is a danger that as many countries develop, suburban sprawl and the takeover of wilderness will march across the world at an unprecedented pace. No one, after all, imagines living in a slum as anything other than a temporary measure en route to better things. Currently, urban areas cover around 2% of the planet's land area, but by 2030, they could sprawl over almost 10%.[25] That would mean losing some 1.2 million square kilometres of other landscapes, many of them rich in biodiversity. The Amazon, for example, is currently experiencing Brazil's most rapid rate of urbanisation and is already home to 25 million people. One of the most important planetary tasks for cities in the Anthropocene is environmental preservation. For while humans, and their landscapes of concrete, glass and industrial sprawl, are kept within the confines of a megacity, the rest of the planet is free to revert to a more natural state. Of course, humans have changed most landscapes irrevocably, so the planet outside of cities will not return to some sort of prehuman state, but ecosystems will nevertheless proliferate and reach their own new equilibria in the absence of continued human pressure. Because of this, cities represent the biggest hope for the survival of other species and ecosystems in the Anthropocene.

Although they seem permanent – some have existed for more than 1,000 years – many cities are unlikely to survive humanity's changes to the planet. Sea-level rise and groundwater extraction are causing major coastal cities to sink into the oceans. The world's capitals, many of the most economically important parts of our planet, already face double the risk of inundation through storm surges – and the greater population of Anthropocene cities puts more lives at risk. Cities from New York to Bangkok have already experienced emergency flood conditions, and many more are to follow. The storm surge associated with Hurricane Sandy in 2012 flooded areas of New York City with more than four metres of water. Located on the Yangtze River delta, Shanghai, which means 'above the sea', is being claimed by the South China Sea – parts of the city have sunk three metres. In response, Chinese authorities have begun pumping 60,000 tonnes of water a year back into wells to reduce the subsidence, built hundreds of kilometres of levees and are planning an emergency floodgate on the river's estuary to protect the nation's most prosperous city and its 20 million inhabitants. Mexico City is suffering a similar fate, with parts of the city subsiding nine metres through over-pumping of groundwater. In the Netherlands, some 50 million cubic metres of sediment has to be moved every year just to maintain current shorelines.

What's the solution? Some cities are investing in new sea walls, dykes and polders, or high-tide gates – like London's Thames Barrier – to hold back high waters. The city of Rotterdam, for example, 90% of which lies below sea level, has incorporated rainwater storage underneath infrastructure such as car parks, and pioneered floating houses. In poorer places, people simply endure the problem until they are forced to abandon their homes. Life is becoming more and more unbearable in many areas of flood-prone cities, such as Ho Chi Minh City, where high tides can cause floods for ten days in a month. Sandbags are often ineffective because the water comes up into houses through the sewage systems. Insurance is already a big problem in many coastal cities – the US government had to underwrite policies for residents of New Orleans after their city was inundated in the aftermath of Hurricane Katrina. But that is a costly and, many would say, doomed enterprise. Coastal cities around the world are likely to have to be abandoned and relocated as the cost of saving lives and repairing infrastructure becomes too great. Even important port cities, like New Orleans on the banks of the mighty Mississippi, will eventually become uninhabitable. And these abandoned cities will leave their marks in the sedimentary layers forming all the time, to be discovered like mythical Atlantis by divers of the far future.

But while some of our cities will drown, others will surely become more magnificent as we innovate new ways to live in the largest, most dense communities our species has known. As we utterly transform our planet humans are no longer just players in this unique production – we are observers too. Humanity can look down, godlike, on our changes with our many cameras – and even from our permanent new home in space, the International Space Station. And yet, for all our observations, for all our technological advances and modelling expertise, the future has never been harder to predict. Our threats are many, including much of what we are bringing on ourselves – but we humans are resourceful, intelligent and endlessly adaptable. As we enter the Anthropocene, our species has never had it so good – more of us than ever are living longer and better. We have the medical and technological knowledge, and logistical ability to improve the lives of poor, hungry and sick people everywhere. And that power doesn't stop with our human world, of course; we could have a similarly positive effect on our wild world. I feel enormously fortunate to have visited so much of this wondrous living planet Earth at the start of the Anthropocene, and to have seen for myself the extraordinary diversity of life created through millions of years of physical, chemical and evolutionary processes. In coming decades, human life on Earth will necessarily become increasingly sophisticated as we learn the consequences of our planetary tweaks and aim to direct our talents to less environmentally destructive ends. Now that we are aware of our impacts, we are the first species to be in a position to choose the future of our planet. I fervently hope we choose a shared future.

EPILOGUE:
THE AGE WE MADE

London, 10 October 2100

Kipp *walks over to his mother's old desk, a glass and steel piece dating from the turn of the twenty-first century made by Ikea, an iconic furniture maker of the time. From a shelf above, he pulls down a book in the classic paper format –* Adventures in the Anthropocene *– written by his mother in the year of his birth. He begins to read.*

Outside, the grapes have ripened on his window-shading vines and the annual heatwave is beginning to wane. He thinks back to the city of his childhood, the sprawl of low-rise brick terraces, the tarmacked roads filled with gasoline-powered motor cars – you'd have to travel to Afghanistan or Somalia to see their like now. London is utterly transformed. The layout of the streets is broadly the same – little different from the medieval layout, in fact – but the rows of little clay-brick houses have been replaced with the big new multi-purpose blocks made from active composites. And between them, the black asphalt road surfaces he remembers from his youth have long since disappeared. Now, carpets of sedges and mosses fill the spaces, interspersed by grasses grazed by capybara, and the planted fig and mango trees, noisy with wild birds. (Many of Britain's native trees are now very rare because of imported diseases.) There are pathways of rubber or concrete lattice for pedestrians and cyclists to use – especially handy during the monsoonal downpours, allowing the sudden deluge to soak away to the rainwater-storage tanks beneath.

It hasn't rained for weeks but the street outside his window is still green, he notices – the vegetation holds the moisture well. He watches his neighbours descend from a pod and approach the building. Larger streets and many smaller ones are strung with electric cablecarpods backed by solar power, there are regular omnipods ('buses') that ply the popular routes for free, and a range of programmable personal pods that can be hired by the minute, hour or day. These include one-rider openpods with a pull-down raincover, familypods for up to six people, and utepods for carrying larger loads. The pods are stationed at junctions and can be 'called up' to your location for a small fee. Some businesses own their own pods, such as construction companies that need to transport their maintenance workers. Pushbikes are also freely available for hire, although for longer journeys, most people pay a small charge to ride electric bikes. The mid-century vogue for remote homeworking has passed and most people now seek the community of local hubs, working remotely but with access to the latest interactive conferencing and other technology. There are specialist hubs, those for administrators, public relations, translators, etc., but most workers use mixed hubs and can walk to their nearest.

As he leafs through the book, Kipp begins to realise what a huge issue energy was at the beginning of the Anthropocene, and how dependent people were on fossil fuels. He remembers his parents fretting about gas bills and petrol prices, but hadn't truly understood how the quest for energy had defined almost every aspect of the beginning of the twenty-first century. As fossil fuels became more and more expensive, through scarcity and taxation, civil unrest grew in countries where the majority had little access. However, by the late 2020s, solar power stations across the world's deserts were supplying around one-third of global power, with nuclear-fission plants (using thorium or uranium) providing another third. Chile led the way with solar reliance, deciding in the end not to exploit the hydropower potential of Patagonia. In Laos, however, several hydrodams were built on the Mekong, benefiting Thailand, China and northern Vietnam with cheap electricity but further impoverishing millions of people in Laos, Cambodia and southern Vietnam, which lost its entire coffee, rice and other agricultural production to salination, soil erosion and storm damage. The impact of the dams on the basin environment caused near collapse of the fisheries (the enormous Mekong catfish went extinct despite attempts to keep some in aquaria), and led to the migration of 3 million people from the lower Mekong region to China, Thailand and beyond. This was nothing compared to the deadly clashes that followed China's decision to dam the Brahmaputra River, however. Most of those dams were later removed because the river flow was so reduced after glacier disappearance that they became inefficient and required continual maintenance.

In 2050, the first full-scale fusion power plant opened in Germany (after successful experiments at ITER, in France, in the 2030s), and by 2065 there were thirty around the world, supplying one-third of global electricity. Now, fusion provides more than half of the world's power, with solar making up around 40% and hydro, wind and waste (biomass) supplying the rest. Electricity is transported across continents using high-voltage direct current lines made with high-temperature superconductors, stored at source in various ways, including in molten-salt banks, hydropotential dams, hydrogen cylinders and efficient chemical batteries.

The effect of cheap, abundant energy brought about the most radical social, technological and environmental change of the century. Transport became almost entirely electrified as the cheapest and most practical solution. Electric cars took off in cities from the US to China. Bolivia, and to a lesser extent Chile and Peru, grew wealthy exporting lithium for electric car batteries. In

Europe, many followed the London model and eliminated private cars from their major cities. Meanwhile, aircraft and some industrial processes were fuelled by hydrogen. In the cities, skies cleared of pollutants, and people began to use their streets for markets and parties. Combined heat and power started to be incorporated into residential and commercial blocks, assisted by ground-source heat pumps. Across the world there was near-universal access to electricity, raising millions from poverty and giving them access to global online resources and community, health care and transport.

Naturally, the cities – and nations, although by mid-century cities had surpassed nations in importance – that were early adopters of electric technology and implementation, gained the most in terms of where the industry was. Thus, Denmark, for example, dominates the market for transit pods, while Britain leads in tidal power, deriving a quarter of its energy from flood-mitigating tidal lagoons built along its coast.

Thinking back on the wars and the blackouts that regularly hit even major cities like London over a particularly grim decade, it's amazing to think that energy is no longer humanity's biggest problem. Instead, fresh water is. The early signs were there in the book, Kipp sees, but no one could have predicted the way water shortages would affect everything. Rice, for example – he remembers eating rice as often as pasta during his childhood but during his twenties it began getting more expensive. Now, it's really only available on the black market or at huge cost as a garnish in upscale restaurants. Even pasta and bread are more costly. A huge amount of energy currently goes towards desalination of seawater and transporting it inland. But it's not enough. Many countries delayed building storage facilities for water, and sea-level rise has caused important aquifers to become fouled across the world. Reading about glaciers, those ancient natural water stores, he realises that he has never seen one outside of a movie. The only natural glaciers on Earth are in Greenland and Antarctica, and, although colleagues of his have visited the thriving cities on Greenland, Kipp has never been there, even for business. There are glaciers outside of the poles but they are artificial, like the ones Norphel built in the book. Nowadays, these seasonal glaciers – and there are *many, made each year in tropical mountains, in strategic places to feed rivers and cropland – need extra help, though. The engineers first have to cool the area with targeted solar radiation management (generally by spraying reflective particles into the stratosphere above the region, although shading satellites have been launched in some places). Then cloud seeding is used to help produce precipitation in the area, although this is still only 60% effective. The resulting glaciers are too precious to be skied on, and most disappear by February in the Himalayas (by August in the Andes). The artificial glaciers are vital, but they're a small drop on to a thirsty planet. Vast covered canal networks are under construction on every continent to bring piped desal water from the coasts inland, and reservoir-building continues. Meanwhile, water scarcity continues to cause regional unrest and has contributed to several conflicts in recent decades, including the ongoing Mexican–US guerrilla war, the 2040–5 civil war in China, the Sinai revolution and the Indian separatist clashes. Domestic water storage, combined with hotter conditions, has also led to waves of insect-borne disease outbreaks, including fatal new strains of malaria and dengue, which have caused epidemics as far north as southern France. Meanwhile, viruses liberated by melting of the permafrost proved just as deadly to humans, animals and crops as before their incarceration. Some, like smallpox and polio, had been successfully eradicated only to be resurrected into an unvaccinated population.*

Kipp flicks ahead in the book and alights on a passage about the Maldives. He leans back in his chair and closes his eyes. The vine leaves filter the sun into a warm amber glow that washes over him, taking him back through the vista of years until he sees a child standing on the shore, holding tightly to the hand of his mother as the waves lap their bare feet. It was 2020, the year he turned seven, and they were about to go snorkelling in the Maldives. It was an introduction to another world – like stepping through the looking glass. Brightly coloured fishes flitted above a rich mosaic of coral. Wafting sea anemones and fans concealed crabs, elusive seahorses, and tiny sparkling fish, while rays glided serenely above. As he snorkelled over the iridescent carpet, he caught a glimpse of a whale shark. The giant from the deep was cruising by looking for plankton. He spent a magical few days in this aqua wonderworld, exploring and delighting in its extraordinary diversity. It was the last time he would see such splendour – coral reefs have all disappeared and, with them, many of the swimming ocean jewels that sparkle only in his memory. Algae is now grown and farmed where people once snorkelled and fished – it loves the warm, carbon-rich water. Researchers have grown and nurtured some sections of coral reef as living museums – there's one in Queensland, Australia – which require regular ocean liming to counteract the acidity, retractable sunshades and massive cooling fans. But they are a poor match for the natural marvels he'd experienced on that trip to the Maldives – a country that itself no longer exists, he thinks, with a start.

He stands up, reflecting on what has been lost. This past century has been brutal, Kipp says aloud as he makes his way to the kitchen. The sea-level rise and storms that swallowed the Maldives also stole large portions of other heavily populated bays and coasts. Hundreds of thousands of Bangladeshis, Indians and Burmese died in the Great Bengal Flood of '69, which also drowned the Sundarbans, the last remaining wild tiger habitat. Globally, there has been a coastal exodus, as cities from Mumbai to New Orleans became uninhabitable. Some, like Miami, were rebuilt inland (although, being far from the beach, New Miami is now unrecognisable from its namesake), while others, like Khulna in Bangladesh, were abandoned. Its former inhabitants became part of the great Bangladeshi diaspora. West of Woolwich, London survived the recent massive floods thanks to the Thames Barrier, although it was a close thing and millions of pounds' worth of damage was done to the estuary cities, with the airport completely washed away. Remarkably, there was no loss of life, because of the early evacuation warning systems, but several thousand people are living in temporary shelters on the Downs. The Thames Barrier fortification programme is under way now.

Kipp opens his electric mealmaker, cracks two eggs into one capsule, adds an arthromeat sausage (made with insect protein), a couple of mockbacon rashers (from lab-grown pork tissue), and some mashed garlic cassava to other capsules and sets the programme. From his vegetable box, he takes out some tomatoes, crispy seaweed, samphire and kale, and makes a salad. Only the kale is local – grown in one of the neighbourhood allotment streets. The samphire and seaweed are grown somewhere in Europe, whereas the tomatoes – like most imported fruits and vegetables – were grown hydroponically in one of the many vast 'sundrops' covering the deserts of north Africa. The first of these had already made it into the book, he sees, an early Australian version of the honeycomb cardboard greenhouse, irrigated by seawater that was desalinated by solar power. Over the past decades the concept grew bigger and even more efficient, so that now, more than half of the world's crops are produced in deserts in this way. The cooling fans counteract the negative effect of the hot outside temperatures, the increased carbon dioxide in the air makes the plants grow faster, and the irrigation system is extremely efficient and independent of weather. Hydroponics allows the green-houses to be many storeys high, growing crops over multiple levels on the same footprint, while the nutrients can be supplied accurately and any surplus recycled. Sundrops have replaced many agricultural fields across China, Australia, India, the US and much of Africa. With the entire process of sowing, growth and harvesting automated, they only use a minimal workforce, so when they were first introduced in the 2030s, there were protests, particularly in India, which delayed the introduction there for nearly a decade. But, with the majority of people living in cities, and the sundrops sited in inhospitable desert (often formerly fertile land that had become desertified), they were soon accepted as the best way to feed people without destroying further wildland for intensive agriculture.

He cracks open a sorghum-millet beer and pours some into a glass. He raises it to his lips as the mealmaker tings its readiness. Kipp ignores it and takes a long slow drink. Ting! Ting! Grunting in irritation, he rises and attends to the machine, retrieving a perfectly cooked meal, while the automaton cleans itself. As he eats, he continues to leaf through the book in the practised manner of one accustomed to eating alone. He reads about the animals that used to roam the planet but which are now extinct in the wild: gorillas and elephants, that went the same way as coral reefs and tigers. He marvels at how many farm animals *there once were. Nowadays, most animal protein comes from insects, with other meats grown synthetically on great cell-culture frames, and fish is entirely farmed.*

His reveries are interrupted by an insistent beeping as a UAV (unmanned aerial vehicle, or 'drone') delivery arrives and tries to access the portal. After several failed attempts, it flies off again. Following an attack by hackers, the portal code has been changed and the delivery company clearly still has the old one. Kipp sighs, he was expecting a couple of new flat-weave algal shirts (once-common cotton is now very pricey), and has already thrown his frayed ones into the biodigester.

Kipp is 87 now, yet his mind is sharp and his body sprightly (thanks to artificial joints, replacement cartilage grown from stem cells, eye and ear parts, and his decadal arterial scrub). He has the lifestyle, he thinks, a 60-year-old might have had in the year he was born: he works two days a week, he plays tennis with virtual and flesh opponents, he meets his virtual and flesh friends and family regularly. When he was born, there were just 7 billion people on the planet; now there are more than 10 billion, fewer than many had predicted. In fact, several city states are trying to incentivise women to have children, and encouraging immigration of minors – families with two or more children get special concessions, as governments worry about an ageing population that is working fewer hours and buying less stuff. So far, though, the schemes have had little success, and population everywhere is falling. The global fertility rate is just 0.7 births per woman, and analysts warn that the human race risks dying out within a century.

As he reads through the book, Kipp becomes newly aware of how much the planet has changed in his lifetime. The Anthropocene predictions have in many cases played out: the world is several degrees warmer and reliant on cooling technologies to make much of it habitable; the ocean is higher and more acidic; the weather everywhere can be extreme and violent; coastlines have changed and maps have been redrawn. Several species have gone extinct in the wild (or altogether) since the book was written. Many places have become unrecognisable – deserts cover vast new swathes of the Earth, and yet the patchworks of agricultural fields he remembers seeing as a child from aeroplanes cover a smaller area. The world's great tropical rainforests have shrunk considerably and are now aided by stratospheric cooling technologies and cloud seeding to protect what is left from drought and forest fires. Meanwhile, great new forests have transformed the top of the world, greening the Arctic Circle and providing new habitat for relocated tropical forest species.

What is ahead, Kipp wonders. How will the Anthropocene unfurl in the twenty-second century? Artificial trees have been planted on every continent to suck carbon dioxide out of the atmosphere, and the new Arctic forests and oceanic algal gardens are helping. But, with the death of so much tropical forestry, and the too-slow relinquishment of fossil fuels over the past century, it is an uphill struggle. Temperatures continue to mount and the great fear is that the trapped methane in the Arctic seabed could escape, melting the last ice sheets in Greenland and Antarctica.

Despite this – perhaps even because of it – the world has become a kinder place, Kipp decides. The terrible wars, the famines, the terrorism, extremism and hate, the drownings and deaths of hundreds of thousands of migrants, the humanitarian crises . . . they seem to be over now. The last two decades have been wonderful years of unparalleled global peace. Billions have been lifted out of poverty – starvation and the type of destitution that existed even fifty years ago no longer exists. The great global mix-up of people that has occurred as a result of climate migration, urbanisation and online networks has produced a new, socially mobile, egalitarian society. The world's giant cities force people to live together in close but diverse communities, and it has gener-ated a spirit of cooperation – a wartime mentality – as people unite against the same external threats of resource limitations, climate disaster and disease. Every year there seems to be another major international agreement to deal with global issues from emissions to overfishing, with governments egging each other on to greater, faster action. Humanity seems to have grown up, he reflects.

Outside his window, children are playing in the green street. Kipp turns the page and begins to read the Epilogue.

ACKNOWLEDGEMENTS

This book – this journey – would never have been possible without the help, patience and many acts of kindness shown to me by so many people around the world. Only a tiny proportion have been named in the book, but to all the others who helped, thank you, I'm so grateful. Some of these inspirational people include Ambica Shrestha of Dwarika's in Kathmandu; Bahadur Khadka of Jungle Cottage in Bardia National Park; Rabi Karmacharya of Open Learning Exchange Nepal, Jora M of Shahi Palace in Jaisalmer; the researchers at TERI, ICIMOD, ICRISAT, Bangladesh Centre for Agricultural Studies; Mrs Kumari who filled my belly so well at her home near Varkala, India; C. B. Ramkumar and Lallita Ramkumar of Our Native Village near Bangalore in India; water expert Tushaar Shah in Anand; Sachin Oza at the Development Support Centre in Anand; Rajesh Shah at the Vikas Centre in Ahmedabad; Paul Roberts who helped in the Maldives; Khidir Box in West Bengal; Blake Ratner in Cambodia; Dan

Cashdan of the Climate Community; Amare G-Egziabher of UNHCR in Addis Ababa; Milkyas Debebe of Gaia Project in Ethiopia; Aliza le Roux; Peter Okoth at the Tropical Soil Biology and Fertility Institute in Nairobi; Yves Rwbutso in Rwanda; Alissa and Josh Ruxin of Heaven in Kigali; Samantha Ludick in Malawi; Tommy Collard in Swakopmund; Steve Phillips in Lismore; Jamie Buchanan-Dunlop of Digital Explorer; Hugo Streeter Cortez in Falda Verde; Effrain Peducasse Castro in Sucre; Vicky Ossio of La Senda Verde in Yolosa; Francis Chauvel of Miraflores guest house in Lima; Carmen Felipe-Morales and Ulises Moreno in Pachacamac; Beth King of Smithsonian Tropical Research Institute in Panama; Sandro Alviani and Encar Garcia of Jaguar Rescue Centre in Playa Chiquita; Alvaro Molino in Ometepe; Tessa de Goede in Antigua, Mike Shanahan of IIED and Jan Zalasiewicz.

A journey like this is expensive and difficult, and I'm enormously grateful to the editors that gave me the work that made it possible, especially Caroline Williams, formerly of *New Scientist*, Michelle Martin at the BBC, John Travis of *Science* magazine, Don Hoyt Gorman, formerly of *Seed* magazine, John Pickrell of *Australian Geographic*, Jon Fildes of BBC Future, and my wonderful editor at Chatto & Windus, Becky Hardie, who believed in the book from its inception, counselled me through the writing process with skill and patience, and made it a far better read. And my copy-editor David Milner, who saved me from several embarrassing errors.

None of this would have been achievable without the support, encouragement and love of friends and family, particularly Emma Young and Jo Marchant who endured my whingeing with empathy and no little stoicism, and took me out to play. My grandparents, Teresa Vince and my much-missed nagypapa Stephen Vince, were a continual lifeline from Sydney; my parents Georgina and Ivan Vince were unfailingly supportive of me and this project from the beginning, despite my regular forays into dangerous places, and provided essential childcare during the writing process; my brother David Vince who welcomed me squatting in his flat for weeks; and my wonderful son Kipp, whose recent arrival delayed the book's delivery in the best way possible. But most of all, my partner Nick, who travelled with me from icy peaks to desert sands, from jungle to ocean, who was my companion on uncomfortable twenty-eight-hour bus rides, suffered far too many cold bucket showers, helped negotiate with scary armed men, carried my backpack when I couldn't, and took all the photos: thank you.

NOTES

Introduction

1. Höning, D. et al., 'Biotic vs abiotic Earth: A model for mantle hydration and continental coverage, Planetary and Space Science', available online 25 October 2013, ISSN 0032–0633, http://dx.doi.org/10.1016/j.pss.2013.10.004.

2. Raup, D. M., & Sepkoski, J. J., 'Periodicity of extinctions in the geologic past', *PNAS* 81 (1984), 801–5.

3. Mora, C., Tittensor, D. P., Adl, S., Simpson, A. G. B., & Worm, B., 'How Many Species Are There on Earth and in the Ocean?', *Public Library of Science Biology* (ed. G. M. Mace), 9(8) (2011), e1001127. doi:10.1371/journal.pbio.1001127.

4. Great Acceleration: IGBP, at www.igbp.net/globalchange/greatacceleration.4.1b8ae20512db692f2a680001630.html.

5. Zalasiewicz, J., Williams, M., Haywood A., & Ellis M., 'The Anthropocene: a new epoch of geological time?', *Philosophical Transactions of the Royal Society* A (2011), 369, 833–4.

6. Gurney, K. R. et al., 'Quantification of Fossil Fuel CO_2 Emissions on the Building/Street Scale for a Large US City', *Environmental Science & Technology* 46 (2012), 12194–202.

7. Crutzen, P. J., 'Geology of mankind', *Nature* 415 (6867) (2002), 23. doi:10.1038/415023a.

8. Subcommission on Quaternary Stratigraphy, ICS Working Groups, at http://quaternary.stratigraphy.org/workinggroups/anthropocene/.

9. TS.2.1.1, 'Changes in Atmospheric Carbon Dioxide, Methane and Nitrous Oxide – AR4 WGI Technical Summary', at www.ipcc.ch/publications_and_data/ar4/wg1/en/tssts-2-1-1.html.

10. Rockström, J., Steffen, W., Noone, K., Persson, Å., Chapin, F. S., Lambin, E. F., Lenton, T. M. et al., 'A safe operating space for humanity', *Nature* 461 (7263) (2009), 472–5. doi:10.1038/461472a. For further reading see Mark Lynas, *The God Species* (2011).

11. AR4 SYR 'Synthesis Report Summary for Policymakers 5 – The long-term perspective', at www.ipcc.ch/publications_and_data/ar4/syr/en/spms5.html ◆ Vuuren, D. P. et al., 'The representative concentration pathways: an overview', *Climatic Change* 109 (2011), 5–31.

12. 'World Population Prospects: The 2012 Revision', UN Department of Economic and Social Affairs (2013), at http://esa.un.org/unpd/wpp/Documentation/pdf/WPP2012_HIGHLIGHTS.pdf.

13. La Rue, F., 'United Nations Human Rights Council Report of the Special Rapporteur on the promotion and protection of the right to freedom of opinion and expression' (2011), at http://www2.ohchr.org/english/bodies/hrcouncil/docs/17session/A.HRC.17.27_en.pdf.

Chapter 1: Atmosphere

1. Srivastava, L., 'Mobile phones and the evolution of social behaviour', *Behaviour & Information Technology* 24(2) (2005), 111–29. doi:10.1080 /01449290512331321910.

2. 'OKR: Poor People Using Mobile Financial Services: Observations on Customer Usage and Impact from M-PESA', at https://openknowledge. worldbank.org/handle/10986/9492.

3. Ray Kurzweil, *How to Create a Mind: The Secret of Human Thought Revealed* (2012).

4. Gibson, C. C., & Long, J. D., 'The presidential and parliamentary elections in Kenya, December 2007', *Electoral Studies* 28 (2009), 497–502.

5. 'Cisco Visual Networking Index: Global Mobile Data Traffic Forecast Update, 2012–2017', at www.cisco.com/en/US/solutions/collateral/ ns341/ns525/ns537/ns705/ns827/white_paper_c11-520862.html.

6. 'UN Millennium Development Goals Report' (2012), at www.un.org/ millenniumgoals/pdf/MDG%20Report%202012.pdf.

7. Myllyvirta, L., 'Silent Killers: Why Europe must replace coal power with green energy' (2013), Greenpeace, at www.greenpeace.org/ international/Global/international/publications/climate/2013/Silent-Killers.pdf.

8. World Bank study, 2007, at http://siteresources.worldbank.org/ INTEAPREGTOPENVIRONMENT/Resources/China_Cost_of_Pollution. pdf ◆ Kahn J., & Yardley J., 'As China Roars, Pollution Reaches Deadly Extremes', *New York Times*, at www.nytimes.com/2007/08/26/world/ asia/26china.html?_r=0 ◆ McGregor, R., '750,000 a year killed by Chinese pollution', *Financial Times* (2007), at www.ft.com/cms/s/8f40e248-28c7-11dc-af78-000b5df10621.Authorised=false.html?_i_location=http%3A%2F%2Fw ww.t.com%2Fcms%2Fs%2F0%2F8f40e248-28c7-11dc-af78-000b5df10621. html%3Fsiteedition%3Duk&siteedition=uk&_i_referer=#axzz2qT1pUxzT.

9. Bates, T. S. et al., 'Measurements of atmospheric aerosol vertical distributions above Svalbard, Norway, using unmanned aerial systems (UAS)', *Atmospheric Measurement Techniques* 6 (2013), 2115–20.

10. Haag, A. L., 'The even darker side of brown clouds', *Nature Reports, Climate Change* (2007), 52–3. doi:10.1038/climate.2007.41.

11. Atmospheric Brown Cloud Regional monitoring and assessment, ICIMOD (2012), http://geoportal.icimod.org/MENRISFactSheets/ Sheets/13icimod-atmospheric_brown_cloud_regional_monitoring_ and_assessment.pdf.

12. Auffhammer, M., Ramanathan, V., & Vincent, J. R., 'Climate change, the monsoon, and rice yield in India', *Climatic Change* 111 (2011), 411–24.

13. Bond, T. C., Doherty, S. J., Fahey, D. W., Forster, P. M., Berntsen, T., DeAngelo, B. J., Flanner, M. G. et al., 'Bounding the role of black carbon in the climate system: A scientific assessment', *Journal of Geophysical Research: Atmospheres* 118(11) (2013), 5380–552. doi:10.1002/ jgrd.50171.

Chapter 2: Mountains

1. Chen, I.-C., Hill, J. K., Ohlemuller, R., Roy, D. B., & Thomas, C. D., 'Rapid Range Shifts of Species Associated with High Levels of Climate Warming', *Science* 333(6045) (2011), 1024–6. doi:10.1126/science.1206432.

2. Pauli, H., Gottfried, M., Dullinger, S., Abdaladze, O., Akhalkatsi, M., Alonso, J. L. B., Coldea, G. et al., 'Recent Plant Diversity Changes on Europe's Mountain Summits', *Science* 336(6079) (2012), 353–5. doi:10.1126/science.1219033.

3. Dyurgerov, M., & Meier M., 'Glaciers and the changing earth system' (2004), at https://instaar.colorado.edu/uploads/occasional-papers/ OP58_dyurgerov_meier.pdf.

4. Radić, V., & Hock, R., 'Regionally differentiated contribution of mountain glaciers and ice caps to future sea-level rise', *Nature Geoscience* 4(2) (2011), 91–4. doi:10.1038/ngeo1052.

5. Campra, P., Garcia, M., Canton, Y., & Palacios-Orueta, A., 'Surface temperature cooling trends and negative radiative forcing due to land use change toward greenhouse farming in southeastern Spain', *Journal of Geophysical Research* 113(D18) (2008). doi:10.1029/2008JD009912.

6. Dutton, E. G., & Christy, J. R., 'Solar radiative forcing at selected locations and evidence for global lower tropospheric cooling following the eruptions of El Chichón and Pinatubo', *Geophysical Research Letters* 19 (1992), 2313–6.

Chapter 3: Rivers

1. Threats to wetlands: WWF, 'Rivers at Risk' report (2004), at wwf.panda. org/about_our_earth/about_freshwater/intro/threats/index.cfm.

2. Syvitski, J. P. M., Kettner, A. J., Overeem, I., Hutton, E. W. H., Hannon, M. T., Brakenridge, G. R., Day, J. et al., 'Sinking deltas due to human activities', *Nature Geoscience* 2(10) (2009), 681–6. doi:10.1038/ngeo629.

3. 'Harrabin's Notes: Safe Assumptions', BBC (2012), at www.bbc.co.uk/ news/science-environment-18020432 ◆ UN, '800 Million People without Drinking Water', Global Research, at www.globalresearch.ca/ un-800-million-people-without-drinking-water/23843.

4. Darwall, W., 'Freshwater Key Biodiversity Areas: work in progress', IUCN Species Programme, at www.unesco.org/mab/doc/iyb/scConf/ Darwall.pdf.

5. Duckworth, J. W., 'Small carnivores in Laos: a status review with notes on ecology, behaviour and conservation', *Small Carnivore Conservation* 16 (1997), 1–21.

6. Dung, V. V., Giao, P. M., Chinh, N. N., Tuoc, D., Arctander, P., & MacKinnon, J., 'A new species of living bovid from Vietnam', *Nature* 363(6428) (1993), 443–5. doi:10.1038/363443a0.

7. Barlow, C., Baran, E., Halls, A. S., & Kshatriya, M., 'How much of the Mekong fish catch is at risk from mainstream dam?', *Catch and Culture* 14 (2008).

8. Fred Pearce, *When the Rivers Run Dry* (2007).

9. Bonheur, N., & Lane, B. D., 'Natural resources management for human security in Cambodia's Tonle Sap Biosphere Reserve', *Environmental Science & Policy* 5 (2002), 33–41.

10. 'Vietnam's rice bowl threatened by rising seas', *Guardian* (2011), at www.theguardian.com/environment/2011/aug/21/vietnam-rice-bowl-threatened-rising-seas.

11. 'Water Services Industry to Double Revenues by 2020 in Efforts to Tackle Scarcity, Says New BofA Merrill Lynch Global Research Report', Bank of America Newsroom, at http://newsroom.bankofamerica.com/press-release/economic-and-industry-outlooks/water-services-industry-double-revenues-2020-efforts-ta.

12. 'Water footprint and virtual water', at www.waterfootprint.org/?page=files/home ◆ Arjen Y. Hoekstra, *The Water Footprint of Modern Consumer Society* (2013).

13. 'Act now to avert a global water crisis', *New Scientist* (24 May 2013), at www.newscientist.com/article/dn23597-act-now-to-avert-a-global-water-crisis.html#.

Chapter 4: Farmlands

1. 'Issues Brief on Desertification, Land Degradation and Drought', led by UNCCD, at http://sustainabledevelopment.un.org/content/documents/1803tstissuesdldd.pdf.

2. Gibbs, H. K. et al., 'Tropical forests were the primary sources of new agricultural land in the 1980s and 1990s', *Proceedings of the National Academy of Sciences* 107 (2010), 16732–7.

3. UN report, 'Water Statistics – Water Use', at www.unwater.org/statistics_use.html.

4. UN report, 'Food Production Must Double by 2050 to Meet Demand from World's Growing Population, Innovative Strategies Needed to

Combat Hunger, Experts Tell Second Committee' (2001), at www.un.org/News/Press/docs/2009/gaef3242.doc.htm.

5. Brown L. R., *Plan B 3.0: Mobilizing to Save Civilization* (2008), Chapter 5, 'Natural Systems Under Stress: Advancing Deserts', at www.earthpolicy.org/books/pb3/PB3ch5_ss5.

6. Jarvis, A., Ramirez-Villegas, J., Herrera Campo, B. V., & Navarro-Racines, C., 'Is Cassava the Answer to African Climate Change Adaptation?', *Tropical Plant Biology* 5(1) (2012), 9–29. doi:10.1007/s12042-012-9096-7.

7. Food and Agriculture Organization of the United Nations, 'The state of food and agriculture; women in agriculture: closing the gender gap for development' (2011), at www.fao.org/docrep/013/i2050e/i2050e00.htm.

8. De Gorter, H., 'Explaining agricultural commodity price increases: The role of biofuel policies', in Oregon State University conference on rising food and energy prices, 'US food policy at a crossroads' (2008), at http://arec.oregonstate.edu/sites/default/files/faculty/perry/degorter.pdf.

9. World Bank, Food Price Watch (February 2011), at www.worldbank.org/foodcrisis/food_price_watch_report_feb2011.htm ◆ Action Aid, 'Meals per gallon: the impact of industrial biofuels on people and global hunger' (2010), at www.actionaid.org.uk/sites/default/files/doc_lib/meals_per_gallon_final.pdf.

10. Food and Agriculture Organization of the United Nations, '2000 World Census of Agriculture Main Results and Metadata by Country (1996–2005)' (2010), at www.fao.org/docrep/013/i1595e/i1595e00.htm.

11. State of the World 2011, *Innovations that Nourish the Planet*, Chapter 6, 'Africa's Soil Fertility Crisis and the Coming Famine', Nourishingtheplanet.org, at http://blogs.worldwatch.org/nourishingtheplanet/wp-content/uploads/2011/02/Chapter-6-Policy-Brief_new.pdf.

12. Montgomery, D. R., 'Soil erosion and agricultural sustainability', *Proceedings of the National Academy of Sciences* 104 (2007), 13268–72.

13. Pimentel, D., 'Soil Erosion: A Food and Environmental Threat', *Environment, Development and Sustainability* 8 (2006), 119–37.

14. 'Effects of Rising Atmospheric Concentrations of Carbon Dioxide on Plants', Learn Science at Scitable, at www.nature.com/scitable/knowledge/library/effects-of-rising-atmospheric-concentrations-of-carbon-13254108.

15. Seufert, V., Ramankutty, N., & Foley, J. A., 'Comparing the yields of organic and conventional agriculture', *Nature* 485(7397) (2012), 229–32. doi:10.1038/nature11069.

16. Ainsworth, E. A., & Ort, D. R., 'How Do We Improve Crop Production in a Warming World?', *Plant Physiology* 154(2) (2010), 526–30. doi:10.1104/pp.110.161349 ◆ Schlenker, W., & Roberts, M. J., 'Nonlinear temperature effects indicate severe damages to US crop yields under climate change', *Proceedings of the National Academy of Sciences* 106(37) (2009), 15594–8. doi:10.1073/pnas.0906865106.

17. Lobell, D. B., Schlenker, W., & Costa-Roberts, J., 'Climate Trends and Global Crop Production Since 1980', *Science* 333(6042) (2011), 616–20. doi:10.1126/science.1204531.

18. 'The use of genetically modified crops in developing countries: a guide to the Nuffield Council on Bioethics Discussion Paper', at www.nuffield-bioethics.org/sites/default/files/GM%20Crops%20short%20version%20FINAL.pdf.

19. Herman, R. A., & Price, W. D., 'Unintended Compositional Changes in Genetically Modified (GM) Crops: 20 Years of Research', *Journal of Agricultural and Food Chemistry* 61 (2013), 11695–701 ◆ DeFrancesco, L., 'How safe does transgenic food need to be?' *Nature Biotechnology* 31 (2013), 794–80 ◆ World Health Organization (Department of Food Safety, Zoonoses and Foodborne Diseases), 'Modern food technology, human health and development: an evidence-based study' (2005).

20. Bond-Lamberty, B., & Thomson, A., 'Temperature-associated increases in the global soil respiration record', *Nature* 464(7288) (2010), 579–82. doi:10.1038/nature08930.

21. 'Save Food: Global Initiative on Food Losses and Waste Reduction: Key Findings', at www.fao.org/save-food/key-findings/en/.

22. Food and Agriculture Organization of the United Nations, 'Food wastage footprint impacts on natural resources: summary report' (2013).

23. Bonhommeau, S., Dubroca, L., Le Pape, O., Barde, J., Kaplan, D. M., Chassot, E., & Nieblas, A.-E., 'Eating up the world's food web and the human trophic level', *Proceedings of the National Academy of Sciences*, 110(51) (2013), 20617–20. doi:10.1073/pnas.1305827110.

24. Tuomisto, H. L., & Teixeira de Mattos, M. J., 'Environmental Impacts of Cultured Meat Production', *Environmental Science & Technology* 45(14) (2011), 6117–23. doi:10.1021/es200130u.

25. Earth Policy Institute, 'Data Highlights: Peak Meat: US Meat Consumption Falling', at www.earth-policy.org/data_highlights/2012/highlights25.

Chapter 5: Oceans

1. Levermann, A., Clark, P. U., Marzeion, B., Milne, G. A., Pollard, D., Radic, V., & Robinson, A., 'The multimillennial sea-level commitment of global warming', *Proceedings of the National Academy of Sciences* 110(34) (2013), 13745–50. doi:10.1073/pnas.1219414110.

2. Peter F. Sale, *Our Dying Planet* (2011), is good on ocean threats.

3. FAQs about ocean acidification. at <http://www.epoca-project.eu/index.php/what-is-ocean-acidification/faq.html.

4. McClanahan, T. R., Graham, N. A. J., MacNeil, M. A., Muthiga, N. A., Cinner, J. E., Bruggemann, J. H., & Wilson, S. K., 'Critical thresholds and tangible targets for ecosystem-based management of coral reef fisheries', *Proceedings of the National Academy of Sciences* 108(41) (2011), 17230–3. doi:10.1073/pnas.1106861108.

5. Jones, A., Berkelmans, R., van Oppen, M. J., Mieog, J., & Sinclair, W., 'A community change in the algal endosymbionts of a scleractinian coral following a natural bleaching event: field evidence of acclimatization', *Proceedings of the Royal Society* B: Biological Sciences 275(1641) (2008), 1359–65. doi:10.1098/rspb.2008.0069.

6. Silverman, J., Lazar, B., Cao, L., Caldeira, K., & Erez, J., 'Coral reefs may start dissolving when atmospheric CO_2 doubles', *Geophysical Research Letters* 36 (2009).

7. Bednaršek, N., Tarling, G. A., Bakker, D. C. E., Fielding, S., Jones, E. M., Venables, H. J., Ward, P. et al., 'Extensive dissolution of live pteropods in the Southern Ocean', *Nature Geoscience* 5(12) (2012), 881–5. doi:10.1038/ngeo1635 ◆ Waldbusser, G. G., Brunner, E. L., Haley, B. A., Hales, B., Langdon, C. J., & Prahl, F. G., 'A developmental and energetic basis linking larval oyster shell formation to acidification sensitivity', *Geophysical Research Letters* 40(10) (2013), 2171–6. doi:10.1002/grl.50449.

8. Bignami, S., Enochs, I. C., Manzello, D. P., Sponaugle, S., & Cowen, R. K., 'Ocean acidification alters the otoliths of a pantropical fish species with implications for sensory function', *Proceedings of the National Academy of Sciences*, 110(18) (2013), 7366–70. doi:10.1073/pnas.1301365110.

9. Nilsson, G. E., Dixson, D. L., Domenici, P., McCormick, M. I., Sørensen, C., Watson, S.-A., & Munday, P. L., 'Near-future carbon dioxide levels alter fish behaviour by interfering with neurotransmitter function', *Nature Climate Change* 2(3) (2012), 201–4. doi:10.1038/nclimate1352.

10. Ilyina, T., Zeebe, R. E., & Brewer, P. G., 'Future ocean increasingly transparent to low-frequency sound owing to carbon dioxide emissions', *Nature Geoscience* 3(1) (2009), 18–22. doi:10.1038/ngeo719.

11. Hall-Spencer, J. M. et al., 'Volcanic carbon dioxide vents show ecosystem effects of ocean acidification', *Nature* 454 (2008), 96–9.

12. Cquestrate: The Idea, at www.cquestrate.com/the-idea.

13. Harvey, L. D. D., 'Mitigating the atmospheric CO_2 increase and ocean acidification by adding limestone powder to upwelling regions', *Journal of Geophysical Research* 113(C4) (2008). doi:10.1029/2007JC004373.

14. World Meteorological Organization, 'Climate, carbon and coral reefs' (2010), at www.wmo.int/pages/prog/wcp/agm/publications/documents/Climate_Carbon_CoralReefs.pdf.

15. Field, I. C., Meekan, M. G., Buckworth, R. C., & Bradshaw, C. J. A., 'Susceptibility of sharks, rays and chimaeras to global extinction', *Advances in Marine Biology* 56 (2009), 275–363.

16. Cowtan, K., & Way, R. G., 'Coverage bias in the HadCRUT4 temperature series and its impact on recent temperature trends', *Quarterly Journal of the Royal Meteorological Society* (2013). doi:10.1002/qj.2297.

17. Colgan, W., Steffen, K., McLamb, W. S., Abdalati, W., Rajaram, H., Motyka, R., Phillips, T. et al., 'An increase in crevasse extent, West Greenland: Hydrologic implications', *Geophysical Research Letters* 38(18) (2011). doi:10.1029/2011GL048491.

18. Perrette, M., Landerer, F., Riva, R., Frieler, K., & Meinshausen, M., 'A scaling approach to project regional sea level rise and its uncertainties', *Earth System Dynamics* 4 (2013), 11–29.

19. Epstein, H. E. et al., 'Arctic Report Card on Vegetation', at www.arctic.noaa.gov/reportcard/vegetation.html.

20. Whiteman, G., Hope, C., & Wadhams, P., 'Climate science: Vast costs of Arctic change', *Nature* 499(7459) (2013), 401–3. doi:10.1038/499401a.

21. Poloczanska, E. S., Brown, C. J., Sydeman, W. J., Kiessling, W., Schoeman, D. S., Moore, P. J., Brander, K. et al., 'Global imprint of climate change on marine life', *Nature Climate Change* 3(10) (2013), 919–25. doi:10.1038/nclimate1958.

22. Jones, B. M., Arp, C. D., Jorgenson, M. T., Hinkel, K. M., Schmutz, J. A., & Flint, P. L., 'Increase in the rate and uniformity of coastline erosion in Arctic Alaska', *Geophysical Research Letters* 36(3) (2009). doi:10.1029/2008GL036205.

23. De Séligny, J. F. P., & Grainger, R., 'The State of World Fisheries and Aquaculture 2010' (2010), at http://41.215.122.106/dspace/handle/0/210.

24. FAO Fisheries & Aquaculture, 'Small-scale and artisanal fisheries', at www.fao.org/fishery/topic/14753/en.

25. De Séligny, J. F. P., & Grainger, R., 'The State of World Fisheries and Aquaculture 2010' (2010), at http://41.215.122.106/dspace/handle/0/210.

26. Halweil, B., *Worldwatch Report: Farming Fish for the Future* (2008).

27. Richardson, A. J., Bakun, A., Hays, G. C., & Gibbons, M. J., 'The jellyfish joyride: causes, consequences and management responses to a more gelatinous future', *Trends in Ecology & Evolution* 24 (2009), 312–22.

28. 'Bird Scaring Lines Protect Seabirds', at www.abcbirds.org/abcprograms/policy/fisheries/tori_lines.html.

29. Pompa, S., Ehrlich, P. R., & Ceballos, G., 'Global distribution and conservation of marine mammals', *Proceedings of the National Academy of Sciences* 108(33) (2011), 13600–5. doi:10.1073/pnas.1101525108.

30. Moore, C. J., Moore, S. L., Leecaster, M. K., & Weisberg, S. B., 'A Comparison of Plastic and Plankton in the North Pacific Central Gyre', *Marine Pollution Bulletin* (2001).

Chapter 6: Deserts

1. Heffernan, O., 'The dry facts', *Nature* 501(7468) (2013), S2–S3. doi:10.1038/501S2a.

2. UN Convention to Combat Desertification, 'Desertification, land degradation and drought, global factsheet', at www.unccd.int/Lists/SiteDocumentLibrary/WDCD/DLDD%20Facts.pdf.

3. McSweeney, C., New, M., & Lizcano, G., 'UNDP Climate Change Country Profiles: Ethiopia', at www.geog.ox.ac.uk/research/climate/projects/undp-cp/UNDP_reports/Ethiopia/Ethiopia.lowres.report.pdf
   • Nicholson, S. E., Nash, D. J., Chase, B. M., Grab, S. W., Shanahan, T. M., Verschuren, D., Asrat, A. et al., 'Temperature variability over

Africa during the last 2,000 years', *Holocene* 23(8) (2013), 1085–94. doi:10.1177/0959683613483618 ◆ UNEP, 'Food Security in the Horn of Africa: The Implications of a Drier, Hotter and More Crowded Future' (2012), at http://na.unep.net/geas/getUNEPPageWithArticleIDScript. php?article_id=72 ◆ Williams, A. P., Funk, C., Michaelsen, J., Rauscher, S. A., Robertson, I., Wils, T. H. G., Koprowski, M. et al., 'Recent summer precipitation trends in the Greater Horn of Africa and the emerging role of Indian Ocean sea surface temperature', *Climate Dynamics* 39(9-10) (2012), 2307–28. doi:10.1007/s00382-011-1222-y.

4. Ngugi M., 'Agro-pastoral systems for dry-land agriculture and good practices for extensive pastoral systems', CAADP Workshop (2011) http://www.caadp.net/pdf/Day%203%20-%20BREAKOUT-Pastoral%20and%20agro-pastoral%20systems_FINAL.pdf.

5. Jacobson, M. Z., & Archer, C. L., 'Saturation wind power potential and its implications for wind energy', *Proceedings of the National Academy of Sciences* 109(39) (2012), 15679–84. doi:10.1073/pnas.1208993109.

6. Marvel, K., Kravitz, B., & Caldeira, K., 'Geophysical limits to global wind power', *Nature Climate Change* 3(2) (2012), 118–21. doi:10.1038/nclimate1683.

Chapter 7: Savannah

1. Tanzania Human Rights Report 2011.

2. Barnosky, A. D., 'Assessing the Causes of Late Pleistocene Extinctions on the Continents', *Science* 306(5693) (2004), 70–5. doi:10.1126/science.1101476.

3. Barnosky, A. D., Matzke, N., Tomiya, S., Wogan, G. O. U., Swartz, B., Quental, T. B., Marshall, C. et al., 'Has the Earth's sixth mass extinction already arrived?', *Nature* 471(7336) (2011), 51–7. doi:10.1038/nature09678.

4. Hof, C., Araújo, M. B., Jetz, W., & Rahbek, C., 'Additive threats from pathogens, climate and land-use change for global amphibian diversity', *Nature* 480 (2011), 516–9. doi:10.1038/nature10650.

5. WWF/Dalberg, 'Fighting Illicit Wildlife Trafficking: a consultation with governments' (2012), at http://awsassets.panda.org/downloads/wwffightingillicitwildlifetrafficking_lr.pdf ◆ Chomel, B. B., Belotto, A., & Meslin, F.-X., 'Wildlife, Exotic Pets, and Emerging Zoonoses', *Emerging Infectious Diseases* 13(1) (2007), 6–11. doi:10.3201/eid1301.060480.

6. 'WWF: Elephant killing continues in Cameroon', *Guardian*, at www.theguardian.com/world/feedarticle/10145985.

7. Turner, W. R., Brandon, K., Brooks, T. M., Gascon, C., Gibbs, H. K., Lawrence, K. S., Mittermeier, R. A. et al., 'Global Biodiversity Conservation and the Alleviation of Poverty', *BioScience* 62(1) (2012), 85–92. doi:10.1525/bio.2012.62.1.13.

8. Smil, V., 'Harvesting the biosphere: The human impact', *Population and Development Review* 37 (2011), 613–36.

9. Dirección del Parque Nacional Galápagos, at www.galapagospark.org/nophprg.php?page=desarrollo_sustentable_especies_invasoras.

10. Hobbs, R. J., Arico, S., Aronson, J., Baron, J. S., Bridgewater, P., Cramer, V. A., Epstein, P. R. et al., 'Novel ecosystems: theoretical and management aspects of the new ecological world order', *Global Ecology and Biogeography* 15(1) (2006), 1–7. doi:10.1111/j.1466-822X.2006.00212.x.

11. Moore, J. L. et al., 'Protecting islands from pest invasion: optimal allocation of biosecurity resources between quarantine and surveillance', *Biological Conservation* 143 (2010), 1068–78.

12. Mora, C., & Sale, P., 'Ongoing global biodiversity loss and the need to move beyond protected areas: a review of the technical and practical shortcomings of protected areas on land and sea', *Marine Ecology Progress Series* 434 (2011), 251–66.

13. 'Frankfurt Zoological Society Statement on The Proposed Serengeti Commercial Road' (2010), at www.zgf.de/download/1131/FZS+statement_+Serengeti+Road.pdf.

14. Folch, J. et al., 'First birth of an animal from an extinct subspecies (*Capra pyrenaica pyrenaica*) by cloning', *Theriogenology* 71(2009), 1026–34.

Chapter 8: Forests

1. Shvidenko, A., & Gonzalez, P., 'Forest and Woodland Systems', at http://pgonzalez.home.igc.org/Shvidenko_et_al_2006.pdf.

2. Van der Werf, G. R. et al., 'CO$_2$ emissions from forest loss', *Nature Geoscience* 2 (2009), 737–8.

3. British Columbia Ministry of Forests, Mines and Lands, 'The State of British Columbia's Forests, Third Edition' (2010), at www.for.gov.bc.ca/hfp/sof/index.htm#2010_report ◆ Ma, Z., Peng, C., Zhu, Q., Chen, H., Yu, G., Li, W., Zhou, X. et al., 'Regional drought-induced reduction in the biomass carbon sink of Canada's boreal forests', *Proceedings of the National Academy of Sciences*, 109(7) (2012), 2423–7. doi:10.1073/pnas.1111576109.

4. Earth Policy Institute, 'Eco-Economy Indicators – Forest Cover – World Forest Area Still on the Decline' (2012), at www.earth-policy.org/indicators/C56/forests_2012.

5. Lindquist, E. J., 'Global forest land-use change 1990–2005', Food and Agriculture Organization of the United Nations (2012).

6. Watts J., 'Environmental activists "being killed at rate of one a week"', *Guardian*, at www.theguardian.com/environment/2012/jun/19/environment-activist-deaths.

7. Ben-Ami, Y. et al., 'Transport of North African dust from the Bodélé depression to the Amazon Basin: a case study', *Atmospheric Chemistry and Physics* 10 (2010), 7533–44.

8. Nellemann, C., Redmond, I., Refisch, J., 'United Nations Environment Programme & GRID–Arendal. The last stand of the gorilla: environmental crime and conflict in the Congo basin', at www.grida.no/_res/site/file/publications/gorilla/GorillaStand_screen.pdf.

9. 'Mountain gorilla numbers rise by 10%', *Guardian*, at www.theguardian.com/environment/2012/nov/13/mountain-gorilla-population-rises.

10. Asner, G. P. et al., 'Condition and fate of logged forests in the Brazilian Amazon', *Proceedings of the National Academy of Sciences* 103(2006),

12947–50 ♦ Ahmed, S. E., Souza, C. M., Riberio, J., & Ewers, R. M., 'Temporal patterns of road network development in the Brazilian Amazon', *Regional Environmental Change* 13 (5) (2013), 927. doi: 10.1007/s10113-012-0397-z ♦ Rosa, I. M. D., Purves, D., Souza, C., & Ewers, R. M., 'Predictive Modelling of Contagious Deforestation in the Brazilian Amazon', *PLOS one* 8(10) (2013), e77231. doi:10.1371/journal.pone.0077231.

11. Fu, R., Yin, L., Li, W., Arias, P. A., Dickinson, R. E., Huang, L., Chakraborty, S. et al., 'Increased dry-season length over southern Amazonia in recent decades and its implication for future climate projection', *Proceedings of the National Academy of Sciences* 110(45) (2013), 18110-18115. doi:10.1073/pnas.1302584110.

12. Cox, P. M., Betts, R. A., Collins, M., Harris, P. P., Huntingford, C., & Jones, C. D., 'Amazonian forest dieback under climate-carbon cycle projections for the 21st century', *Theoretical and Applied Climatology* 78(1–3) (2004). doi:10.1007/s00704-004-0049-4.

13. Betts, R. A., Cox, P. M., Collins, M., Harris, P. P., Huntingford, C., & Jones, C. D., 'The role of ecosystem-atmosphere interactions in simu-lated Amazonian precipitation decrease and forest dieback under global climate warming', *Theoretical and Applied Climatology* 78(1–3) (2004). doi:10.1007/s00704-004-0050-y.

14. Asner, G. P., Llactayo, W., Tupayachi, R., & Luna, E. R., 'Elevated rates of gold mining in the Amazon revealed through high-resolution monitoring', *Proceedings of the National Academy of Sciences* 110 (2013), 18454-9.

15. Tollefson, J., 'Climate: Counting carbon in the Amazon', *Nature* 461(7267) (2009), 1048–52. doi:10.1038/4611048a.

16. Laurance, W., 'China's appetite for wood takes a heavy toll', *Timber and Forestry E-News 2012* (2012), 12–13.

17. WWF, 'Deforestation: Threats', at http://worldwildlife.org/threats/deforestation.

18. Wearn, O. R., Reuman, D. C., & Ewers, R. M., 'Extinction Debt and Windows of Conservation Opportunity in the Brazilian Amazon', *Science* 337(6091) (2012), 228–32. doi:10.1126/science.1219013.

19. 'California's Redwoods May be Benefiting From Climate Change', Science | KQED Public Media for Northern CA, at http://blogs.kqed.org/science/2013/08/14/californias-redwoods-may-be-benefiting-from-climate-change/.

20. Pross, J. et al., 'Persistent near-tropical warmth on the Antarctic continent during the early Eocene epoch', *Nature* 488 (2012), 73–7.

21. Cao, L., & Caldeira, K., 'Atmospheric carbon dioxide removal: long-term consequences and commitment', *Environmental Research Letters* 5 (2010), 024011.

Chapter 9: Rocks

1. Fischer-Kowalski, M. et al., 'Decoupling natural resource use and environmental impacts from economic growth', United Nations Environment Programme (2011).

2. Syvitski, J. P. M., 'Impact of Humans on the Flux of Terrestrial Sediment to the Global Coastal Ocean', *Science* 308(5720) (2005), 376–80. doi:10.1126/science.1109454.

3. Syvitski J. et al., 'Changing the History of the Earth: The Role of Water in the Anthropocene', GWSP Conference (2013), at www.gwsp.org/fileadmin/Conference_2013/water_anthropocene_A3.pdf.

4. Gregory M., 'Why are China's mines so dangerous?', BBC, at www.bbc.co.uk/news/business-11497070.

5. Schwarzer, S., De Bono, A., Giuliani, G., Kluser, S., and Peduzzi, P., 'E-waste, the hidden side of IT equipment's manufacturing and use', *UNEP DEWA/GRID–Europe Environment Alert Bulletin* 5 (2005).

6. 'E-waste: Annual gold, silver "deposits" in new high-tech goods worth $21b; less than 15% recovered', at www.sciencedaily.com/releases/2012/07/120706164159.htm ♦ US EPA, O. 'Statistics on the

Management of Used and End-of-Life Electronics', at www.epa.gov/epawaste/conserve/materials/ecycling/manage.htm.

7. See Cosima Dannortizer's documentary *The Light Bulb Conspiracy*.

8. WWF, 'Living Planet Report 2008', at http://awsassets.panda.org/downloads/living_planet_report_2008.pdf.

9. International Energy Agency, World Energy Outlook 2012, at http://www.iea.org/publications/freepublications/publication/English.pdf.

10. Evans, L., Maio, G. R., Corner, A., Hodgetts, C. J., Ahmed, S., & Hahn, U., 'Self-interest and pro-environmental behaviour', *Nature Climate Change* 3(2) (2012), 122–5. doi:10.1038/nclimate1662.

11. OECD, 'Environmental Outlook to 2050', Climate Change chapter, at www.oecd.org/env/cc/49082173.pdf.

12. 'Plastic to Oil Fantastic', at www.youtube.com/watch?v=R-Lg_kvLaAM.

13. WHO, 'Air quality and health', at www.who.int/mediacentre/factsheets/fs313/en/.

14. Fischetti M., 'The Human Cost of Energy', *Scientific American* (2011), at www.scientificamerican.com/article.cfm?id=the-human-cost-of-energy.

Chapter 10: Cities

1. 'Urbanization, Biodiversity and Ecosystem Services: Challenges and Opportunities – a SpringerOpen book', at www.springer.com/life+sciences/ecology/book/978-94-007-7087-4.

2. 'Colombia's murder rate down 45% during Uribe: Police', Colombia Reports, at http://colombiareports.co/colombias-murder-rate-down-45-during-uribe-police/.

3. Bettencourt, L. M. A., Lobo, J., Helbing, D., Kuhnert, C., & West, G. B., 'Growth, innovation, scaling, and the pace of life in cities', *Proceedings of the National Academy of Sciences* 104(17) (2007), 7301–6. doi:10.1073/pnas.0610172104.

4. Hamilton, M. J., Milne, B. T., Walker, R. S., & Brown, J. H., 'Nonlinear scaling of space use in human hunter-gatherers', *Proceedings of the*

National Academy of Sciences 104(11) (2007), 4765–9. doi:10.1073/pnas.0611197104.

5. UN, 'The Millennium Development Goals Report 2013', at www.un.org/millenniumgoals/pdf/report-2013/mdg-report-2013-english.pdf.

6. Hoornweg, D., Sugar, L., & Trejos Gomez, C. L., 'Cities and greenhouse gas emissions: moving forward', *Environment and Urbanization* 23(1) (2011), 207–27. doi:10.1177/0956247810392270.

7. From P. D. Smith, *City: A Guidebook for the Urban Age* (2012).

8. Li, W., Maduro, G., & Begier, E. M., 'Life Expectancy in New York City: What Accounts for the Gains?', New York City Department of Health and Mental Hygiene: Epi Research Report, March 2013, 1–12, at www.nyc.gov/html/doh/downloads/pdf/epi/epiresearch-lifeexpectancy.pdf.

9. McGranahan, G., & Satterthwaite, D., 'Better Health for the Uncounted Urban Masses', *Scientific American* 305(3) (2011), 62. doi:10.1038/scientificamerican0911-62 ◆ Patel, R. B., & Burke, T. F., 'Urbanization – An Emerging Humanitarian Disaster', *New England Journal of Medicine* 361(8) (2009), 741–3. doi:10.1056/NEJMp0810878.

10. OECD, 'Is Informal Normal? Towards More and Better Jobs in Developing Countries', OECD Development Centre Studies (2009), ISBN 978-92-64- 05923-8.

11. McKinsey & Company, 'Urban world: Mapping the economic power of cities', at www.mckinsey.com/insights/urbanization/urban_world.

12. Nourishing the Planet, 'Urban Harvest', at http://blogs.worldwatch.org/nourishingtheplanet/tag/urban-harvest/ .

13. 'Brazilian Army Caves in to Favela's Drug Dealers', Brazzilmag.com, at http://brazzilmag.com/component/content/article/34/5790-brazilian-army-caves-in-to-favelas-drug-dealers.html.

14. Acuto M., & Khanna P., 'Nations are no longer driving globalization – cities are', Quartz (2013), at http://qz.com/80657/the- return-of-the-city-state/.

15. UN–Habitat report, 'State of the World's Cities 2010/2011', at www. unhabitat.org/content.asp?cid=8051&catid=7&typeid=46.

16. The Revolutionary Optimists, 'Map Your World', at http://revolution-aryoptimists.org/map-your-world.

17. Rick Smolan and Jennifer Erwitt, *The Human Face of Big Data* (2012).

18. 'How Target Figured Out A Teen Girl Was Pregnant Before Her Father Did', *Forbes*, at www.forbes.com/sites/kashmirhill/2012/02/16/how-target-figured-out-a-teen-girl-was-pregnant-before-her-father-did/.

19. World Bank Group President Robert B. Zoellick opening remarks at the C40 Large Cities Climate Summit, at http://web.worldbank.org/WBSITE/EXTERNAL/NEWS/0,,contentMDK:22928910~pagePK:6425 7043~piPK:437376~theSitePK:4607,00.html.

20. 'Vanity Height: the Empty Space in Today's Tallest', *CTBUH Journal*, Issue III (2013), 42, at www.ctbuh.org/LinkClick. aspx?fileticket=k%2bK%2fgAvA7YU%3d&tabid=5837&language=en-US.

21. Sivak, M., 'Has motorization in the US peaked? (2013), at http://141.213.232.243/handle/2027.42/98098.'

22. Vidal, J., & Pathak, S., 'How urban heat islands are making India hotter', *Guardian*, at www.theguardian.com/global-development/poverty-matters/2013/jan/09/delhi-mumbai-urban-heat-islands-india.

23. Proppe, D. S., Sturdy, C. B., & St Clair, C. C., 'Anthropogenic noise decreases urban songbird diversity and may contribute to homogenization', *Global Change Biology* 19(4) (2013), 1075–84. doi:10.1111/gcb.12098 ◆ Miranda, A. C., Schielzeth, H., Sonntag, T., & Partecke, J., 'Urbanization and its effects on personality traits: a result of microevolution or phenotypic plasticity?', *Global Change Biology* 19(9) (2013), 2634–4. doi:10.1111/gcb.12258 ◆ Snell-Rood, E. C., & Wick, N., 'Anthropogenic environments exert variable selection on cranial capacity in mammals', *Proceedings of the Royal Society* B: Biological Sciences 280(1769) (2013), 20131384. doi:10.1098/rspb.2013.1384 ◆ Harris, S. E., Munshi-South, J., Obergfell, C., & O'Neill, R., 'Signatures of rapid evolution in urban and rural

transcriptomes of white-footed mice (*Peromyscus leucopus*) in the New York metropolitan area' (2013), PeerJ PrePrints 1:e13v2, at http://dx.doi. org/10.7287/peerj.preprints.13v2.

24. Lederbogen, F., Kirsch, P., Haddad, L., Streit, F., Tost, H., Schuch, P., Wüst, S. et al., 'City living and urban upbringing affect neural social stress processing in humans', *Nature* 474(7352) (2011), 498–501. doi:10.1038/nature10190 ◆ Pedersen, C. B., & Mortensen, P. B., 'Evidence of a dose-response relationship between urbanicity during upbringing and schizophrenia risk', *Archives of General Psychiatry* 58 (2001), 1039–46.

25. Seto, K. C., Guneralp, B., & Hutyra, L. R., 'Global forecasts of urban expansion to 2030 and direct impacts on biodiversity and carbon pools', *Proceedings of the National Academy of Sciences* 109(40) (2012), 16083–8. doi:10.1073/pnas.1211658109.

INDEX